ELDRIDGE
TIDE AND PILOT BOOK
2009

One Hundred Thirty-fifth Year of Continuous Publication

CONTENTS

Publishers: Robert Eldridge White, Jr. and Linda Foster White
P.O. Box 775, Medfield, MA 02052
ebb2flood@gmail.com

Tel. 617-482-8460
Fax: 617-482-8304

Printed in U.S.A.   ISBN 978-1-883465-15-5   **PRICE: $14.00**

THE ORIGIN OF ELDRIDGE

In 1854 George Eldridge of Chatham, a celebrated cartographer, published Eldridge's Pilot for Vineyard Sound and Monomoy Shoals. The book had 32 pages, a grey paper cover and no recorded price. Its pages were devoted to "Dangers," embellished with his personal observations, and to Compass Courses and Distances, etc. This volume was the precursor of the Tide and Pilot Book, which followed 21 years later.

In 1870 George Eldridge published another small book, called the "Compass Test," and asked his son, George W. Eldridge, to go to Vineyard Haven and sell it for him, along with the charts he produced.

Son George W. Eldridge was dynamic, restless, and inventive. He was glad to move to the Vineyard, for Vineyard Haven was at that time an important harbor for large vessels. The number of ships passing through Vineyard and Nantucket Sounds was second only to those plying the English Channel. As the ships came into the harbor (frequently as many as 100 schooners would anchor to await a fair current), George W. would go out to them in his catboat to sell his father's charts and the "Compass Test." He was constantly asked by mariners what time the current TURNED to run East or West in the Sound. He then began making observations, and one day, while in the ship chandlery of Charles Holmes, made the first draft of a current table. Shortly after, with the help of his father, he worked out the tables for places other than Vineyard Sound, and in 1875 the first Tide Book was published. It did not take long for mariners to realize the value of this information, and it soon became an indispensable book to all who sailed the Atlantic Coast from New York east. Gradually George W. added more important information, such as his explanation of the unusual currents which caused so many vessels to founder in the "Graveyard."

Captain George W. Eldridge based the tables on his own observations. In later years, knowing that the government's scientific calculations are the most accurate obtainable, the publishers have made use of them; some tables are directly taken from government figures and others, which the government does not give in daily schedules, are computed by the publishers from government predictions. Since the Captain's day there have been many changes and additions in the book to keep abreast of modern navigational aids.

In 1910 Captain George W. Eldridge transferred the management of the book to the next generation of his family, as he was interested in developing his chart business and inventing aids to navigation. At his death in 1914, his son-in-law became Publisher. Wilfrid O. White, an expert in marine navigation and President of Wilfrid O. White & Sons Co., compass manufacturers, served as Publisher until his death at 1955. Wilfrid's son Robert Eldridge White then became publisher, and with great help from his wife Molly he expanded the coverage of the book and significantly increased its readership. On Bob's death in 1990, Molly continued to expand the book's scope and circulation, with valuable assistance from her son Ridge and daughter-in-law Linda. On Molly's passing in 2004 the book moved once again into the hands of the next (fourth) generation. At every generational transfer the new Publisher was well prepared, each having apprenticed for years.

Whether new to ELDRIDGE or a longtime reader, we welcome you aboard! Please continue to offer your suggestions and, where necessary, corrections. Your sharp eyes keep us on course. We hope, as did Captain George W. Eldridge, that this book might ensure for you a "Fair Tide" and the safety of your ship.

Robert Eldridge White, Jr.
Linda Foster White
Publishers

Yours for a fair tide

Geo. W. Eldridge

About the 2009 ELDRIDGE, our 135th edition

☆ **NOTE:** *The information in this volume has been compiled from U. S. Government sources and others, and carefully checked. The Publishers cannot assume any liability for errors, omissions, or changes.*

Daylight Saving Time begins at 2 a.m., Sunday, March 8, 2009
Eastern Standard Time begins at 2 a.m., Sunday, November 1, 2009

p. 10 **Caution on Using Tide and Current Data**
    A very important note explaining why your harbor may have been dropped from NOAA's listings (and ours), and our warning about using the data for substations that have been relisted, with an asterisk*.

New Articles

pp. 164-165 **The Shore Angler and Alongshore Currents** - an expert's secrets
p. 201 **Rule of Twelfths** - secrets of the rising and falling tide
p. 203 **Two Ways to Adjust Your Compass** - make your compass trustworthy
p. 204 **Time at Sea** - a brief review of timekeeping, from hourglass to atomic time
p. 205 **What Your GPS Can and Cannot Do** - the genie has limitations
p. 236 **Daily Moon Phase Calendar** - 365 moon phases for quick reference
p. 246 **Keys to Predicting Weather** - we condense Fitzroy to the essentials
p. 248 **Humidity and Dewpoint at Sea** - fog, dew, and dampness explained
p. 252 **Story Contest** - a first for ELDRIDGE! Got a factual or fictional tale?
p. 258 **Vessel Safety Check, US Coast Guard Boardings** - what you should know

We extend a very warm welcome to a number of new advertisers this year. We hope you will support them and say that you saw their ad in ELDRIDGE.

Publishers: Robert Eldridge White, Jr. - Linda Foster White
Editorial and Advertising: Tel.: 617-482-8460, Fax: 617-482-8304
Email: ebb2flood@gmail.com, **Web:** www.eldridgetide.com
Mailing Address: P.O. Box 775, Medfield, MA 02052

FREE SUPPLEMENT ON REQUEST - available after May 15, 2009
listing changes and updates through May 1, 2009
To obtain a free Supplement:
    1) Download a pdf from: www.eldridgetide.com *or*
        www.robertwhite.com (click on Eldridge)
    2) Mail a stamped, self-addressed envelope, with your request, to:

ELDRIDGE TIDE & PILOT BOOK
P.O. Box 775
Medfield, MA 02052

I enclose a **stamped, self-addressed envelope**. Please send me a free Supplement, updating data through May 1, 2009.

Name   _____

Address   _____

City, State, Zip   _____

INLAND NAVIGATION RULES

Good Seamanship Rule (Rule 7): Every vessel shall use all available means appropriate to the prevailing circumstances and conditions to determine if risk of collision exists. If there is any doubt, such risk shall be deemed to exist.

General Prudential Rule (Rule 2b): Due regard shall be had to all dangers of navigation and collision, and to any special circumstances, including the limitations of the vessels involved, which may make a departure from these Rules necessary to avoid immediate danger.

General right-of-way (Rule 18): Vessel categories are listed in <u>decreasing</u> order of having the right-of-way:
- Vessel not under command (most right of way)
- Vessel restricted in ability to maneuver, in a narrow fairway or channel
- Vessel engaged in fishing with nets, lines, or trawls (but not trolling lines)
- Sailing vessel (sails only)
- Power-driven vessel (least right of way)

Vessels Under Power

Overtaking (Rule 13): A vessel overtaking another is the "give-way" vessel and must stay clear of the overtaken or "stand-on" vessel. The overtaking vessel is to sound one short blast if it intends to pass on the other vessel's starboard side, and two short blasts if it intends to pass on the other's port side. The overtaken vessel must respond with the identical sound signal if it agrees, and must maintain course and speed during the passing situation.

Meeting head-on (Rule 14): When two vessels are meeting approximately head-on, neither has right-of-way. Unless it is otherwise agreed, each vessel should turn to starboard and pass port to port.

Crossing (Rule 15): When two vessels approaching each other are neither in an overtaking or meeting situation, they are deemed to be crossing. The power vessel which has the other on its starboard side is the give-way vessel and must change course, slow down, or stop. The vessel which is on the right, is in the right.

Vessels Under Sail

Port-Starboard (Rule 12): A vessel on the port tack shall keep clear of one on the starboard tack.

Windward-Leeward (Rule 12): When both vessels are on the same tack, the vessel to windward shall keep clear of a vessel to leeward.

Sail vs. Power (Rule 18): Generally, a sailboat has right of way over a powerboat. However: (1) a sailboat overtaking a powerboat must keep clear; (2) sailboats operating in a narrow channel shall keep clear of a power vessel which can safely navigate only within a narrow channel; (3) sailboats must give way to a vessel which is fishing, a vessel restricted in its ability to maneuver, and a vessel not under command.

FEDERAL SAFETY EQUIPMENT REQUIREMENTS
(Minimum)
some states require additional equipment

Sound Signaling Devices

Under 39.4' or 12 meters:
> Must have some means of making an efficient sound signal

Over 39.4' or 12 meters:
> Whistle or horn, audible for 1/2 mile, and a bell at least 8" dia.

Visual Distress Signals (all with approval number)

Under 16' or 5 meters:
> Night: 1 electric SOS flashlight or 3 day/night red flares

Over 16' or 5 meters:
> Day only: 1 orange flag, 3 floating or hand-held orange smoke signals
>
> Day and night: 3 hand-held, or 3 pistol, or 3 hand-held rocket, or 3 pyrotechnic red flares

The following signals indicate distress or need of assistance

A gun or other explosive signal fired at intervals of about 1 minute

A continuous sounding with any fog-signaling apparatus

Rockets or shells, fired one at a time or at short intervals

SOS transmitted by any signaling method

"Mayday" on the radiotelephone (channel 16)

International Code Signal flags "NC"

An orange square flag with a black square over a black ball

Flames on the vessel

Rocket parachute flare or hand-held flare

Orange colored smoke

Slowly and repeatedly raising and lowering outstretched arms

Signals transmitted by EPIRB

High intensity white light flashing 50-70 times per minute

Personal Flotation Devices (must be USCG approved)

Under 16' or 5 meters:
> 1 Type I, II, III, or V per person, USCG approved

Over 16' or 5 meters:
> 1 Type I, II, III, or V per person, and 1 Type IV per boat, USCG approved

Portable Fire Extinguishers (approved)

under 26' 1 B-I, if no fixed extinguisher system in machinery space.
> (Not required on out-boards built so that vapor entrapment cannot occur.)

26-39' 2 B-I or 1 B-II & 1 B-I, if no fixed exting. system; 1 B-I, with a fixed exting. system.

40-65' 3 B-I or 1 B-II & 1 B-I, if no fixed exting. system; 2 B-I or 1 B-II with a fixed exting. system.

Back-Fire Flame Arrestor

One approved device per carburetor of all inboard gasoline engines.

At least 2 ventilator ducts fitted with cowls or their equivalent to ventilate efficiently the bilges of every engine and fuel tank compartment of boats using gasoline or other fuel with a flashpoint less than 110°F.

NAVIGATION LIGHTS
Definition of Lights

Masthead Light — a white light fixed over the vessel's centerline showing an unbroken light over an arc of the horizon of 225°, from dead ahead to 22.5° abaft the beam on either side.

Sidelights — a green light on the starboard side and a red light on the port side showing an unbroken light over an arc of the horizon of 112.5°, from dead ahead to 22.5° abaft the beam on either side.

Sternlight — a white light placed as nearly as practicable at the stern showing an unbroken light over an arc of the horizon of 135°, 67.5° from dead aft to each side of the vessel.

All-round Light — an unbroken light over an arc of the horizon of 360°.

Towing Light — a yellow light with same characteristics as the sternlight.

Note: R . and Y. Flashing Lights are now authorized for vessels assigned to Traffic Control, Medical Emergencies, Search and Rescue, Fire-Fighting, Salvage and Disabled Vessels.

When under way, in all weathers from sunset to sunrise, every vessel shall carry and exhibit the following lights

When Under Power Alone or When Under Power and Sail Combined

Under 39.4' or 12 meters:
   Masthead light visible 2 miles
   Sidelights visible 1 mile
   Stern light visible 2 miles (may be combined with masthead light)

Over 39.4' or 12 meters to 65' or 20 meters::
   Masthead light visible 3 miles
   Sidelights visible 2 miles
   Stern light visible 2 miles (or in lieu of separate masthead and stern lights, an all-round white light visible 2 miles)

Sailing Vessels Under Way (Sail Only)

Under 22' or 7 meters:
   Either the lights listed below for sailing vessels under 65'; or a white light to be exhibited (for example, by shining it on the sail) in sufficient time to prevent collision
Under 65' or 20 meters:
   Sidelights visible 2 miles
   Stern light visible 2 miles
   These lights may be combined in one tricolor light carried near the top of the mast

At Anchor

Vessels under 50 meters (165') must show an all-round white light where it can best be seen and visible 2 miles.

Vessels under 7 meters (22') need no light unless they are near a channel, a fairway, an anchorage or area where other vessels navigate.

Fishing

Vessels Trawling shall show, in addition to the appropriate lights above, 2 all-round lights in a vertical line, the upper green and the lower white.

Vessels Fishing (other than trawling) shall show, in addition to the appropriate lights above, 2 all-round lights, the upper red and the lower white.

When Towing or Being Towed

Towing Vessel: 2 masthead lights (if tow is less than 200 meters); 3 masthead lights in a vertical line forward (if tow exceeds 200 meters); sidelights; sternlight; a yellow tow light in vertical line above sternlight; a diamond shape where it can best be seen (if tow exceeds 200 meters).

Vessel Being Towed: sidelights; sternlight; a diamond shape where it can best be seen (if tow exceeds 200 meters).

SOUND SIGNALS FOR FOG

Ask your Chart Dealer for the latest Navigation Rules—Inland/International

All signals prescribed by this article for vessels under way shall be given:

>**First:** By Power-driven Vessels – On the Whistle or Siren.
>
>**Second:** By Sailing Vessels or Vessels being Towed – On the Fog Horn.

A prolonged blast shall mean a blast of 4 to 6 seconds' duration.

A power-driven vessel having way upon her shall sound at intervals of no more than 2 minutes, a prolonged blast.

A power-driven vessel under way, but stopped and having no way upon her, shall sound at intervals of no more than 2 minutes, 2 prolonged blasts with about 2 seconds between them.

A sailing vessel under way, shall sound at intervals of not more than 2 minutes, 1 prolonged blast followed by 2 short blasts, regardless of tack.

A fishing vessel or a power-driven vessel towing or pushing another vessel, shall sound every 2 minutes, 1 prolonged blast followed by 2 short blasts. A vessel being towed shall sound 1 prolonged blast followed by 3 short blasts.

A vessel at anchor shall ring a bell rapidly for about 5 seconds at intervals of not more than 1 minute and may in addition sound 3 blasts, 1 short, 1 prolonged, 1 short, to give warning of her position to an approaching vessel. Vessels under 20 meters (65') shall not be required to sound these signals when anchored in a special anchorage area.

A vessel aground shall give the bell signal and shall, in addition, give 3 separate and distinct strokes of the bell.

PASSING SIGNALS

Inland Rules:

>1 short blast: I intend to direct my course to Starboard.
>
>2 short blasts: I intend to direct my course to Port.

International Rules:

>1 short blast: I am directing my course to Starboard.
>
>2 short blasts: I am directing my course to Port.
>
>2 long, 1 short blast: I am overtaking you on your Starboard side.
>
>2 long, 2 short blasts: I am overtaking you on your Port side.

Response: long blast, short blast, long blast, short blast if agreeable.

Frequently, in fog, small sail or power boats cannot be heard or picked up by other vessels' radar. The Coast Guard strongly recommends that, to avoid collisions, all vessels carry Radar Reflectors mounted as high as possible.

Editors' note:

Lovers of maritime tradition relinquish only reluctantly some of the old lost ways of life at sea: heaving the lead for soundings, taking sextant sights, winding the chronometer, signaling other vessels with flags, and using a taffrail log. Modern instruments and procedures have taken their place aboard ship.

In a move toward modernity the Publishers eliminated one small piece we had printed in ELDRIDGE for countless years. But when a seasoned sea-farer came into our shop a while back and, with a nostalgic smile, recited all four verses perfectly, we were struck by the desire to restore this piece of the past. For him and for those others who lament the loss of tradition at sea, we herewith return it to our pages, because it is still useful, worth knowing, and good poetry besides.

The Rule of the Road

When all three lights I see ahead,
I turn to **Starboard** and show my **Red:**
Green to Green, Red to Red,
Perfect Safety – **Go Ahead.**

But if to **Starboard Red appear,**
It is my duty to keep clear –
To act as judgment says is proper:
To **Port** or **Starboard, Back** or **Stop** her.

And if upon my **Port** is seen
A Steamer's **Starboard** light of **Green,**
I hold my course and watch to see *
That **Green** to **Port** keeps Clear of me.

Both in safety and in doubt
Always keep a good look out.
In danger, with no room to turn,
Ease her, **Stop** her, **Go Astern.**

* "There's nought for me to do but see" is the original version

A Cautionary Word on Using Tide and Current Data

The Budget Crunch at NOAA/NOS, and How It Affects You

As some of our readers may have noticed, starting several years ago, a number of subordinate stations for tides and currents have been omitted from the usual listing beginning on p. 12. The reason for this is that in 2000, NOAA/NOS began a complete review of substations to check the times of change for tides and currents, the heights of tides, and strength of currents. Some of the data they found were over 100 years old, and deemed not reliable. As a result, a number of the tide and current substations have recently been dropped from government tables. And years ago, for the same reason, the Government ceased publication of current chart diagrams.

What NOAA/NOS requires, in order to put a tide or current substation back on the list, is a minimum of a one-month-long stream of data from that location, collected some time in the past 60 years. It is costly to set up a tidal station, and especially so for a current station. Given severe budgetary restraints, NOAA/NOS says that it is not likely that the substations recently dropped will be reintroduced anytime soon.

Reacting to this move, we at ELDRIDGE, in the interest of caution, decided to follow NOAA's lead. Our listings of the **Time of High Water**, pp. 12-20, and **Time of Current Change**, pp. 22-29, have diminished accordingly. We hoped that readers would find harbors close by the place they were interested in, and use that information to make an estimate. A number of readers called, asking why their favorite harbor didn't appear. We told them the story of NOAA/NOS, and suggested that they could continue to use the old figures, as inaccurate information is sometimes better than none, but only with the clear warning to be especially cautious and not to rely on it.

Recognizing that old, somewhat suspect data may still be of some use to the mariner, the Publishers have decided to reintroduce a number of the substations we dropped in the last few years. These locations are clearly identified by an asterisk: * We do so with the obvious cautions that (**1**) *no single source of information, even this publication, should be relied upon completely*, (**2**) *vigilance and prudence should govern the decisions of the navigator, and* (**3**) *that experience shows that tide and current predictions are approximate to begin with.*

We warn our readers to note the *asterisk in our list of substations for tides and currents (p. 12 and following), and to exercise a greater than usual degree of caution in relying on the data which is given for these relisted stations.

HOW TO USE THE TIDE AND CURRENT TABLES AND CURRENT CHARTS

High Water – In addition to presenting tide tables for nine reference ports, from Portland to Miami, we show the approximate time of High Water and mean (average) height at some 350 substations, pp. 12-20. In those pages, find your harbor or the nearest listed harbor to it, and note the time difference shown. Apply this to the referenced table for that date. Low Water will follow by an average of 6 hours, 12 minutes. When the height of High Water in the reference table is higher or lower than its average, it will be correspondingly higher or lower at your harbor. The heights listed in these tables are the number of feet above Mean Lower Low Water (the same datum as the soundings on U.S. Government charts). We urge you to read the article on p. 10, "A Cautionary Word on Using Tide and Current Data."

Currents – There are eight current tables covering from Massachusetts to the Chesapeake. At over 300 other points, pp. 22-29, we show the approximate time of Current Change, with directions of Flood and Ebb and average maximum velocities. Find the place you are concerned with, or the listed place nearest to it, note the time difference shown and apply this to the referenced table for that date. When the velocity of the current in the referenced table exceeds the average maximum, the current in your area will also exceed the average maximum.

Note: *As the terms Ebb and Flood can sometimes confuse, we recommend using the direction as the name of the current. For instance, it is more helpful to refer to an Easterly Current, which means that it is Eastbound or runs toward the East, than to an Ebb or Flood Current, which leaves the listener guessing its direction.*

Current Charts – Complete current Charts in the book are for: BUZZARDS BAY, VINEYARD & NANTUCKET SOUNDS, LONG ISLAND & BLOCK ISLAND SOUNDS, and NEW YORK HARBOR. (There are also Current Diagrams for BOSTON HARBOR, NARRAGANSETT BAY and CHESAPEAKE BAY, but these are simplifications of government current charts.) Find the Current Chart covering the area you are concerned with and note the Table to which the Chart is referenced. Turn to this Table, which tells the time of Start of Flood and Start of Ebb, and take the Time of the Start of the helpful current for the day in question. Go back to the Current Charts, which are labeled for each hour of the Flood and Ebb, and apply these hours to the time of the helpful current, which you have taken from the reference table, and give yourself these free hours of fair tide. The difference between having a fair tide or a head tide literally means hours of difference to the auxiliary and considerable quantities of fuel, even for the 20-knot cruiser.

Effect of the Moon – Tides are higher and currents stronger at the times of Full Moon and at New Moon. When either Full Moon or New Moon occurs at Perigee, there will be rather spectacular tides and currents. When the Moon is on the Equator, the a.m. and p.m. tides will be about equal in height, and the a.m. and p.m. currents about equal in velocity. When the Moon is at its most Southerly or Northerly declination, however, heights of Highs and Lows and velocities of Floods and Ebbs will be markedly different. (See p. 234 for Phases of the Moon and p. 235, Tides and the Moon, for more information.)

TIME OF HIGH WATER

Time figures shown are the *average* differences throughout the year. Rise in feet is mean range.
(Low Water times are given *only* when they vary more than 20 min. from High Water times.)

NOTE: *Asterisk indicates that NOAA has recently dropped these substations from its listing because the data are judged to be of questionable accuracy. We have published NOAA's most recently available figures with this warning: Mariners are cautioned that the starred information is only approximate and not supported by NOAA or the Publishers of Eldridge.

see NOAA Article p. 10, *A Cautionary Word on Using Tide and Current Data.*

---

*For **Canadian Ports**, if your watch is set for Atlantic Time, use the time differences listed here; if your watch is set for Eastern Time, subtract one hour from these time differences.*

	H.M.			Rise in feet
NOVA SCOTIA, Outer Coast				
Guysborough	3 05	before	PORTLAND	3.8
Whitehaven Harbour	3 20	"	"	3.7
Liscomb Harbour	3 20	"	"	4.2
Sheet Harbor	3 20	"	"	4.2
Ship Harbor	3 20	"	"	4.2
Jeddore Harbor	3 15	"	"	4.3
Halifax	3 10	"	"	4.4
Sable Island, North	3 15	"	"	2.6
Sable Island, South	3 10	"	"	3.9
Chester, Mahone Bay	3 10	"	"	4.4
Mahone Harbour, Mahone Bay	3 10	"	"	4.5
Lunenburg	3 05	"	"	4.2
Riverport, La Have River	3 00	"	"	4.5
Liverpool Bay	2 55	"	"	4.3
Lockeport	2 40	"	"	4.6
Shelburne	2 40	"	"	4.8
NOVA SCOTIA & NEW BRUNSWICK, Bay of Fundy				
Lower E. Pubnico	1 10	before	PORTLAND	8.7
Yarmouth Harbour	0 20	"	"	11.5
Annapolis Royal, Annapolis R.	0 50	after	"	22.6
Parrsboro, Minas Basin, Partridge Is.	1 35	"	"	34.4
Burntcoat Head, Minas Basin	1 50	"	"	38.4
Amherst Point, Cumberland Basin	1 20	"	"	35.6
Grindstone Is, Petitcodiac River	1 05	"	"	31.1
Hopewell Cape, Petitcodiac River	1 00	"	"	33.2
Saint John	0 45	"	"	20.8
Indiantown, Saint John River	2 15	"	"	1.2
L'Etang Harbor	0 45	"	"	18.4

REVERSING FALLS....SAINT JOHN, N.B.

The most turbulence in the gorge occurs on days when the tides are largest. On largest tides the outward fall is between 15 and 16 1/2 feet and is accompanied by a greater turbulence than the inward fall which is between 11 and 12 1/2 feet. The outward fall is at its greatest between two hours before and one hour after low water at St. John; the inwaard fall is greater just before the time of high water. For complete tidal information of Canadian ports see Tide Tables of the Atlantic Coast of Canada. (Purchase tables from nautical dealers in Canadian ports or from the Queen's Printer, Department of Public Printing, Ottawa.

PORTLAND Tables, pp. 30-35

When a high tide exceeds av. ht., the *following* low tide will be lower than av.

*Times and Hts. are approximate. *Important*: see NOTE, top p. 12, and NOAA Article, p. 10.

TIME OF HIGH WATER

Time figures shown are the *average* differences throughout the year. Rise in feet is mean range.
(Low Water times are given *only* when they vary more than 20 min. from High Water times.)

U.S. ATLANTIC COAST, from Maine southward

	H.M.			Rise in feet
MAINE				
Eastport	0 15	before	PORTLAND	18.4
Cutler, Little River..	0 25	"	"	13.5
Shoppee Pt. Englishman Bay	0 20	"	"	12.1
Steele Harbor Island	0 30	"	"	11.6
*Jonesport..	0 20	"	"	11.5
Green Island, Petit Manan Bar	0 30	"	"	10.6
Prospect Harbor..	0 25	"	"	10.5
Winter Harbor, Frenchman Bay..........................	0 20	"	"	10.1
Bar Harbor, Mt. Desert Island............................	0 20	"	"	10.6
Southwest Harbor, Mt. Desert Island	0 20	"	"	10.2
Bass Harbor................................. **high 0 15** *before, low*..0 45		"	"	9.9
Blue Hill Harbor, Blue Hill Bay	0 10	"	"	10.1
Burnt Coat Harbor, Swans Island.......................	0 20	"	"	9.5
Penobscot Bay				
Center Harbor, Eggemoggin Reach.....................	0 10	"	"	10.1
Little Deer Isle, Eggemoggin Reach	0 05	"	"	10.0
Isle Au Haut..	0 20	"	"	9.3
Stonington, Deer Isle..	0 10	"	"	9.7
Matinicus Harbor, Wheaton Is.	0 15	"	"	9.0
Vinalhaven..	0 10	"	"	9.3
North Haven..	0 05	"	"	9.7
Pulpit Harbor, North Haven Is.	0 10	"	"	9.9
Castine..	0 05	"	"	10.1
Bucksport, Penobscott River...............................	0 25	"	"	10.8
Bangor, Penobscot River.............. **high 0 25** *before, low..same as*			"	13.4
Belfast..	0 10	before	"	10.2
*Camden..	0 10	"	"	9.6
Rockland..	0 10	"	"	9.8
MAINE, Outer Coast				
Tenants Harbor	0 10	before	PORTLAND	9.3
Monhegan Island	0 15	"	"	8.8
Port Clyde, St. George River................................	0 10	"	"	8.9
Thomaston, St. George River...............................	0 05	"	"	9.4
New Harbor, Muscongus Bay...............................	0 10	"	"	8.8
Friendship Harbor	0 20	"	"	9.0
Waldoboro, Medomak River.................................	0 15	"	"	9.5
East Boothbay, Damariscotta River .... same as			"	8.9
Boothbay Harbor...	0 05	before	"	8.8
Wiscasset, Sheepscot River.................................	0 15	after	"	9.4
Robinhood, Sasanoa River...................................	0 15	"	"	8.8
Phippsburg, Kennebec River	0 25	"	"	8.0
Bath, Kennebec River	1 00	"	"	6.4
Casco Bay				
*Small Point Harbor..	0 10	before	"	8.8
Cundy Harbor, New Meadows River .... same as			"	8.9
South Harpswell, Potts Harbor .... same as			"	8.9
South Freeport..	0 10	after	"	9.0

PORTLAND Tables, pp. 30-35

When a high tide exceeds av. ht., the *following* low tide will be lower than av.
*Times and Hts. are approximate. *Important*: see NOTE, top p. 12, and NOAA Article, p. 10.

TIDE STATIONS

TIME OF HIGH WATER

Time figures shown are the *average* differences throughout the year. Rise in feet is mean range.
(Low Water times are given *only* when they vary more than 20 min. from High Water times.)

	H.M.			Rise in feet
MAINE, Cont.				
Falmouth Foreside	same as		PORTLAND	9.1
Great Chebeague Island	same as		"	9.1
Portland Head Light	same as		"	8.9
Cape Porpoise	0 10	after	"	8.7
Kennebunkport	0 05	"	"	8.8
York Harbor	0 05	"	"	8.6
NEW HAMPSHIRE				
Portsmouth	0 20	after	PORTLAND	7.8
Gosport Harbor, Isles of Shoals	same as		"	8.5
Hampton Harbor	0 15	after	"	8.3
MASSACHUSETTS, Outer Coast				
Newburyport, Merrimack River.. **high 0 30** after, low 1 10		after	PORTLAND	7.8
Plum Island Sound, S. End .... **high 0 10** after, low 0 35		"	"	8.6
Annisquam, Lobster Cove	0 10	"	"	8.8
Rockport	0 05	"	"	8.7
Gloucester Harbor	same as		BOSTON	8.8
*Manchester	same as		"	8.8
Salem	same as		"	8.9
*Marblehead	same as		"	9.1
Lynn, Lynn Harbor	same as		"	9.2
Neponset, Neponset R.	same as		"	9.5
Weymouth, Fore River Bridge	0 10	after	"	9.5
Hingham	0 10	"	"	9.5
Hull	0 05	"	"	9.3
Cohasset Harbor (White Head)	0 05	"	"	8.8
Scituate, Scituate Harbor	0 05	"	"	8.9
Cape Cod Bay				
Duxbury Hbr. .... **high 0 05** after, low 0 35		"	"	9.9
Plymouth	0 05	"	"	9.8
Cape Cod Canal, East Entrance	same as		"	8.7
Barnstable Harbor, Beach Point	0 10	after	"	9.5
Wellfleet	0 15	"	"	10.0
Provincetown	0 15	"	"	9.1
Cape Cod				
Chatham Hbr, Aunt Lydias Cove **high 1 10** after, low 1 55		"	"	4.6
Pleasant Bay.............................. **high 2 30** after, low 3 25		"	"	3.2
Nantucket Sound				
Stage Harbor	1 00	"	"	3.9
Wychmere Harbor....................... **high 0 50** after, low 0 25		"	"	3.7
Dennisport .... **high 1 05** after, low 0 40		"	"	3.4
South Yarmouth, Bass River	1 50	"	"	2.8
Hyannis Port............................... **high 1 05** after, low 0 30		"	"	3.1
Cotuit Highlands........................ **high 1 15** after, low 0 45		"	"	2.5
Falmouth Heights	0 15	before	"	1.3
Nantucket Island				
Nantucket	1 05	after	"	3.0
Great Point	0 45	"	"	3.1
Muskeget Island, North side	0 25	"	"	2.0

PORTLAND Tables, pp. 30-35, BOSTON Tables, pp. 38-43
When a high tide exceeds av. ht., the *following* low tide will be lower than av.
*Times and Hts. are approximate. *Important*: see NOTE, top p. 12, and NOAA Article, p. 10.

TIME OF HIGH WATER

Time figures shown are the *average* differences throughout the year. Rise in feet is mean range.
(Low Water times are given *only* when they vary more than 20 min. from High Water times.)

	H.M.			Rise in feet
MASSACHUSETTS, Martha's Vineyard				
Edgartown (Caution: new data needed but not yet available)				
Oak Bluffs...................**high 0 30** *after, low* ...0 10		*before*	*BOSTON*	1.7
Vineyard Haven.......................**high 0 25** *after, low* .. *same as*			"	1.7
Lake Tashmoo (inside).....................2 30		*before*	"	2.0
Cedar Tree Neck ....**high 0 10** *after, low* 1 30		*after*	*NEWPORT*	2.2
Menemsha Bight........................**high** *same as, low* 0 35		"	"	2.7
Gay Head.....................................**high 0 05** *before, low*..0 45		"	"	2.9
Squibnocket Point.......................**high 0 45** *before, low* *same as*			"	2.9
Wasque Point ....**high 2 00** *after, low* 3 20		*after*	"	1.1
Nomans Land ....**high 0 20** *before, low*..0 20		"	"	3.0
Vineyard Sound North Side				
Little Hbr., Woods Hole................**high 0 30** *after, low* 2 20		"	"	1.4
Quick's Hole, N. side 0 10		*before*	"	3.5
Cuttyhunk...1 20		*after*	"	3.4
Buzzards Bay				
Cuttyhunk Pond Entr................................*same as*			"	3.4
W. Falmouth Harbor, Chappaquoit Pt.............................0 05		*after*	"	3.8
Pocasset Hbr., Barlows Landing 0 25		"	"	4.0
Monument Beach...0 15		"	"	4.0
Wareham River...0 20		"	"	4.1
Great Hill..0 10		"	"	4.0
Marion, Sippican Harbor............................0 10		"	"	4.0
Mattapoisett Harbor....................................0 10		"	"	3.9
Clarks Point...0 15		"	"	3.6
New Bedford..0 05		"	"	3.7
South Dartmouth 0 25		"	"	3.7
Westport Harbor.........................**high 0 10** *after, low* 0 35		*after*	"	3.0
RHODE ISLAND & MASS, Narragansett Bay				
Sakonnet..0 10		*before*	*NEWPORT*	3.2
Beavertail Point..0 05		"	"	3.3
Conanicut Point..0 05		*after*	"	3.8
Prudence Island (south end)................................0 10		"	"	3.8
Bristol Harbor 0 15		"	"	4.1
Fall River, MA...0 20		"	"	4.4
Providence, State Pier no. 1...............................0 15		"	"	4.4
Pawtucket, Seekonk River 0 20		"	"	4.6
East Greenwich 0 15		"	"	4.0
Wickford..0 05		"	"	3.7
Narragansett Pier.......................**high 0 10** *before, low*..0 10		*after*	"	3.2
RHODE ISLAND, Outer Coast				
Pt. Judith, Harbor of Refuge.......**high** *same as, low*........0 35		*after*	*NEWPORT*	3.0
Block Island, Old Harbor............**high 0 15** *before, low*..0 15		"	"	2.9
Watch Hill Pt. ....**high 0 40** *after, low* 1 15		"	"	2.6
CONNECTICUT, L.I. Sound				
Stonington 2 15		*before*	*BRIDGEPORT*	2.7
Noank...2 05		"	"	2.3
New London, Thames River.................................1 45		"	"	2.6
Norwich, Thames River..1 20		"	"	3.0

NEWPORT Tables, pp. 78-83, BRIDGEPORT Tables, pp. 98-103

When a high tide exceeds av. ht., the *following* low tide will be lower than av.

*Times and Hts. are approximate. *Important*: see NOTE, top p. 12, and NOAA Article, p. 10.

TIDE STATIONS

TIME OF HIGH WATER

Time figures shown are the *average* differences throughout the year. Rise in feet is mean range.
(Low Water times are given *only* when they vary more than 20 min. from High Water times.)

	H.M.			Rise in feet
CONNECTICUT, L.I. Sound, Cont.				
Saybrook Jetty, Connecticut River	0 35	before	BRIDGEPORT	3.5
Essex, Connecticut River	0 05	"	"	3.0
Madison	0 20	"	"	4.9
Branford	0 05	"	"	5.9
New Haven Harbor, New Haven Reach	same as		"	6.2
Milford Harbor	same as		"	6.3
Stratford, Housatonic River, Sniffens Point.	0 10	after	"	6.4
South Norwalk	0 10	"	"	7.1
Stamford	0 05	"	"	7.2
Cos Cob Harbor	0 05	"	"	7.2
*Greenwich	same as		"	7.4
NEW YORK, Long Island Sound, North Side				
Rye Beach	0 20	before	KINGS POINT	7.3
New Rochelle	0 15	"	"	7.3
Throgs Neck	0 10	after	"	7.0
Whitestone, East River	0 05	"	"	7.1
College Point, Flushing Bay	0 15	"	"	6.8
Hunts Point, East River	0 15	"	"	7.0
North Brother Island, East River	0 20	"	"	6.6
Port Morris, Stony Pt., East River	0 05	"	"	6.2
NEW YORK, Long Island, Long Island Sound				
Willets Point	same as		KINGS POINT	7.2
Port Washington, Manhasset Bay	0 10	before	"	7.3
Glen Cove, Hempstead Harbor	0 20	"	"	7.3
Oyster Bay Harbor, Oyster Bay	0 05	after	BRIDGEPORT	7.3
Cold Spring Harbor, Oyster Bay	0 05	before	"	7.3
Eatons Neck Point	same as		"	7.1
Lloyd Harbor, Huntington Bay	same as		"	7.0
Northport, Northport Bay	0 05	before	"	7.3
Port Jefferson Harbor Entrance	same as		"	6.6
Mattituck Inlet	0 05	after	"	5.2
Shelter Island Sound				
Orient Point	1 10	before	"	2.5
Greenport	0 40	"	"	2.4
Southold	same as		"	2.3
Sag Harbor	0 45	before	"	2.5
New Suffolk	0 40	after	"	2.6
South Jamesport, Peconic Bay	0 40	"	"	2.8
Three Mile Harbor	1 05	before	"	2.5
Montauk Harbor Entr.	2 10	"	"	2.0
Long Island, South Shore				
Shinnecock Inlet, Ocean........ **high 0 40** before, low..1 05		before	SANDY HOOK	3.3
Moriches Inlet	1 00	"	"	2.9
Democrat Point, Fire Island Inlet	0 40	"	"	2.6
Patchogue, Great South Bay	3 15	after	"	1.1
Bay Shore, Watchogue Creek Entrance	2 15	"	"	1.0
Jones Inlet (Point Lookout)	0 20	before	"	3.6
Bellmore, Hempstead Bay.......... **high 1 30** after, low 2 00		after	"	2.0

BRIDGEPORT Tables, pp. 98-103, KINGS POINT Tables, pp. 104-109, SANDY HOOK Tables, pp. 134-139

When a high tide exceeds av. ht., the *following* low tide will be lower than av.
*Times and Hts. are approximate. *Important*: see NOTE, top p. 12, and NOAA Article, p. 10.

TIME OF HIGH WATER

Time figures shown are the *average* differences throughout the year. Rise in feet is mean range.
(Low Water times are given *only* when they vary more than 20 min. from High Water times.)

	H.M.			Rise in feet
NEW YORK, Long Island, South Shore, Cont.				
Freeport, Baldwin Bay	0 40	after	SANDY HOOK	3.0
E. Rockaway Inlet	0 10	before	"	4.1
Barren Is., Rockaway Inlet, Jamaica Bay	same as		"	5.0
NEW YORK & NEW JERSEY				
New York Harbor				
Coney Island	0 05	before	SANDY HOOK	4.7
Fort Hamilton, The Narrows	same as		"	4.7
Tarrytown, Hudson River	1 50	after	BATTERY	3.2
Poughkeepsie, Hudson River	4 35	"	"	3.1
Kingston, Hudson River	5 20	"	"	3.7
NY & NJ, the Kills and Newark Bay				
Constable Hook, Kill Van Kull	0 20	before	"	4.6
Port Elizabeth	same as		"	5.1
Bellville, Passaic River .... **high 0 10** *after, low*	0 50	after	"	5.6
Kearny Pt., Hackensack River.	0 10	"	"	5.2
Hackensack, Hackensack River	1 05	"	"	6.0
Lower NY Bay, Raritan Bay				
South Amboy, Raritan River	0 05	before	SANDY HOOK	5.1
Great Kills Harbor	0 05	after	"	4.7
New Brunswick, Raritan River	0 30	"	"	5.7
Keyport	0 05	before	"	5.0
Atlantic Highlands, Sandy Hook Bay	0 10	"	"	4.7
Highlands, Shrewsbury R., Rte. 36 bridge, Sandy Hook	0 15	after	"	4.2
Red Bank, Navesink River, Sandy Hook Bay	1 20	"	"	3.5
Sea Bright, Shrewsbury River, Sandy Hook Bay	1 15	"	"	3.2
NEW JERSEY, Outer Coast				
Shark River, R.R. Bridge, Shark River Island	0 15	before	"	4.3
Manasquan Inlet, USCG Station	0 10	"	"	4.0
Brielle, Rte. 35 bridge, Manasquan River	0 05	"	"	3.9
Barnegat Inlet,USCG Station, Barnegat Bay	0 10	"	"	2.2
Manahawkin Drawbridge .... **high 2 50** *after, low*	3 40	after	"	1.3
Beach Haven, USCG Station, Little Egg Harbor	1 20	"	"	2.2
Absecon Creek, Rte. 30 bridge	1 05	"	"	3.9
Atlantic City, Ocean	0 25	before	"	4.0
Great Egg Hbr. Bay, Beesleys Pt. .. **high 0 30** *after, low*	1 05	after	"	3.6
Townsends Inlet, Ocean Dr. bridge	0 10	"	"	3.9
Hereford Inlet, Stone Harbor, Great Channel	0 35	"	"	4.0
Cape May Harbor, Cape May Inlet	0 10	"	"	4.5
NEW JERSEY & DELAWARE BAY				
Delaware Bay, Eastern Shore				
Brandywine Shoal Light .... **high 0 20** *after, low*	1 10	after	BATTERY	5.1
Cape May Point, Sunset Beach	0 15	"	"	4.8
Dennis Creek Entr., 2.5 mi. above .... **high 1 15** *after, low*	2 05	"	"	5.2
Mauricetown, Maurice R.	2 40	"	"	4.4
Millville, Maurice R.	3 55	"	"	5.0

BATTERY Tables, pp. 122-127, SANDY HOOK Tables, pp. 134-139

When a high tide exceeds av. ht., the *following* low tide will be lower than av.
*Times and Hts. are approximate. *Important*: see NOTE, top p. 12, and NOAA Article, p. 10.

TIDE STATIONS

TIME OF HIGH WATER

Time figures shown are the *average* differences throughout the year. Rise in feet is mean range.
(Low Water times are given *only* when they vary more than 20 min. from High Water times.)

	H.M.			Rise in feet
NEW JERSEY & DELAWARE BAY, Cont.				
Delaware Bay, Western Shore				
*Cape Henlopen..0 10	after	BATTERY		4.1
Lewes (Breakwater Harbor)........ **high 0 20** *after, low* 0 45	"	"		4.1
*St. Jones River Ent................... **high 1 10** *after, low* 1 55	"	"		4.8
Delaware River				
*Liston Point, Delaware 2 05	"	"		5.7
Salem, Salem River, NJ 3 55	"	"		4.2
Reedy Point, Delaware................. **high 3 05** *after, low* 3 25	"	"		5.3
C&D Summit Bridge, Delaware 2 35	after	"		3.5
Chesapeake City, MD 2 20	"	"		2.9
New Castle, Deleware.................. **high 3 35** *after, low* 4 05	"	"		5.2
Wilmington Marine Terminal..... **high 3 55** *after, low* 4 30	"	"		5.3
DELAWARE, MARYLAND & VIRGINIA				
Indian River Inlet, USCG Station, Delaware 0 55	after	SANDY HOOK		2.5
Ocean City Fishing Pier.......................................0 20	before	"		3.4
Harbor of Refuge, Chincoteague Bay 0 10	after	"		2.4
Chincoteague Channel, south end 0 20	"	"		2.2
Chincoteague Island, USCG Station....................0 40	"	"		1.6
Metomkin Inlet 0 40	"	"		3.6
Wachapreague Channel 0 50	"	"		4.0
*Quinby Inlet Entrance......................................0 05	"	"		4.0
Great Machipongo Inlet, inside 0 45	"	"		3.9
Chesapeake Bay, Eastern Shore				
Cape Charles Harbor 0 40	after	BATTERY		2.3
Crisfield, Little Annemessex River 4 30	"	"		1.9
Salisbury, Wicomico River 7 15	"	"		3.0
Hooper Island, middle 4 40	before	BALTIMORE		1.5
Taylors Island, Little Choptank River................3 15	"	"		1.3
*Sharps Is. Lt. 3 50	"	"		1.3
Cambridge, Choptank River 2 40	"	"		1.6
Dover Bridge, Choptank River............................0 20	"	"		1.7
Oxford, Tred Avon River......................................2 50	"	"		1.4
Easton Pt., Tred Avon River................................2 45	"	"		1.6
St. Michaels, Miles River.....................................2 10	"	"		1.4
Kent Island Narrows..1 30	"	"		1.2
*Bloody Pt. Bar Lt..2 40	"	"		1.1
Worton Creek Entrance 1 20	after	"		1.3
Town Point Wharf, Elk River 3 20	"	"		2.2
Chesapeake Bay, Western Shore				
Havre de Grace, Susquehanna River......................3 15	after	"		1.9
*Pooles Is..0 55	"	"		1.2
Annapolis, Severn River (US Naval Academy) 1 30	before	"		1.0
*Sandy Point 1 20	"	"		0.8
Thomas Pt. Shoal Lt...1 55	"	"		0.9
*Drum Point, Pawtuxent River 4 50	"	"		1.2
Solomons Island, Pawtuxent River......................4 40	"	"		1.2
Point Lookout..5 30	"	"		1.2
Sunnybank, Little Wicomico River 6 35	after	BATTERY		0.8
Glebe Point, Great Wicomico River 4 10	"	"		1.2

BATTERY Tables, pp. 122-127, SANDY HOOK Tables, pp. 134-139, BALTIMORE Tables, pp. 156-159

When a high tide exceeds av. ht., the *following* low tide will be lower than av.

*Times and Hts. are approximate. *Important*: see NOTE, top p. 12, and NOAA Article, p. 10.

TIME OF HIGH WATER

Time figures shown are the *average* differences throughout the year. Rise in feet is mean range.
(Low Water times are given *only* when they vary more than 20 min. from High Water times.)

	H.M.			Rise in feet
DELAWARE, MARYLAND & VIRGINIA, Cont.				
Windmill Point, Rappahannock River	2 45	after	BATTERY	1.2
*Orchard Point, Rappahannock River	3 20	"	"	1.4
*New Point Comfort, Mobjack Bay	0 45	"	"	2.3
Tue Marshes Light, York River	0 55	"	"	2.2
*Perrin River, York River	1 05	"	"	2.3
Goodwin Neck, York River	1 10	"	"	2.2
Hampton Roads	**high 0 50** after, low 0 40	"	"	2.4
Norfolk, Elizabeth River	1 15	"	"	2.8
Newport News, James River	1 20	"	"	2.6
Jamestown Is., James River	**high 3 55** after, low 4 15	"	"	2.0
*Windmill Pt., James River	**high 6 15** after, low 6 30	"	"	2.3
Chesapeake Bay Br. Tunnel	**high 0 05** before, low 0 20	before	"	2.6
Cape Henry	0 05	"	"	2.8
NORTH CAROLINA				
Roanoke Sound Channel	1 10	after	BATTERY	0.5
Oregon Inlet Marina	**high 0 20** before, low 0 10	before	"	0.9
Oregon Inlet Channel	0 30	"	"	1.2
Oregon Inlet, USCG Station	**high 0 40** before, low 1 00	"	"	1.9
Cape Hatteras Fishing Pier	1 00	"	"	3.0
Hatteras Inlet	0 50	"	"	2.0
Ocracoke Inlet	0 50	"	"	1.9
Beaufort Inlet Channel Range	0 55	"	"	3.2
Morehead City	0 35	"	"	3.1
Bogue Inlet	0 45	"	"	2.2
New River Inlet	0 45	"	"	3.0
New Topsail Inlet	**high 0 40** before, low 0 10	"	"	3.0
Masonboro Inlet	0 35	"	"	3.8
Bald Head, Cape Fear River	1 15	"	"	4.5
Wilmington	**high 1 20** after, low 1 45	after	"	4.3
Lockwoods Folly Inlet	0 55	"	"	4.2
SOUTH CAROLINA				
Little River Neck, north end	1 20	after	BATTERY	4.7
Hog Inlet Pier	0 45	"	"	5.0
Myrtle Beach, Springmaid Pier	0 50	"	"	5.0
Pawleys Island Pier (ocean)	0 55	"	"	4.9
Winyah Bay Entrance, south jetty	0 45	before	"	4.6
South Island Plantation, C.G. Station	0 10	after	"	3.8
Georgetown, Sampit River	**high 1 00** after, low 1 40	"	"	3.7
North Santee River Inlet	0 35	before	"	4.5
Charleston (Custom House)	0 25	"	"	5.2
Folly River, north, Folly Island	**high** same as, low 0 35	"	"	5.4
North Edisto River, Rockville, Bohicket Creek	0 05	"	"	5.8
South Edisto River Marina, Big Bay Creek entr.	0 20	"	"	6.0
Harbor River Bridge, St. Helena Sound	0 10	"	"	6.1
Hutchinson Island, Ashepoo River, St. Helena Sound	0 15	after	"	6.0
Fripps Inlet, Hunting Island Bridge, St. Helena Sound	0 25	before	"	6.1
Hilton Head Is., Port Royal Plantation	0 15	"	"	6.1
Battery Creek, Port Royal Sound, 4 mi. above entr.	**high 1 00** after, low 0 15	after	"	7.7

BATTERY Tables, pp. 122-127

When a high tide exceeds av. ht., the *following* low tide will be lower than av.

*Times and Hts. are approximate. *Important*: see NOTE, top p. 12, and NOAA Article, p. 10.

TIDE STATIONS

19

TIME OF HIGH WATER

Time figures shown are the *average* differences throughout the year. Rise in feet is mean range.
(Low Water times are given *only* when they vary more than 20 min. from High Water times.)

	H.M.			Rise in feet
SOUTH CAROLINA, Cont.				
Beaufort, Beaufort River .... **high 0 55** *after, low* 0 30	after	BATTERY	7.4
Braddock Point, Hilton Head Island................................0 10	before	"	6.7
GEORGIA			
Savannah River Entrance, Fort Polasky 0 15	before	BATTERY	6.9
Tybee Creek Entrance..0 25	"	"	6.8
Wilmington River, north entrance..................................0 25	after	"	7.6
Isle of Hope, Skidaway River ...... **high 0 35** *after, low* 0 10	"	"	7.8
Egg Islands, Ossabaw Sound 0 10	before	"	7.2
Walburg Creek Entr., St. Catherines Sd...........................same as	"	7.1
Blackbeard Island..0 05	after	"	6.9
Blackbeard Creek, Blackbeard Island..............................0 05	"	"	6.5
Sapelo Island, Doboy Sound, Old Tower. ....same as	"	6.8
Threemile Cut Entrance, Darien River.............................0 30	after	"	7.1
St. Simons Sound Bar...0 15	before	"	6.5
Frederica River, St. Simons Sound................................0 35	after	"	7.2
Brunswick, East River, St. Simons Sound 0 45	"	"	7.2
Jekyll Is. Marina, Jekyll Creek, St. Andrew Sound 0 30	"	"	6.8
Cumberland Wharf, Cumberland River 0 30	after	"	6.8
FLORIDA, East Coast			
St. Mary's Entrance, n. jetty......................................0 10	before	BATTERY	5.8
Fernandina Beach, Amelia R...... **high 0 30** *after, low* 0 05	after	"	6.0
Nassau River Entrance............... **high 0 10** *after, low* 0 50	"	"	5.1
Amelia City, South Amelia River...................................0 50	"	"	5.4
Mayport....................................... **high 0 15** *after, low* 0 15	before	"	4.6
St. Augustine, City Dock...0 10	after	"	4.5
Ponce Inlet, Halifax River .... **high 0 05** *after, low* 0 30	after	MIAMI	2.8
Cape Canaveral......................... **high 1 05** *before, low*..0 45	before	"	3.5
Port Canaveral, Trident Pier......................................same as	"	3.5
Sebastian Inlet bridge................ **high 0 50** *before, low*..0 25	before	"	2.2
St. Lucie....................................... **high 0 40** *after, low* 1 45	after	"	1.0
Vero Beach, ocean 0 55	before	"	3.4
Fort Pierce Inlet, south jetty....................................0 30	"	"	2.6
Stuart, St. Lucie River .... **high 2 15** *after, low* 3 30	after	"	0.9
Jupiter Inlet, south jetty..0 10	before	"	2.5
North Palm Beach, Lake Worth .. **high 0 20** *before, low*..0 15	after	"	2.8
Port of Palm Beach, Lake Worth . **high 0 20** *before, low*..0 05	"	"	2.7
Lake Worth Pier, ocean .... **high 0 45** *before, low*..0 20	before	"	2.7
Hillsboro Inlet, C.G. Station 0 15	"	"	2.5
Hillsboro Inlet Marina................ **high 0 05** *before, low*..0 25	after	"	2.5
Lauderdale-by-the-Sea, fish pier **high 0 35** *before, low*....0 15	before	"	2.6
Bahia Mar, Yt. Club .... **high 0 05** *before, low*..0 35	after	"	2.4
Port Everglades, turning basin ... **high 0 30** *before, low*..0 10	before	"	2.5
North Miami Beach, fishing pier **high 0 20** *before, low*..same as	"	2.5
Miami Harbor Entrance 0 20	"	"	2.5
Miami Marina, Biscayne Bay..... **high 0 20** *after, low* 0 50	after	"	2.2
Dinner Key Marina, *Biscayne Bay* **high 0 55** *after, low* 1 50	"	"	1.9
Key Biscayne Yt. Club, *Biscayne B* **high 0 45** *after low*....1 30	"	"	2.0
Ocean Reef Hbr., Key Largo .... **high 0 10** *before, low*..0 15	"	"	2.3
Tavernier Harbor, Hawk Ch........ **high 0 05** *after, low* 0 25	"	"	2.0
Key West...0 50	before	BOSTON	1.3

BOSTON Tables, pp. 38-43, BATTERY Tables, pp. 122-127, MIAMI Tables, pp. 160-163

When a high tide exceeds av. ht., the *following* low tide will be lower than av.
*Times and Hts. are approximate. *Important*: see NOTE, top p. 12, and NOAA Article, p. 10.

Why Tides and Currents Often Behave Differently
Frequently Asked Questions

We are often asked such questions as, **"Why are the times of high water and current change not the same?"** Shouldn't an ebb current begin right after a high tide? Although tides (vertical height of water) and currents (horizontal movement) are inextricably related, they often behave rather differently.

If the Earth had a uniform seabed and no land masses, it is likely that a high tide at one point would occur simultaneously with a change in the current direction. However, the existence of continents, a sea bottom which is anything but uniform, and the great ocean currents and different prevailing winds around the world, make the picture extremely complex.

As one example of how a time of high tide can differ greatly from the time of a current change, see the Relationship of High Water and Ebb Current, p. 155. Picture a fjord or long indentation into the coastline, with a narrow opening to the ocean. When a flood current is reaching its peak, or the tide is high outside the mouth of this fjord, the fjord is still filling, unable to keep pace with conditions on the outer coast.

Why do the heights of tides differ so much from one place to the next? Turn to Time of High Water at various ports, pp. 12-20, and compare the Rise in Feet of tides for Nova Scotia's outer coast (2.6 to 4.8 feet) to those for the Bay of Fundy (just below), with a range of up to 35.6 feet. Why the difference? The answer is geography, both above and below water. Tidal ranges of points out on the edge of an outer coast (Nantucket, for instance) tend to be moderate, while estuaries and deep bays with narrowing contours often experience a funneling effect which exaggerates the tidal range. Another explanation is proximity to the continental shelf: the closer a port is to the shelf, the more likely it is to experience a lower tidal range; the farther from the shelf, the more likely it is that a harbor is subject to surges, as when a wave crest hits the shallow water at a beach.

There are other anomalies between tides and currents. **Do stronger currents indicate higher tides?** Woods Hole, MA often has very strong currents through its narrow passage, sometimes as much as 7 knots, but the tidal range is less than 2 feet. Conversely, Boston Harbor has a mean tidal range of about 9.6 feet, but the average currents at the opening, between Deer Island and Hull, do not exceed 2 knots. There is no necessary correlation between current strength and range of tide.

Why did the tidal or current prediction in ELDRIDGE differ from what I saw? Unless there was an error in the Government tables we take our data from, the answer is either (1) weather-related, as when a storm either retards or advances a tidal event, or (2) the discrepancy is small enough to be explained by the approximate nature of tide and current predictions, and figures are sometimes rounded off. We appreciate hearing from readers of any observed discrepancies or errors. Call us at 617-482-8460, Monday to Friday, 9 a.m. to 5 p.m.

TIME OF CURRENT CHANGE
(See Note at bottom of Boston Tables, pp. 38-43: Rule-of-Thumb for Current Velocities.)

> NOTE: NOAA has recently dropped many substations from its listing because the data are judged to be of questionable accuracy.
>
> see NOAA Article p. 10, *A Cautionary Word on Using Tide and Current Data.*

CURRENTS IN THE GULF OF MAINE - In the Gulf of Maine, on the western side, the Flood Current splits at Cape Ann, Mass., and floods north and east along the shore towards the Bay of Fundy. At the same time, on the eastern side of the Gulf, at the southern tip of Nova Scotia, the Flood Current runs to the west and then north and eastwards along the shore into the Bay of Fundy. The Ebb Current is just the reverse. In addition to these large principal currents, along the Maine Coast, at least at the mouths of principal bays, there is a shoreward set during the Flood and an offshore set during the Ebb, although this set is of considerably less velocity.

West of Mount Desert, the average along-shore current is rarely more than a knot but the farther east one goes, the greater are the average velocities to be expected, up to 2 knots or more. When heading west, therefore, start off at the time shown for High Water in your area (see p. 13) and have a fair Ebb current for 6 hours. Headed east, start at the time for Low Water in your area (about 6 ½ hours after High Water) and carry the beneficial Flood current. East of Schoodic Point, the average currents are up to 2 knots and taking advantage of them will save considerable time and fuel.

Off shore, in the Gulf of Maine, unlike the along-shore currents that come to dead slack and _reverse,_ there are so-called _rotary_ currents. These currents constantly change direction in a clockwise flow completing the circle in about 12 ½ hours. The maximum currents are when it is flooding in the northeasterly direction or ebbing in a southwesterly direction; minimum currents occur halfway between. There is no slack water.

Entering the Bay of Fundy through Grand Manan Channel, one finds that the average velocities are from 1-2 ½ knots, although in the narrower channels off the Bay, velocities are higher (Friar Roads at Eastport has average velocities of 3 knots of more). The Current in the Bay Floods to the Northeast and Ebbs to the Southwest.

In using this table, bear in mind that **actual times of Slack or Maximum occasionally differ from the predicted times** by as much as half an hour and in rare instances as much as an hour. Referring the Time of Current Change at the subordinate stations listed below, to the predicted Current Change at the reference station gives the _approximate_ time only. Therefore, to make make sure of getting the full advantage of a favorable current or slack water, the navigator should reach the entrance or strait at least half an hour before the predicted time. (This is basically the same precautionary note found in the U.S. Tidal Currents Table Book.)

Figures shown below are **average maximum** velocities in knots. We have omitted places having an average maximum velocity of less than 1 knot. To find the Time of Current Change (Start of Flood and Start of Ebb) at a selected point, refer to the table heading that particular section (in bold type) and add or subtract the time listed.

	TIME DIFFERENCES Flood Starts; Ebb Starts H.M.	MAXIMUM FLOOD Dir.(true) in degrees	Av. Max. in knots	MAXIMUM EBB Dir.(true) in degrees	Av. Max. in knots
MAINE COAST – based on Boston, pp. 38-43					
(Fl. starts at Low Water; Ebb starts at High Water)					
Isle Au Haut, 0.8 mi. E of Richs Pt.	-0 05	335	1.4	140	1.5
Damariscotta R., off Cavis Pt.	F+1 15, E+0 05	350	0.6	215	1.0
Sheepscot R., off Barter Is.	F+1 15, E+0 15	005	0.8	200	1.1
Lowe Pt., NE of, Sasanoa R.	F+1 15, E+0 45	325	1.7	150	1.8
Lower Hell Gate, Knubble Bay*	F+1 40, E+0 45	290	3.0	155	3.5

* Velocities up to 9.0 kts. have been observed in the vicinity of the Boilers.

Important: see NOTE, bottom p. 29 and NOAA Article, p. 10.

TIME OF CURRENT CHANGE
(See Note at bottom of Boston Tables, pp. 38-43: Rule-of-Thumb for Current Velocities.)

	TIME DIFFERENCES Flood Starts; Ebb Starts H.M.	MAXIMUM FLOOD Dir.(true) in degrees	Av. Max. in knots	MAXIMUM EBB Dir.(true) in degrees	Av. Max. in knots
KENNEBEC RIVER – based on Boston, pp. 38-43					
(Fl. starts at Low Water; Ebb starts at High Water)					
Hunniwell Pt., NE of	F+2 10, E+1 35	330	2.4	150	2.9
Bald Head, 0.3 mi. SW of	F+2 30, E+1 25	320	1.6	155	2.3
Bluff Head, W of	F+2 40, E+1 55	015	2.3	185	3.4
Fiddler Ledge, N of	F+2 50, E+1 50	265	1.9	115	2.6
Doubling Pt., S of	F+1 45, E+1 55	300	2.6	125	3.0
Lincoln Ledge, E. of	F+2 40, E+1 55	000	1.9	175	2.8
Bath, 0.2 mi. S of bridge	F+2 35, E+2 15	005	1.0	175	1.5
CASCO BAY – based on Boston, pp. 38-43					
(Fl. starts at Low Water; Ebb starts at High Water)					
Broad Sound, W. of Eagle Is.	F+0 50, E+0 05	010	0.9	170	1.3
Hussey Sound, SW of Overset Is.	+0 25	310	1.1	150	1.1
Portland Hbr. Ent., SW of Cushing Is.	+0 15	320	1.0	155	1.1
Portland Bridge, Center of draw	+0 25	220	0.6	045	0.5
PORTSMOUTH HARBOR – based on Boston, pp. 38-43					
(Fl. starts at Low Water; Ebb starts at High Water)					
Kitts Rocks, 0.2 mi. W of	F+2 05, E+1 30	325	0.8	175	1.6
Portsmouth Hbr. entr., off Wood Is.	F+2 05, E+1 30	355	1.2	195	1.8
Fort Point	F+2 10, E+1 30	350	1.5	130	2.0
Seavey Is., N of	F+2 20, E+1 45	260	1.4	080	1.8
Clarks Is., S of	+2 15	270	1.6	085	2.3
Seavey Is., S of	+2 20	260	3.0	090	3.8
Henderson Pt., W of	F+2 35, E+2 00	340	2.6	170	2.3
Off Gangway Rock	F+2 35, E+2 00	280	2.1	110	3.0
Badgers Is., SW of	F+2 35, E+2 00	330	3.3	125	3.7
MASSACHUSETTS COAST – based on Boston, pp. 38-43					
(Fl. starts at Low Water; Ebb starts at High Water)					
Merrimac River entr.	+1 35	285	2.2	105	1.4
Newburyport	F+1 25, E+2 05	290	1.5	100	1.4
Plum Is. Sound entr.	F+0 30, E+1 10	315	1.6	185	1.5
Gloucester Hbr., Blynman Canal entr.	-0 05	310	3.0	130	3.3
Hypocrite Channel	F+0 10, E+1 10	260	0.9	070	1.0
BOSTON HARBOR – based on Boston, pp. 38-43					
(Fl. starts at Low Water; Ebb starts at High Water)					
Pt. Allerton, 0.4 mi. NW.	F-0 15, E+0 35	265	0.7	080	0.8
Deer Island Lt.	F-0 05, E+0 18	255	1.1	110	1.2
Nantasket Rds					
Hull Gut	+0 15	165	1.2	350	1.8
West Head (West Gut) 0.2mi. SW	F-0 10, E+1 25	165	1.4	322	1.4
Weir R. Ent., Worlds End, N of	F+0 15, E+1 05	075	0.7	270	0.8
Bumkin Is., 0.4mi. W. of	F-0 20, E+0 45	195	0.5	305	0.3
Weymouth Back R., betw. Grape I. and Lower Neck	F-0 20, E+0 30	095	0.7	280	0.9
CAPE COD BAY – based on Boston, pp. 38-43					
(Fl. starts at Low Water; Ebb starts at High Water)					
Race Point, 7 mi. N of	F-0 05, E+0 20	290	1.5	–	1.5
Race Point, 1 mi. NW of	F-0 10, E+0 15	225	1.0	060	0.9
Barnstable Harbor	F+0 15, E+0 40	190	1.2	005	1.4
Manomet Point	F+0 00, E+0 25	155	1.1	010	0.9
Gurnet Point, 1 mi. E of	F-0 10, E+0 15	250	1.4	–	1.0
Farnham Rock, 1 mi. E of	F-0 25, E-0 00	180	1.1	010	0.9

Important: see NOTE, bottom p. 29 and NOAA Article, p. 10.

CURRENT STATIONS

TIME OF CURRENT CHANGE

(See Note at bottom of Boston Tables, pp. 38-43: Rule-of-Thumb for Current Velocities.)

	TIME DIFFERENCES Flood Starts; Ebb Starts H.M.	MAXIMUM FLOOD Dir.(true) in degrees	Av. Max. in knots	MAXIMUM EBB Dir.(true) in degrees	Av. Max. in knots
NANTUCKET SOUND – based on Pollock Rip Channel, pp. 60-65					
Pollock Rip Channel, E end	-0 20	055	2.0	210	1.8
*POLLOCK RIP CHANNEL at Butler Hole - See table, pp. 60-65					
Monomoy Point, 0.2 mi. W of	+0 10	170	1.7	345	2.0
Halfmoon Shoal, 3.5 mi. E of	+1 10	090	1.1	295	1.0
Great Point, 0.5 mi. W of	F+0 25, E+1 15	030	1.1	195	1.2
Tuckernuck Shoal, off E end	+1 15	115	0.9	285	0.9
Nantucket Hbr. entr. chan.	F+3 20, E+2 45	170	1.2	350	1.5
Muskeget Is. chan., 1 mi. NE of	F+1 30, E+1 00	110	1.1	295	1.5
Muskeget Rock, 1.3 mi. SW of	+1 05	025	1.3	190	1.0
Muskeget Channel	+1 35	020	3.8	200	3.3
Betw. Long Shoal-Norton Shoal	+1 30	100	1.4	260	1.1
Cape Poge Lt., 1.7 mi. SSE of	+0 55	025	1.6	215	1.3
Cross Rip Channel	+1 50	090	1.3	270	0.9
Cape Poge, 3.2 mi. NE of	+2 35	095	1.6	300	1.2
Betw. Broken Gr.-Horseshoe Sh.	F+1 45, E+1 15	105	1.1	275	0.9
Point Gammon, 1.2 mi. S of	+1 10	105	1.1	260	1.0
Lewis Bay entr. chan.	+2 45	005	0.9	185	1.3
Betw. Wreck Shoal-Eldridge Shoal	+1 45	060	1.7	245	1.4
Hedge Fence Lighted Gong Buoy 22	+2 45	110	1.4	270	1.2
Betw. E. Chop-Squash Meadow	F+2 10, E+1 45	130	1.4	330	1.8
East Chop, 1 mi. N of	F+2 40, E+2 20	115	2.2	295	2.2
West Chop, 0.8 mi. N of	F+2 50, E+2 20	095	3.1	280	3.0
Betw. Hedge Fence-L'hommedieu Shoal	+2 15	105	2.1	275	2.2
Waquoit Bay Entr.	F+3 20, E+3 40	350	1.5	205	1.4
L'hommedieu Shoal, N of W end	+2 20	080	2.3	270	2.3
Nobska Point, 1.8 mi. E of	+2 05	065	2.3	240	1.7
VINEYARD SOUND – based on Pollock Rip Channel, pp. 60-65					
West Chop, 0.2 mi. W of	F+1 20, E+1 50	060	2.7	240	1.4
Nobska Point, 1 mi. SE of	+2 30	070	2.6	260	2.4
Norton Point, 0.5 mi. N of	+2 00	050	3.4	240	2.4
Tarpaulin Cove, 1.5 mi. E of	F+2 50, E+2 10	055	1.9	230	2.3
Robinson's Hole, 1.2 mi. SE of	F+2 30, E+2 10	060	1.9	240	2.1
Gay Head, 3 mi. N of	+2 05	075	1.1	255	1.2
Gay Head, 1.5 mi. NW of	+1 35	010	2.0	250	2.0
VINEYARD SOUND-BUZZARDS BAY – based on Cape Cod Canal, pp. 46-51					
WOODS HOLE - table, pp. 52-57					
Robinsons Hole, S end	+1 15	160	0.8	340	1.0
Robinsons Hole, Middle	F+1 30, E+1 00	145	2.8	315	2.9
Robinsons Hole, N end	F+1 55, E+0 50	160	1.0	340	1.2
Quicks Hole, S end	F+2 20, E+1 15	140	1.9	300	2.0
Quicks Hole, Middle	F+2 20, E+1 25	165	2.5	340	2.2
Quicks Hole, N end	F+2 40, E+1 45	165	2.0	000	2.6
Canapitsit Channel	F+2 05, E+1 00	155	2.6	310	1.7
BUZZARDS BAY – based on Pollock Rip Channel, pp. 60-65					
Westport River Entr.	F+0 10, E-0 25	290	2.2	110	2.5
Gooseberry Nk., 2 mi. SSE of Buzz Bay Entr.- *rotary current, no slack water. Av. max. 0.6 kts, approx. dir. 52° true at 3 1/2 hrs. after Fl. starts at Poll. Rip. Ave. max. 0.5 kts, approx. dir. 232° true 2 1/2 hrs. after Ebb starts at Poll. Rip.*					
Betw. Ribbon Reef-Sow &Pigs Rf.	F-0 20, E-2 45	060	0.8	235	1.2
Penikese Is., 0.8 mi. NW of	F-1 40, E-0 55	050	1.2	255	1.1
Betw. Gull Is.-Nashawena Is.	-2 10	090	0.9	245	1.1
Abiels Ledge, 0.4 mi. S of	F+0 25, E-0 05	035	0.8	215	1.0
Dumpling Rocks, 0.2 mi. SE of	-1 40	065	0.8	190	1.1
CAPE COD CANAL - table, pp. 46-51		070	4.0	250	4.5

*see Tidal Current Chart Buzzards Bay, Vineyard and Nantucket Sounds, pp. 66-77

Important: **see NOTE, bottom p. 29 and NOAA Article, p. 10.**

TIME OF CURRENT CHANGE

(See Note at bottom of Boston Tables, pp. 38-43: Rule-of-Thumb for Current Velocities.)

	TIME DIFFERENCES Flood Starts; Ebb Starts H.M.	MAXIMUM FLOOD Dir.(true) in degrees	Av. Max. in knots	MAXIMUM EBB Dir.(true) in degrees	Av. Max. in knots
**NARRAGANSETT BAY – based on Pollock Rip Channel, pp. 60-65					
Tiverton, Stone Bridge	F-3 00, E-2 25	010	2.7	190	2.7
Tiverton, RR Bridge	F-3 25, E-2 50	000	2.3	180	2.4
Castle Hill, W of East Passage	F-0 05, E-1 05	015	0.7	235	1.2
Bull Point, E of	-1 10	000	1.2	205	1.5
Rose Is., NE of	F-1 55, E-1 15	310	0.8	125	1.0
Rose Is., W of	F-0 40, E-1 20	000	0.7	170	1.0
Dyer Is., W of	-1 00	025	0.8	215	1.0
Mount Hope Bridge	-1 15	045	1.1	230	1.4
Kickamuit R., Mt. Hope Bay	F-2 05, E-1 20	000	1.4	190	1.7
Beavertail Point, 0.8 mi NW of	F-0 10, E-1 30	005	0.5	190	1.0
Betw. Dutch Is.-Beaver Head	-1 55	030	1.0	235	1.0
Dutch Is., W of	-1 25	015	1.3	205	1.2
Warren R., Warren	-0 20	000	1.0	170	0.9
India Pt. RR Bridge, Seekonk R.	-1 40	020	1.0	180	1.4
BLOCK ISLAND SOUND – based on The Race, pp. 86-91					
Pt. Judith Pond entr.	-3 05	350	1.8	185	1.5
Sandy Pt., Block Is. 1.5 mi N of	F-0 10, E-1 10	315	1.9	065	2.1
Lewis Pt., 1.0 mi. SW of	F-1 15, E-0 25	300	1.9	135	1.8
Lewis Pt., 1.5 mi. W of	F-1 20, E-0 50	320	1.4	170	1.7
Southwest Ledge	-0 20	320	1.5	140	2.1
Watch Hill Pt., 2.2 mi. E of	F-0 15, E+0 45	260	1.2	085	0.7
Montauk Pt., 1.2 mi. E of	F-1 10, E-0 40	345	2.8	160	2.8
Montauk Pt., 1 mi. NE of	F-1 50, E-1 15	355	2.4	145	1.9
Betw. Shagwong Reef-Cerberus Shoal	-0 20	240	1.9	055	1.8
Betw. Cerberus Sh.-Fishers Is.	F-0 45, E+0 05	265	1.3	095	1.3
Gardiners Is., 3 mi. NE of	-0 30	305	0.9	140	1.0
GARDINERS BAY etc. – based on The Race, pp. 86-91					
Goff Point, 0.4 mi. NW of	-1 30	225	1.2	010	1.6
Acabonack Hbr. entr., 0.6 mi. ESE of	-1 10	345	1.4	140	1.2
Gardiners Pt. Ruins, 1.1 mi. N of	-0 05	270	1.2	065	1.8
Betw. Gardiners Point-Plum Is.	F-0 05, E-0 35	288	1.4	100	1.6
Jennings Pt., 0.2 mi. NNW of	+0 40	290	1.6	055	1.5
Cedar Pt., 0.2 mi. W of	F+0 00, E+0 30	195	1.8	005	1.6
North Haven Peninsula, N of	+0 30	230	2.4	035	2.1
Paradise Pt., 0.4 mi. E of	+0 40	145	1.5	345	1.5
Little Peconic Bay entr.	+0 50	240	1.6	015	1.5
Robins Is., 0.5 mi. S of	+0 50	245	1.7	065	0.6
FISHERS ISLAND SOUND – based on The Race, pp. 86-91					
Napatree Point, 0.7 mi. SW of	-0 40	285	1.7	115	2.2
Little Narragansett Bay entr.	F-1 45, E-2 15	090	1.3	270	1.3
Ram Island Reef, S of	-0 45	255	1.3	090	1.6
LONG ISLAND SOUND – based on The Race, pp. 86-91					
*THE RACE (near Valiant Rock) – See pp. 86-91		300	2.7	110	3.0
Race Point, 0.4 mi. SW of	F-0 05, E-0 35	290	2.6	135	3.5
Little Gull Is., 0.5 mi. NE of	-0 20	000	3.3	105	3.1
Little Gull Is., 1.1 mi. ENE of	+0 10	300	4.0	130	4.7
Little Gull Is., 0.8 mi. NNW of	F+0 40, E-2 20	260	1.9	045	2.9
Great Gull Is., 0.7 mi. WSW of	-0 25	300	2.6	135	3.2
Goshen Pt., 1.9 mi. SSE of	-0 50	285	1.2	060	1.6
Bartlett Reef, 0.2 mi. S of	F-1 50, E-1 05	255	1.4	090	1.3
Twotree Is. Channel	-0 40	265	1.2	100	1.6

*see Tidal Current Chart Long Is. and Block Is. Sounds, pp. 92-97
**Floods somewhat unstable. Flood currents differing from predicted should be expected.

Important: see NOTE, bottom p. 29 and NOAA Article, p. 10.

CURRENT STATIONS

TIME OF CURRENT CHANGE
(See Note at bottom of Boston Tables, pp. 38-43: Rule-of-Thumb for Current Velocities.)

	TIME DIFFERENCES Flood Starts; Ebb Starts H.M.	MAXIMUM FLOOD Dir.(true) in degrees	Av. Max. in knots	MAXIMUM EBB Dir.(true) in degrees	Av. Max. in knots
LONG ISLAND SOUND – based on The Race, pp. 86-91					
Black Point, 0.8 mi. S of	-0 20	260	1.2	075	1.4
Betw. Black Pt.-Plum Is.	+0 40	235	2.1	075	2.4
Plum Is., 0.8 mi. NNW of	F+0 25,E-1 05	245	1.7	065	2.4
Plum Gut	-1 10	305	1.9	120	3.2
Hatchett Pt., 1.1 mi. WSW of	F-2 15, E-0 45	240	1.3	045	1.2
Saybrook Bkwtr., 1.5 mi. SE of	F-1 10, E-0 45	260	1.9	070	2.0
Conn. River RR Drawbridge	+0 55	000	1.0	200	1.0
Mulford Pt., 3.1 mi. NW of	+0 10	270	1.9	065	2.3
Cornfield Point, 3 mi. S of	F-0 45, E-0 10	255	2.0	095	1.7
Cornfield Point, 1.1 mi. S of	-0 45	295	1.4	110	1.6
Kelsey Point, 1 mi. S of	-1 15	250	2.0	120	1.5
Six Mile Reef, 2 mi. E of	-0 10	235	1.6	040	2.1
Sachem Head, 1 mi. SSE of	-0 20	255	1.1	065	1.0
New Haven Harbor entr.	-0 50	320	1.4	150	0.9
Housatonic R., Milford Pt., 0.2 mi. W of	+0 20	330	1.2	135	1.2
Point No Point, 2.1 mi. S of	-0 05	250	1.3	075	1.2
Port Jefferson Harbor entr.	+0 25	150	2.6	325	1.9
Crane Neck Point, 0.5 mi. NW of	F-0 35, E-1 45	255	1.3	015	1.5
Eatons Neck Pt., 1.3 mi. N of	F+0 40, E+0 15	285	1.4	075	1.4
Lloyd Point, 1.3 mi. NNW of	+1 30	255	1.0	055	0.9
EAST RIVER – based on Hell Gate, pp. 110-115					
Cryders Pt., 0.4 mi. NNW of	-0 30	110	1.3	285	1.1
College Pt. Rf., .25 mi. NW of	-0 30	075	1.5	260	1.4
Rikers Is. Chann. off La Guardia Field	+0 05	090	1.1	260	1.3
Hunts Point, SW of	0 00	110	1.7	280	1.3
S. Brother Is. NW of	-0 10	055	1.5	250	1.2
Off Winthrop Ave., Astoria	0 00	040	3.4	220	2.5
Mill Rock, NE of	-0 25	105	2.3	290	0.6
Mill Rock, W of	F-0 25, E-0 00	000	1.2	180	1.0
HELL GATE (off Mill Rock) – table, pp. 110-115		050	3.4	230	4.6
Roosevelt Is., W of, off 75th St.	-0 05	035	3.8	215	4.7
Roosevelt Is., E of, off 36th Ave.	-0 10	030	3.5	210	3.4
Roosevelt Is., W of, off 67th St.	+0 10	010	3.6	230	4.0
Off 19th St. (Pier 67)	-0 10	355	1.8	180	1.9
Williamsburg Br., 0.3 mi. N of	-0 05	020	2.7	220	2.9
Brooklyn Bridge, 0.1 mi. SW of	-0 10	045	2.9	220	3.5
*Buttermilk Channel	F-0 30, E+0 05	050	1.8	220	2.6
LONG ISLAND, South Coast – based on The Narrows, pp. 116-121					
Shinnecock Inlet	F+0 05, E-0 40	350	2.5	170	2.3
Fire Is. Inlet, 0.5 mi. S. of Oak Bch.	+0 15	080	2.4	245	2.4
Jones Inlet	-1 05	035	3.1	215	2.6
East Rockaway Inlet	F-1 35, E-1 10	040	2.2	225	2.3
JAMAICA BAY – based on The Narrows, pp. 116-121					
Rockaway Inlet entr.	-1 45	085	1.8	245	2.7
Barren Is., E of	F-1 50, E-0 10	005	1.2	190	1.7
Beach Channel (bridge)	F-1 40, E-0 05	060	1.9	225	2.0
Grass Hassock Channel	-1 10	050	1.0	230	1.0

* **Caution-** During the first two hours of flood in the channel north of Governers Island, the current in the Hudson River is still ebbing while during the first 1 1/2 hours of ebb in this channel, the current in the Hudson River is still flooding.

Important: see NOTE, bottom p. 29 and NOAA Article, p. 10.

TIME OF CURRENT CHANGE

(See Note at bottom of Boston Tables, pp. 38-43: Rule-of-Thumb for Current Velocities.)

	TIME DIFFERENCES Flood Starts; Ebb Starts H.M.	MAXIMUM FLOOD Dir.(true) in degrees	Av. Max. in knots	MAXIMUM EBB Dir.(true) in degrees	Av. Max. in knots
NEW YORK HARBOR ENTRANCE – based on The Narrows, pp. 116-121					
Ambrose Channel	-0 40	305	1.6	125	1.7
Norton Pt., WSW of	+0 10	340	1.0	165	1.2
THE NARROWS (mid-ch.) – table, pp. 116-121		335	1.6	165	1.9
NEW YORK HARBOR, Upper Bay – based on The Narrows, pp. 116-121					
Bay Ridge, W of	F+0 00, E+0 35	355	1.4	185	1.5
Red Hook Channel	F-0 55, E-0 15	355	1.0	170	0.7
Robbins Reef Light, E of	F+0 25, E-0 05	015	1.3	205	1.6
Red Hook, 1 mi. W of	+0 45	025	1.3	205	2.3
Statue of Liberty, E of	+0 55	030	1.4	205	1.9
HUDSON RIVER, Midchannel – based on The Narrows, pp. 116-121					
George Washington Bridge	+1 35	010	1.8	205	2.5
Spuyten Duyvil	+1 35	020	1.6	–	2.1
Riverdale	F+2 25, E+1 50	015	1.4	200	2.0
Dobbs Ferry	F+2 45, E+2 10	010	1.3	–	1.7
Tarrytown	+2 40	000	1.1	–	1.5
West Point, off Duck Is.	+3 40	010	1.0	–	1.1
NEW YORK HARBOR, Lower Bay – based on The Narrows, pp. 116-121					
Sandy Hook Channel	-1 20	285	1.6	095	1.9
Sandy Hook Channel, 0.4 mi. W of N. tip	-1 40	235	2.0	050	1.6
Coney Is. Lt., 1.5 mi. SSE of	-1 10	310	1.1	125	1.3
Rockaway Inlet Jetty, 1 mi. SW of	F-2 05, E-1 35	285	1.2	140	1.4
Coney Is. Channel, W end	F-1 15, E-0 30	295	1.1	100	1.2
SANDY HOOK BAY – based on The Narrows, pp. 116-121					
Highlands Bridge, Shrewsbury R.	+0 25	170	2.6	–	2.5
Sea Bright Br., Shrewsbury R.	F+1 05, E+0 45	185	1.4	–	1.7
RARITAN RIVER – based on The Narrows, pp. 116-121					
Washington Canal, N entr.	F-1 00, E-1 40	240	1.5	060	1.5
South River entr.	F-1 45, E-0 35	180	1.1	000	1.0
*ARTHUR KILL & KILL VAN KULL – based on The Narrows, pp. 116-121					
Tottenville, Arthur Kill River	-0 50	025	1.0	210	1.1
Tufts Pt.-Smoking Pt.	-0 35	110	1.2	265	1.2
Elizabethport	+0 20	090	1.4	260	1.1
Bergen Pt., East Reach	-1 35	275	1.1	095	1.2
New Brighton	-1 35	260	1.3	070	1.9
NEW JERSEY COAST – based on Del. Bay Entr., pp. 140-145					
Manasquan Inlet	F-0 45, E-1 10	300	1.7	120	1.8
Manasquan R. Hy. Br. Main Ch.	F-0 40, E-1 15	230	2.2	050	2.1
**Pt. Pleasant Canal, north bridge	F+1 45, E+0 50	170	1.8	350	2.0
Barnegat Inlet	F+1 00, E+0 15	270	2.2	090	2.5
Manahawkin Drawbridge	+2 30	030	1.1	210	0.9
McCrie Shoal	-0 35	280	1.3	100	1.4
Cape May Harbor entr.	-1 35	325	1.6	140	1.7
Cape May Canal, E end	-1 50	310	1.9	130	1.9

* Tidal flow erratic due to dredging.

**Waters are extremely turbulent. Currents of 6 to 7 knots have been reported near the bridges.

Important: **see NOTE, bottom p. 29 and NOAA Article, p. 10.**

TIME OF CURRENT CHANGE
(See Note at bottom of Boston Tables, pp. 38-43: Rule-of-Thumb for Current Velocities.)

	TIME DIFFERENCES Flood Starts; Ebb Starts H.M.	MAXIMUM FLOOD Dir.(true) in degrees	Av. Max. in knots	MAXIMUM EBB Dir.(true) in degrees	Av. Max. in knots
DELAWARE BAY & RIVER – based on Del. Bay Entr., pp. 140-145					
Cape May Channel .... -1 10		305	1.5	150	2.3
DELAWARE BAY ENTR. – table, pp. 140-145		325	1.4	145	1.3
Cape Henlopen, 0.7 mi. ESE of ...F- 0 05, E- 040		330	1.8	140	2.4
Cape Henlopen, 2 mi. NE of........F+0 20, E-0 05		315	2.0	145	2.3
Cape Henlopen, 5 mi. N of .... +0 30		345	2.0	175	1.9
Mispillion River Mouth.............. F+2 35, E+1 50		025	1.5	190	1.0
Bay Shore chan., City of Town Bank.......... - 0 40		005	0.9	185	1.0
Fourteen Ft. Bk., Lt., 1.2 mi. E of.................. +0 10		340	1.3	175	1.5
Maurice River entr. .... +1 00		010	1.1	190	1.0
Kelly Island, 1.5 mi. E of .... +0 50		350	0.9	165	1.2
Miah Maull rge. at Cross Ledge rge. .... +1 25		335	1.5	160	1.8
False Egg Is. Pt., 2 mi. off.............................. +0 20		340	1.1	160	1.3
Ben Davis Pt. Shoal., SW of .... +1 40		320	1.8	145	1.9
Cohansey R., 0.5 mi. above entr. .... +1 30		075	1.2	255	1.4
Arnold Point, 2.2 mi. SW of........................ +2 25		325	2.1	145	1.9
Smyrna River entr. .... +1 55		250	1.2	070	1.5
Stony Point chan., W of............... F+3 25, E+2 30		325	1.5	150	1.9
Appoquinimink R. entr. .... +2 25		230	1.0	050	1.2
Reedy Is., off end of pier .... +2 55		025	2.4	195	2.6
Alloway Creek entr., 0.2 mi. above .... +2 15		130	2.1	325	2.1
Reedy Point, 0.85 mi. NE of........ F+3 35, E+2 50		340	1.6	165	2.2
Salem River entr. .... +3 40		060	1.5	245	1.6
Bulkhead Sh. chan., off Del. City F+3 15, E+2 55		310	2.1	140	2.1
Pea Patch Is., chan., E of............................. +3 30		320	2.3	150	2.3
New Castle, chan., abreast of ..... F+3 35, E+3 00		050	1.9	230	2.4
CHESAPEAKE BAY – based on The Race, pp. 86-91					
(over 90% correlation within 15 min. throughout year)					
Cape Henry Light, 2.0 mi. N of .... +0 40		290	1.2	110	1.1
Chesapeake Bay entr. .... F+0 20, E-0 20		300	0.8	130	1.2
Cape Henry Light, 4.6 mi. N of F-0 25, E-0 05		295	1.3	105	1.3
Cape Henry Light, 8.3 mi. NW of .... +0 30		330	1.0	135	1.1
Tail of the Horseshoe .... F+0 20, E-0 05		300	0.9	110	1.0
Chesapeake Channel (Bridge Tunnel)...... +0 15		335	1.8	145	1.5
Fisherman Is., 1.7 mi. S of F-0 00, E-0 35		300	1.0	125	1.4
Old Plantation Flats Lt., 0.5 mi. W of .... +1 40		005	1.2	175	1.3
York Spit Channel N buoy "26" . F+1 50, E+1 05		010	0.8	195	1.1
Wolf Trap Lt., 0.5 mi. W of .... F+2 00, E+1 15		015	1.0	190	1.2
Stingray Point, 5.5 mi. E of .... +2 50		345	1.0	180	0.9
Stingray Point, 12.5 mi. E of .... F+2 35, E+1 50		030	1.0	175	0.8
Smith Point Lt., 6.0 mi. N of .... +4 45		350	0.4	135	1.0

Cove Point - See Chesapeake Bay Current Diagram, p. 154
Pooles Island - See Chesapeake Bay Current Diagram, p. 154
Worton Point - See Chesapeake Bay Current Diagram, p. 154
CHESAPEAKE & DELAWARE CANAL - table, pp. 148-153

	TIME DIFFERENCES	MAXIMUM FLOOD		MAXIMUM EBB	
C& D CANAL POINTS – based on C&D Canal, pp. 148-153					
Back Creek, 0.3 mi. W of Sandy Pt.............. -0 05		055	1.2	245	1.4
Reedy Point Radio Tower, S of......F-1 00, E-0 05		080	1.9	265	1.3

HAMPTON ROADS – based on The Race, pp. 86-91
(over 90% correlation within 15 min. throughout year)

	TIME DIFFERENCES	MAXIMUM FLOOD		MAXIMUM EBB	
Thimble Shoal Channel (West End) .... +0 05		295	0.9	115	1.2
Old Point Comfort, 0.2 mi. S of F-0 20, E-1 15		240	1.7	075	1.4
Willoughby Spit, 0.8 mi. NW of ...F-1 15, E-2 00		260	0.7	040	1.0
Sewells Point, chan., W of F-0 25, E-1 50		195	0.9	000	1.2
Newport News, chan., middle ....F-0 25, E -0 32		245	1.1	075	1.1

Important: see NOTE, bottom p. 29 and NOAA Article, p. 10.

TIME OF CURRENT CHANGE

(See Note at bottom of Boston Tables, pp. 38-43: Rule-of-Thumb for Current Velocities.)

	TIME DIFFERENCES Flood Starts; Ebb Starts H.M.	MAXIMUM FLOOD Dir.(true) in degrees	Av. Max. in knots	MAXIMUM EBB Dir.(true) in degrees	Av. Max. in knots
VA, NC, SC, GA & FL, outer coast – based on Hell Gate, pp. 110-115					
(over 90% correlation within 15 min. throughout year)					
Hatteras Inlet	F+1 00, E+0 40	305	2.1	150	2.0
Ocracoke Inlet chan. entr.	F+1 10, E+0 45	000	1.7	145	2.4
Beaufort Inlet Approach	F+0 25, E-1 00	000	0.3	160	1.4
Cape Fear R. Bald Head	F-1 25, E-1 30	035	2.2	190	2.9
Winyah Bay entr.	+0 05	320	1.9	140	2.0
North Santee R. entr.	F-0 40, E-1 35	010	1.5	165	1.8
South Santee R. entr.	F-1 20, E-1 15	045	1.5	240	1.6
Charleston Hbr. entr., betw. jetties	-1 40	320	1.8	120	1.8
Charleston Hbr., off Ft. Sumter	-1 40	315	1.7	130	2.0
Charleston Hbr. S. ch. 0.8 mi. ENE of Ft. Johnson	F-0 55, E-1 40	275	0.8	115	2.6
Charleston Hbr., Drum Is. (bridge)	-1 20	020	1.2	185	2.0
North Edisto River entr.	-0 35	330	2.9	140	3.7
South Edisto River entr.	F-1 20, E-1 50	350	1.8	145	2.2
Ashepoo R. off Jefford Cr. entr.	-0 40	015	1.5	195	1.6
Port Royal Sd., SE chan. entr.	F-2 10, E-1 50	310	1.3	150	1.6
Hilton Head	-1 20	325	1.8	145	1.8
Beaufort River entr.	-1 20	010	1.3	195	1.4
Savannah River entr.	F-1 00, E-0 50	285	2.0	110	2.0
Vernon R. 1.2 mi. S of Possum Pt.	F-1 25, E-1 00	325	1.1	165	1.7
Raccoon Key & Egg Is. Shoal bet.	F-0 40, E -1 15	255	1.6	130	2.0
St. Catherines Sound entr.	F-1 40, E-0 35	290	1.8	125	1.7
Sapelo Sound entr.	F-1 30, E-0 55	290	1.7	120	2.2
Doboy Sound entr.	-1 25	290	1.6	105	1.8
Altamaha Sd., 1 mi. SE of Onemile Cut	F-0 15, E-2 00	270	1.0	090	1.9
St. Simons Sound Bar Channel	F-1 15, E-0 40	305	0.8	120	1.7
St. Andrews Sound entr.	F-1 20, E-0 50	270	2.1	105	2.2
Cumberland Sd., St. Mary's River entr., Ft. Clinch, 0.3 mi. N	F-1 25, E-1 00	275	1.4	085	1.6
Drum Point Is., rge. D chan	-0 50	350	1.1	170	1.5
Nassau Sd., 1 mi. N of Sawpit Cr. entr.	F-0 15, E-0 40	310	1.7	135	1.7
FLORIDA EAST COAST – based on The Narrows, pp. 116-121					
(over 90% correlation within 15 min. throughout year)					
St. Johns R. Entr. betw. jetties	+0 20	260	2.0	080	2.0
Mayport	+0 30	210	2.2	025	3.3
St. Johns Bluff	F+0 50, E+0 05	245	1.6	060	2.4
FLORIDA EAST COAST – based on Hell Gate, pp. 110-115					
(over 90% correlation within 15 min. throughout year)					
Fort Pierce Inlet	F+1 05, E+0 25	250	2.6	070	3.1
Lake Worth Inlet, betw. jetties	F+0 00, E-0 25	275	2.4	095	3.6
Miami Hbr., Bakers Haulover Cut	-0 10	270	2.9	090	2.5
Miami Hbr. Entr.	-0 20	295	1.8	110	1.6

CURRENT STATIONS

NOTE: Velocities shown are from U.S. Gov't. figures. It is obvious, however, to local mariners and other observers, that coastal inlets may have far greater velocities than indicated here. Strong winds and opposing tides can cause even more dangerous conditions, and great caution should be used. Separate times for Flood and Ebb are given only when the times are more than 20 minutes apart.

Important: **see NOAA Article, p. 10.**

2009 HIGH & LOW WATER
PORTLAND, ME
43°39.6'N, 70°14.8'W

		Standard Time JANUARY								Standard Time FEBRUARY					
D A Y O F M O N T H	D A Y O F W E E K	HIGH				LOW		D A Y O F M O N T H	D A Y O F W E E K	HIGH				LOW	
		a.m.	Ht.	p.m.	Ht.	a.m.	p.m.			a.m.	Ht.	p.m.	Ht.	a.m.	p.m.
1	T	1 45	8.6	1 50	9.2	7 41	8 08	1	S	2 31	9.4	2 54	8.7	8 45	9 01
2	F	2 24	8.7	2 34	8.9	8 25	8 49	2	M	3 18	9.5	3 49	8.4	9 39	9 52
3	S	3 06	8.8	3 22	8.7	9 13	9 34	3	T	4 11	9.5	4 50	8.1	10 39	10 51
4	S	3 52	9.0	4 16	8.4	10 07	10 24	4	W	5 12	9.6	5 59	8.0	11 46	11 57
5	M	4 44	9.3	5 16	8.3	11 06	11 20	5	T	6 19	9.8	7 11	8.2	...	12 56
6	T	5 41	9.5	6 21	8.3	...	12 10	6	F	7 27	10.1	8 18	8.6	1 05	2 03
7	W	6 41	9.9	7 27	8.4	12 20	1 15	7	S	8 32	10.5	9 18	9.2	2 12	3 03
8	T	7 42	10.4	8 31	8.8	1 22	2 17	8	S	9 32	10.9	10 12	9.7	3 13	3 58
9	F	8 44	10.8	9 31	9.2	2 24	3 16	9	M	10 28	11.2	11 04	10.1	4 09	4 49
10	S	9 41	11.2	10 26	9.6	3 23	4 12	10	T	11 18	11.2	11 51	10.4	5 03	5 37
11	S	10 37	11.5	11 19	9.9	4 19	5 05	11	W	...	...	12 08	11.0	5 54	6 23
12	M	11 31	11.5	...	...	5 14	5 56	12	T	12 37	10.4	12 57	10.5	6 44	7 08
13	T	12 11	10.1	12 24	11.3	6 08	6 46	13	F	1 23	10.3	1 46	9.9	7 34	7 54
14	W	1 02	10.2	1 17	10.8	7 03	7 36	14	S	2 09	10.0	2 36	9.2	8 24	8 41
15	T	1 53	10.1	2 11	10.2	7 57	8 26	15	S	2 57	9.6	3 29	8.5	9 17	9 30
16	F	2 45	9.9	3 06	9.5	8 54	9 18	16	M	3 47	9.1	4 26	7.9	10 14	10 23
17	S	3 37	9.6	4 04	8.8	9 53	10 11	17	T	4 42	8.7	5 28	7.5	11 15	11 22
18	S	4 32	9.3	5 05	8.2	10 54	11 07	18	W	5 42	8.5	6 32	7.3	...	12 19
19	M	5 29	9.0	6 09	7.8	11 57	...	19	T	6 43	8.4	7 33	7.4	12 24	1 20
20	T	6 26	8.9	7 11	7.7	12 06	12 59	20	F	7 41	8.6	8 26	7.6	1 24	2 14
21	W	7 23	8.8	8 09	7.7	1 04	1 57	21	S	8 33	8.8	9 12	8.0	2 17	3 01
22	T	8 16	9.0	9 00	7.8	1 59	2 48	22	S	9 18	9.1	9 52	8.3	3 04	3 42
23	F	9 04	9.1	9 44	8.0	2 48	3 33	23	M	9 58	9.4	10 27	8.7	3 45	4 17
24	S	9 47	9.3	10 24	8.2	3 32	4 14	24	T	10 35	9.6	11 00	9.0	4 23	4 50
25	S	10 25	9.5	11 00	8.4	4 12	4 50	25	W	11 10	9.7	11 32	9.3	4 59	5 22
26	M	11 01	9.6	11 33	8.6	4 49	5 23	26	T	11 46	9.7	...	...	5 34	5 54
27	T	11 36	9.7	...	...	5 24	5 54	27	F	12 05	9.6	12 23	9.6	6 12	6 29
28	W	12 05	8.8	12 10	9.6	5 59	6 26	28	S	12 40	9.8	1 03	9.4	6 52	7 07
29	T	12 38	8.9	12 46	9.5	6 36	6 59								
30	F	1 12	9.1	1 25	9.3	7 15	7 36								
31	S	1 49	9.3	2 07	9.0	7 58	8 16								

Dates when Ht. of **Low** Water is below Mean Lower Low with Ht. of lowest given for each period and Date of lowest in ( ):

8th - 15th: -1.8' (11th - 12th)

6th - 14th: -1.6' (9th - 10th)
28th: -0.2'

Average Rise and Fall 9.1 ft.

When a high tide exceeds av. ht., the *following* low tide will be lower than av.

2009 HIGH & LOW WATER
PORTLAND, ME
43°39.6'N, 70°14.8'W

Daylight Time starts March 8 at 2 a.m.   **Daylight Saving Time**

DAY OF MONTH	DAY OF WEEK	MARCH HIGH a.m.	Ht.	HIGH p.m.	Ht.	LOW a.m.	LOW p.m.	DAY OF MONTH	DAY OF WEEK	APRIL HIGH a.m.	Ht.	HIGH p.m.	Ht.	LOW a.m.	LOW p.m.
1	S	1 19	9.9	1 46	9.1	7 36	7 49	1	W	3 36	10.2	4 24	8.6	10 07	10 19
2	M	2 03	9.9	2 36	8.8	8 24	8 37	2	T	4 37	9.9	5 30	8.5	11 11	11 26
3	T	2 52	9.9	3 32	8.4	9 19	9 31	3	F	5 45	9.7	6 40	8.5	...	12 19
4	W	3 49	9.7	4 36	8.2	10 21	10 34	4	S	6 58	9.6	7 49	8.8	12 38	1 28
5	T	4 54	9.6	5 48	8.1	11 31	11 44	5	S	8 08	9.7	8 51	9.3	1 50	2 32
6	F	6 06	9.6	7 00	8.3	...	12 42	6	M	9 12	9.9	9 46	9.8	2 56	3 29
7	S	7 18	9.9	8 06	8.8	12 57	1 49	7	T	10 09	10.1	10 36	10.2	3 54	4 20
8	S	*9 23	10.2	*10 04	9.4	*3 04	*3 48	8	W	11 01	10.1	11 20	10.5	4 46	5 06
9	M	10 23	10.5	10 57	10.0	4 05	4 41	9	T	11 49	10.1	...	...	5 34	5 49
10	T	11 15	10.7	11 43	10.4	4 59	5 29	10	F	12 02	10.6	12 32	9.8	6 19	6 31
11	W	...	...	12 04	10.7	5 49	6 14	11	S	12 42	10.5	1 15	9.5	7 02	7 11
12	T	12 27	10.6	12 51	10.5	6 37	6 57	12	S	1 21	10.3	1 57	9.1	7 44	7 51
13	F	1 10	10.6	1 36	10.1	7 23	7 39	13	M	2 01	9.9	2 40	8.7	8 26	8 32
14	S	1 51	10.3	2 21	9.5	8 08	8 21	14	T	2 42	9.5	3 25	8.3	9 10	9 16
15	S	2 33	10.0	3 07	8.9	8 54	9 04	15	W	3 27	9.1	4 13	8.0	9 57	10 03
16	M	3 17	9.5	3 55	8.4	9 42	9 50	16	T	4 16	8.8	5 05	7.7	10 48	10 56
17	T	4 04	9.1	4 48	7.9	10 33	10 41	17	F	5 10	8.5	6 01	7.7	11 42	11 54
18	W	4 56	8.7	5 46	7.5	11 30	11 38	18	S	6 08	8.3	6 57	7.8	...	12 38
19	T	5 54	8.4	6 47	7.4	...	12 31	19	S	7 07	8.3	7 50	8.0	12 54	1 32
20	F	6 57	8.3	7 48	7.4	12 40	1 33	20	M	8 03	8.5	8 37	8.5	1 52	2 21
21	S	7 58	8.3	8 42	7.7	1 42	2 29	21	T	8 55	8.7	9 21	9.0	2 44	3 05
22	S	8 53	8.6	9 29	8.1	2 38	3 17	22	W	9 42	9.0	10 01	9.5	3 31	3 47
23	M	9 41	8.9	10 10	8.6	3 28	3 58	23	T	10 27	9.2	10 41	10.1	4 16	4 28
24	T	10 23	9.2	10 47	9.1	4 11	4 35	24	F	11 11	9.5	11 21	10.5	4 59	5 09
25	W	11 03	9.4	11 22	9.5	4 51	5 10	25	S	11 55	9.6	...	...	5 42	5 51
26	T	11 41	9.6	11 56	9.9	5 30	5 45	26	S	12 03	10.9	12 41	9.6	6 27	6 35
27	F	...	...	12 20	9.7	6 08	6 22	27	M	12 48	11.0	1 29	9.6	7 15	7 23
28	S	12 33	10.3	1 01	9.6	6 49	7 00	28	T	1 36	11.0	2 21	9.4	8 05	8 14
29	S	1 12	10.5	1 44	9.5	7 32	7 42	29	W	2 29	10.8	3 17	9.2	8 59	9 11
30	M	1 55	10.5	2 32	9.2	8 18	8 29	30	T	3 26	10.5	4 18	9.1	9 58	10 13
31	T	2 42	10.4	3 24	8.9	9 10	9 21								

*Daylight Saving Time starts

Dates when Ht. of **Low** Water is below Mean Lower Low with Ht. of lowest given for each period and Date of lowest in ( ):

1st - 2nd: -0.3' (1st)
7th - 15th: -1.1' (10th, 12th)
27th - 31st: -0.7' (29th - 30th)

1st: -0.2'
6th - 12th: -0.8' (10th)
24th - 30th: -1.1' (27th)

Average Rise and Fall 9.1 ft.

When a high tide exceeds av. ht., the *following* low tide will be lower than av.

2009 HIGH & LOW WATER
PORTLAND, ME
43°39.6'N, 70°14.8'W

		MAY								JUNE					
		Daylight Saving Time								Daylight Saving Time					
DAY OF MONTH	DAY OF WEEK	HIGH				LOW		DAY OF MONTH	DAY OF WEEK	HIGH				LOW	
		a.m.	Ht.	p.m.	Ht.	a.m.	p.m.			a.m.	Ht.	p.m.	Ht.	a.m.	p.m.
1	F	4 29	10.2	5 22	9.0	11 00	11 20	1	M	6 29	9.5	7 07	9.8	12 17	12 44
2	S	5 37	9.8	6 28	9.1	...	12 05	2	T	7 34	9.2	8 04	9.9	1 23	1 42
3	S	6 47	9.6	7 32	9.4	12 31	1 09	3	W	8 36	9.0	8 58	10.0	2 25	2 37
4	M	7 54	9.5	8 30	9.7	1 40	2 10	4	T	9 34	8.9	9 47	10.1	3 21	3 29
5	T	8 57	9.5	9 24	10.1	2 43	3 05	5	F	10 26	8.8	10 33	10.1	4 13	4 17
6	W	9 53	9.5	10 12	10.3	3 40	3 55	6	S	11 13	8.8	11 15	10.1	5 00	5 01
7	T	10 44	9.5	10 56	10.4	4 31	4 42	7	S	11 56	8.7	11 55	10.0	5 43	5 43
8	F	11 31	9.4	11 38	10.4	5 18	5 25	8	M	...	...	12 37	8.6	6 24	6 22
9	S	...	...	12 15	9.2	6 01	6 05	9	T	12 34	9.9	1 16	8.5	7 02	7 00
10	S	12 17	10.3	12 56	9.0	6 42	6 45	10	W	1 11	9.8	1 53	8.5	7 40	7 39
11	M	12 55	10.1	1 36	8.8	7 22	7 24	11	T	1 50	9.6	2 31	8.4	8 17	8 19
12	T	1 33	9.8	2 16	8.5	8 02	8 04	12	F	2 29	9.4	3 11	8.4	8 55	9 01
13	W	2 14	9.6	2 58	8.3	8 43	8 46	13	S	3 11	9.2	3 51	8.5	9 33	9 45
14	T	2 56	9.3	3 42	8.2	9 25	9 31	14	S	3 54	9.0	4 33	8.6	10 13	10 32
15	F	3 42	9.0	4 28	8.1	10 09	10 19	15	M	4 40	8.7	5 17	8.8	10 56	11 23
16	S	4 30	8.8	5 16	8.1	10 56	11 12	16	T	5 30	8.6	6 03	9.0	11 42	...
17	S	5 22	8.6	6 05	8.3	11 44	...	17	W	6 24	8.4	6 52	9.3	12 17	12 31
18	M	6 16	8.4	6 54	8.5	12 07	12 33	18	T	7 21	8.4	7 44	9.8	1 14	1 23
19	T	7 11	8.4	7 42	8.9	1 02	1 22	19	F	8 20	8.5	8 37	10.2	2 11	2 18
20	W	8 06	8.5	8 30	9.4	1 57	2 11	20	S	9 19	8.7	9 31	10.7	3 08	3 13
21	T	8 59	8.7	9 16	9.9	2 50	2 59	21	S	10 16	9.0	10 26	11.1	4 04	4 08
22	F	9 51	9.0	10 03	10.5	3 40	3 47	22	M	11 12	9.4	11 20	11.5	4 58	5 03
23	S	10 41	9.2	10 50	10.9	4 29	4 35	23	T	...	...	12 07	9.6	5 52	5 58
24	S	11 32	9.5	11 39	11.3	5 19	5 24	24	W	12 15	11.6	1 01	9.9	6 45	6 54
25	M	...	...	12 23	9.6	6 09	6 15	25	T	1 11	11.6	1 56	10.0	7 39	7 51
26	T	12 29	11.4	1 16	9.7	7 00	7 07	26	F	2 07	11.3	2 51	10.1	8 32	8 49
27	W	1 23	11.4	2 10	9.7	7 53	8 03	27	S	3 04	10.9	3 47	10.1	9 27	9 50
28	T	2 19	11.2	3 07	9.6	8 48	9 02	28	S	4 03	10.4	4 43	10.1	10 22	10 52
29	F	3 17	10.8	4 07	9.6	9 46	10 04	29	M	5 04	9.8	5 40	10.0	11 17	11 55
30	S	4 19	10.4	5 07	9.6	10 44	11 10	30	T	6 07	9.2	6 37	9.9	...	12 14
31	S	5 24	9.9	6 08	9.7	11 44	...								

Dates when Ht. of **Low** Water is below Mean Lower Low with Ht. of lowest given for each period and Date of lowest in ( ):

1st: -0.2'
7th - 10th: -0.4' (8th - 9th)
23rd - 30th: -1.3' (26th)

21st - 28th: -1.4' (24th - 25th)

Average Rise and Fall 9.1 ft.

When a high tide exceeds av. ht., the *following* low tide will be lower than av.

2009 HIGH & LOW WATER
PORTLAND, ME
43°39.6'N, 70°14.8'W

Daylight Saving Time   Daylight Saving Time

D A Y O F M O N T H	D A Y O F W E E K	JULY HIGH a.m.	Ht.	p.m.	Ht.	LOW a.m.	p.m.	D A Y O F M O N T H	D A Y O F W E E K	AUGUST HIGH a.m.	Ht.	p.m.	Ht.	LOW a.m.	p.m.
1	W	7 10	8.8	7 34	9.8	12 59	1 12	1	S	8 44	7.9	8 55	9.3	2 33	2 36
2	T	8 12	8.5	8 30	9.7	2 01	2 08	2	S	9 39	8.0	9 46	9.4	3 27	3 28
3	F	9 11	8.3	9 22	9.7	2 59	3 03	3	M	10 27	8.1	10 31	9.5	4 16	4 16
4	S	10 04	8.3	10 10	9.7	3 52	3 53	4	T	11 09	8.3	11 12	9.6	4 59	4 58
5	S	10 52	8.3	10 54	9.7	4 40	4 39	5	W	11 47	8.5	11 50	9.7	5 37	5 37
6	M	11 35	8.4	11 35	9.8	5 24	5 21	6	T	...	...	12 22	8.7	6 12	6 13
7	T	...	...	12 15	8.4	6 04	6 00	7	F	12 26	9.7	12 55	8.8	6 44	6 49
8	W	12 13	9.8	12 52	8.5	6 40	6 38	8	S	1 00	9.6	1 27	9.0	7 15	7 25
9	T	12 51	9.7	1 28	8.6	7 15	7 15	9	S	1 36	9.5	2 01	9.2	7 47	8 02
10	F	1 26	9.6	2 02	8.6	7 48	7 52	10	M	2 12	9.3	2 35	9.3	8 21	8 42
11	S	2 03	9.5	2 37	8.7	8 22	8 31	11	T	2 52	9.1	3 13	9.5	8 59	9 26
12	S	2 41	9.3	3 13	8.9	8 57	9 13	12	W	3 35	8.9	3 57	9.6	9 40	10 15
13	M	3 21	9.1	3 51	9.0	9 34	9 57	13	T	4 24	8.6	4 46	9.6	10 27	11 10
14	T	4 05	8.8	4 33	9.2	10 15	10 46	14	F	5 20	8.3	5 41	9.7	11 21	...
15	W	4 53	8.6	5 20	9.4	11 00	11 39	15	S	6 23	8.2	6 43	9.9	12 12	12 22
16	T	5 47	8.4	6 12	9.6	11 51	...	16	S	7 31	8.3	7 49	10.2	1 18	1 27
17	F	6 47	8.3	7 08	9.9	12 38	12 47	17	M	8 38	8.6	8 54	10.5	2 25	2 33
18	S	7 50	8.4	8 08	10.3	1 40	1 47	18	T	9 41	9.1	9 56	10.9	3 27	3 37
19	S	8 54	8.6	9 09	10.7	2 43	2 48	19	W	10 38	9.7	10 54	11.3	4 24	4 36
20	M	9 56	9.0	10 08	11.1	3 44	3 49	20	T	11 31	10.2	11 49	11.4	5 18	5 32
21	T	10 54	9.4	11 06	11.4	4 41	4 48	21	F	...	...	12 22	10.6	6 08	6 26
22	W	11 50	9.8	...	...	5 36	5 45	22	S	12 41	11.3	1 11	10.7	6 56	7 18
23	T	12 02	11.6	12 43	10.2	6 28	6 41	23	S	1 33	10.9	1 59	10.7	7 44	8 10
24	F	12 57	11.6	1 35	10.4	7 20	7 36	24	M	2 24	10.4	2 48	10.5	8 32	9 03
25	S	1 52	11.3	2 27	10.5	8 11	8 32	25	T	3 17	9.7	3 38	10.1	9 21	9 58
26	S	2 46	10.8	3 20	10.4	9 01	9 29	26	W	4 11	9.1	4 30	9.7	10 12	10 56
27	M	3 42	10.1	4 13	10.2	9 53	10 27	27	T	5 09	8.5	5 26	9.3	11 07	11 56
28	T	4 39	9.5	5 07	9.9	10 46	11 28	28	F	6 10	8.0	6 25	9.0	...	12 05
29	W	5 39	8.8	6 03	9.6	11 42	...	29	S	7 12	7.8	7 26	8.9	12 59	1 06
30	T	6 42	8.3	7 01	9.4	12 30	12 39	30	S	8 12	7.8	8 24	8.9	2 00	2 06
31	F	7 45	8.0	7 59	9.3	1 33	1 38	31	M	9 06	7.9	9 16	9.1	2 55	3 00

Dates when Ht. of **Low** Water is below Mean Lower Low with Ht. of lowest given for each period and Date of lowest in ( ):

20th - 27th: -1.5' (23rd - 24th)   18th - 24th: -1.3' (21st - 22nd)

Average Rise and Fall 9.1 ft.

When a high tide exceeds av. ht., the *following* low tide will be lower than av.

2009 HIGH & LOW WATER
PORTLAND, ME
43°39.6'N, 70°14.8'W

Daylight Saving Time · Daylight Saving Time

D A Y O F M O N T H	D A Y O F W E E K	SEPTEMBER HIGH a.m.	Ht.	p.m.	Ht.	LOW a.m.	p.m.	D A Y O F M O N T H	D A Y O F W E E K	OCTOBER HIGH a.m.	Ht.	p.m.	Ht.	LOW a.m.	p.m.
1	T	9 54	8.2	10 03	9.3	3 44	3 48	1	T	9 54	8.7	10 09	9.2	3 43	3 57
2	W	10 35	8.5	10 44	9.5	4 26	4 30	2	F	10 31	9.1	10 48	9.4	4 20	4 37
3	T	11 12	8.8	11 22	9.6	5 03	5 09	3	S	11 05	9.5	11 26	9.5	4 55	5 14
4	F	11 46	9.0	11 57	9.6	5 36	5 45	4	S	11 39	9.8	...	...	5 28	5 51
5	S	...	...	12 18	9.3	6 08	6 21	5	M	12 03	9.5	12 13	10.1	6 03	6 30
6	S	12 32	9.6	12 50	9.5	6 39	6 57	6	T	12 42	9.4	12 50	10.2	6 39	7 10
7	M	1 08	9.5	1 23	9.7	7 12	7 34	7	W	1 22	9.3	1 31	10.3	7 19	7 54
8	T	1 45	9.3	2 00	9.8	7 48	8 16	8	T	2 07	9.1	2 16	10.3	8 03	8 43
9	W	2 28	9.1	2 42	9.9	8 28	9 01	9	F	2 58	8.9	3 08	10.2	8 53	9 37
10	T	3 13	8.8	3 27	9.9	9 13	9 52	10	S	3 52	8.7	4 05	10.0	9 48	10 37
11	F	4 05	8.6	4 21	9.8	10 04	10 50	11	S	4 55	8.5	5 09	9.8	10 51	11 42
12	S	5 04	8.4	5 21	9.8	11 02	11 55	12	M	6 02	8.6	6 19	9.7	11 59	...
13	S	6 10	8.3	6 28	9.8	...	12 07	13	T	7 09	8.9	7 29	9.8	12 49	1 11
14	M	7 19	8.5	7 38	10.0	1 03	1 17	14	W	8 13	9.3	8 35	10.0	1 54	2 18
15	T	8 26	8.9	8 45	10.3	2 10	2 25	15	T	9 10	9.9	9 35	10.2	2 53	3 19
16	W	9 26	9.5	9 46	10.6	3 11	3 28	16	F	10 02	10.4	10 29	10.3	3 46	4 15
17	T	10 21	10.1	10 42	10.9	4 06	4 25	17	S	10 50	10.7	11 20	10.2	4 35	5 06
18	F	11 11	10.5	11 34	10.9	4 57	5 19	18	S	11 35	10.9	...	...	5 22	5 54
19	S	11 58	10.8	...	...	5 45	6 10	19	M	12 07	10.1	12 18	10.8	6 06	6 40
20	S	12 24	10.7	12 44	10.9	6 31	6 59	20	T	12 54	9.8	1 01	10.6	6 49	7 25
21	M	1 13	10.4	1 29	10.7	7 16	7 47	21	W	1 39	9.4	1 44	10.2	7 33	8 10
22	T	2 01	9.9	2 15	10.4	8 01	8 36	22	T	2 25	8.9	2 28	9.8	8 17	8 57
23	W	2 51	9.3	3 02	10.0	8 48	9 27	23	F	3 12	8.5	3 15	9.3	9 03	9 46
24	T	3 42	8.8	3 51	9.5	9 37	10 21	24	S	4 02	8.2	4 06	8.9	9 53	10 38
25	F	4 36	8.3	4 46	9.0	10 30	11 18	25	S	4 55	7.9	5 01	8.6	10 48	11 32
26	S	5 34	7.9	5 45	8.7	11 28	...	26	M	5 50	7.8	5 59	8.4	11 46	...
27	S	6 35	7.7	6 46	8.6	12 19	12 29	27	T	6 45	7.9	6 57	8.4	12 28	12 45
28	M	7 33	7.8	7 45	8.6	1 19	1 29	28	W	7 37	8.2	7 52	8.5	1 20	1 41
29	T	8 26	8.0	8 39	8.8	2 13	2 25	29	T	8 24	8.5	8 43	8.6	2 09	2 32
30	W	9 13	8.3	9 26	9.0	3 01	3 14	30	F	9 07	8.9	9 29	8.8	2 52	3 19
								31	S	9 46	9.4	10 12	9.0	3 33	4 01

Dates when Ht. of **Low** Water is below Mean Lower Low with Ht. of lowest given for each period and Date of lowest in ( ):

16st - 22nd: -1.1' (19th - 20th)

5th - 8th: -0.3' (6th - 7th)
16th - 21st: -1.0' (18th)

Average Rise and Fall 9.1 ft.

When a high tide exceeds av. ht., the *following* low tide will be lower than av.

34

2009 HIGH & LOW WATER
PORTLAND, ME
43°39.6'N, 70°14.8'W

Standard Time starts Nov. 1 at 2 a.m.　　　　　　　Standard Time

D A Y O F M O N T H	D A Y O F W E E K	NOVEMBER						D A Y O F M O N T H	D A Y O F W E E K	DECEMBER					
		HIGH				LOW				HIGH				LOW	
		a.m.	Ht.	p.m.	Ht.	a.m.	p.m.			a.m.	Ht.	p.m.	Ht.	a.m.	p.m.
1	S	*9 24	9.8	*9 54	9.2	*3 11	*3 42	1	T	9 30	10.5	10 11	9.1	3 15	3 58
2	M	10 02	10.2	10 35	9.3	3 50	4 23	2	W	10 16	10.9	10 58	9.3	4 02	4 45
3	T	10 42	10.5	11 18	9.4	4 30	5 05	3	T	11 04	11.1	11 47	9.4	4 50	5 33
4	W	11 24	10.7	...	...	5 12	5 50	4	F	11 54	11.2	...	...	5 39	6 23
5	T	12 04	9.3	12 09	10.8	5 57	6 37	5	S	12 39	9.5	12 47	11.0	6 32	7 16
6	F	12 52	9.2	12 59	10.7	6 45	7 29	6	S	1 33	9.5	1 43	10.7	7 28	8 10
7	S	1 45	9.1	1 54	10.5	7 39	8 24	7	M	2 29	9.5	2 43	10.3	8 28	9 08
8	S	2 43	9.0	2 54	10.1	8 38	9 24	8	T	3 29	9.5	3 46	9.8	9 32	10 07
9	M	3 46	9.0	4 01	9.8	9 43	10 28	9	W	4 31	9.5	4 54	9.4	10 39	11 08
10	T	4 50	9.1	5 09	9.6	10 52	11 32	10	T	5 31	9.6	6 01	9.1	11 47	...
11	W	5 54	9.3	6 17	9.5	...	12 02	11	F	6 32	9.8	7 06	8.9	12 09	12 53
12	T	6 55	9.7	7 22	9.5	12 33	1 08	12	S	7 29	9.9	8 08	8.8	1 08	1 54
13	F	7 51	10.1	8 22	9.5	1 31	2 08	13	S	8 23	10.1	9 03	8.8	2 04	2 49
14	S	8 42	10.4	9 16	9.5	2 25	3 03	14	M	9 12	10.1	9 53	8.8	2 55	3 39
15	S	9 30	10.6	10 06	9.5	3 14	3 52	15	T	9 57	10.2	10 38	8.8	3 43	4 25
16	M	10 14	10.6	10 52	9.4	4 01	4 39	16	W	10 40	10.1	11 20	8.7	4 27	5 07
17	T	10 57	10.5	11 36	9.2	4 44	5 23	17	T	11 19	10.0	11 59	8.7	5 08	5 47
18	W	11 37	10.3	...	...	5 26	6 05	18	F	11 58	9.8	...	...	5 47	6 25
19	T	12 19	8.9	12 18	10.0	6 08	6 46	19	S	12 37	8.6	12 36	9.6	6 26	7 02
20	F	1 01	8.7	1 00	9.7	6 49	7 28	20	S	1 15	8.5	1 15	9.4	7 05	7 39
21	S	1 43	8.4	1 43	9.3	7 32	8 12	21	M	1 53	8.4	1 55	9.1	7 45	8 16
22	S	2 27	8.2	2 29	9.0	8 18	8 56	22	T	2 32	8.4	2 38	8.8	8 29	8 56
23	M	3 14	8.1	3 18	8.7	9 07	9 43	23	W	3 13	8.4	3 24	8.5	9 15	9 38
24	T	4 02	8.1	4 10	8.4	9 59	10 31	24	T	3 57	8.5	4 13	8.2	10 05	10 23
25	W	4 51	8.1	5 04	8.2	10 55	11 20	25	F	4 44	8.6	5 07	8.0	10 59	11 11
26	T	5 41	8.3	5 59	8.2	11 50	...	26	S	5 33	8.8	6 04	7.9	11 55	...
27	F	6 29	8.7	6 53	8.2	12 09	12 45	27	S	6 25	9.1	7 03	8.0	12 04	12 53
28	S	7 15	9.1	7 46	8.4	12 57	1 36	28	M	7 18	9.5	8 00	8.2	12 58	1 49
29	S	8 00	9.5	8 35	8.6	1 44	2 25	29	T	8 11	10.0	8 56	8.6	1 53	2 43
30	M	8 45	10.0	9 23	8.9	2 30	3 11	30	W	9 04	10.6	9 49	9.0	2 47	3 36
								31	T	9 57	11.0	10 40	9.4	3 40	4 27

*Standard Time starts

Dates when Ht. of **Low** Water is below Mean Lower Low with Ht. of lowest given for
each period and Date of lowest in ( ):

2nd - 7th: -0.8' (4th - 5th)　　　　　　1st - 8th: -1.2' (3rd - 4th)
14th - 18th: -0.7' (15th - 16th　　　　13th - 17th: -0.4' (14th - 15th)
　　　　　　　　　　　　　　　　　　29th - 31st: -1.2' (31st)

Average Rise and Fall 9.1 ft.

When a high tide exceeds av. ht., the *following* low tide will be lower than av.

Coping with Currents

When going directly with or against a current, our piloting problems are simple: there is no change in course and our speed over the bottom is easily figured, but we tend to guess a bit when the current is at some other angle. Where these currents are strong, as between New York and Nantucket, it will be vital to figure the factors carefully, especially in haze or fog.

The Table below tells 1) how many degrees to change your course; 2) by what percent your speed is decreased, with the current off the Bow; 3) or by what percent it is increased, with the current off the Stern.

First: estimate your boat's speed through the water. Then refer to the appropriate TIDAL CURRENT CHART (see pp. 66-77 or pp. 92-97) and estimate the current's speed. Put these two in the form of a ratio, for example: boat speed is 8 kts, current 2 kts; ratio is 4 to 1.

Second: using the same CURRENT CHART, estimate the relative direction of the current to the nearest 15°. Example: your desired course is 60°, the current is from the East, or a relative angle of 30 on your starboard bow.

Third: Enter the Tables under Ratio of 4.0; drop down to the 30° block of numbers (indicated in the left margin). The top figure in the block shows you must change your course 7°, always toward the current, and in this example, to 67°. The middle figure, 22%, is the amount by which your speed over the bottom will be decreased if the current is off your bow, i.e. from 8 kts down to 6.25 kts. Had the figure been 30% off your stern, instead of your bow, you would apply the third figure, 21%, adding it to your 8 kts, making your true speed about 9.7 kts.

RATIOS OF BOAT SPEED TO CURRENT SPEED

Relative Angle of Current		2	2½	3	3½	4	5	6	7	8	10	12	15	20
0° from	°	0	0	0	0	0	0	0	0	0	0	0	0	0
Bow	−%	50	40	33	29	25	20	17	14	12	10	8.3	6.7	5.0
Stern	+%	50	40	33	29	25	20	17	14	12	10	8.3	6.7	5.0
15° from	°	7.0	6.0	5.0	4.0	3.5	3.0	2.5	2.0	1.5	1.5	1.0	1.0	0.5
Bow	−%	49	39	33	28	24	20	16	14	12	10	8.0	6.4	4.8
Stern	+%	48	38	32	27	24	19	16	14	12	10	8.0	6.4	4.8
30° from	°	14	11	9.5	8.0	7.0	5.5	4.5	4.0	3.0	2.5	2.0	2.0	1.0
Bow	−%	46	36	30	26	22	18	15	13	11	8.8	7.3	5.9	4.3
Stern	+%	40	33	28	24	21	17	14	12	11	8.6	7.1	5.7	4.3
45° from	°	20	16	13	11	10	8.0	7.0	5.5	5.0	4.0	3.0	2.5	1.5
Bow	−%	42	32	26	22	19	15	12	11	9.2	7.4	6.1	4.9	3.6
Stern	+%	29	24	21	18	16	13	11	10	8.4	6.8	5.7	4.5	3.4
60° from	°	25	20	16	14	12	9.5	8.0	7.0	6.0	4.5	3.5	3.0	2.0
Bow	−%	34	26	21	18	15	11	9.3	7.8	6.7	5.4	4.4	3.5	2.6
Stern	+%	16	14	13	11	10	8.6	7.3	6.4	5.7	4.6	3.8	3.1	2.4
75° from	°	29	23	18	16	14	10	9.0	7.5	6.5	5.0	4.0	3.5	2.5
Bow	−%	25	18	14	11	9.1	6.8	5.5	4.5	3.8	3.0	2.5	1.9	1.4
Stern	+%	0.8	2.5	3.7	3.8	3.6	3.6	3.1	2.9	2.6	2.2	1.9	1.5	1.2
90°	°	30	24	19	17	14	11	9.5	8.0	7.0	5.5	4.5	3.5	2.5
Abeam	−%	13	8.6	5.4	4.1	3.0	1.8	1.4	1.0	0.7	0.4	0.3	0.2	0.1

Note: In general, while rounding a headland where head current is strong, hug the shore as far as safety will permit or go well out. (Current is usually apt to be strongest between these two points.)

Boston Harbor Currents

The diagram below shows the direction of the Flood Currents in Boston Harbor at the Maximum Flood Current, generally 3.5 hours after Low Water at Boston. Generally speaking the Ebb Currents flow in precisely the opposite direction (note one exception, shown by dotted arrow east of Winthrop), and reach these maximum velocities about 4 hours after high Water at Boston.

Basically, the velocities of the Ebb Currents are about the same as those of the Flood Currents. Where the Ebb Current differs by .2 kts., the velocity of the Ebb is shown in parentheses.

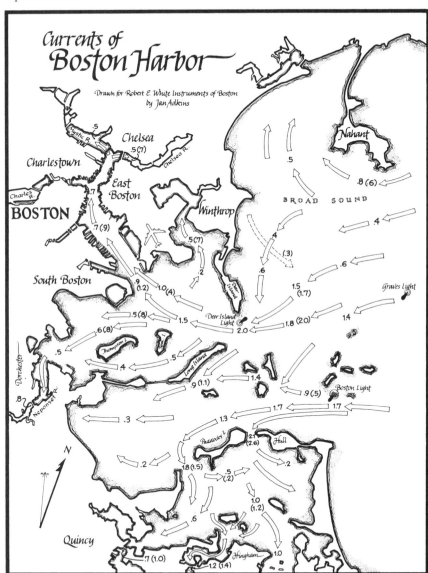

The Velocities shown on this Current Diagram are the maximums normally encountered each month at Full Moon and at New Moon. At other times the velocities will be smaller. As a rule of thumb, the velocities shown are those found on days when High Water at Boston is 11.0 ft. to 11.5' (see Boston High Water Tables pp. 38-43). When the height of High Water is 10.5', subtract 10% from the velocities shown; at 10.0', subtract 20%; at 9.0', 30%; at 8.0', 40%; below 7.5', 50%.

2009 HIGH & LOW WATER
BOSTON, MA
42°21.3'N, 71°03.1'W

Standard Time | Standard Time

DAY OF MONTH	DAY OF WEEK	JANUARY HIGH a.m.	Ht.	HIGH p.m.	Ht.	LOW a.m.	LOW p.m.	DAY OF MONTH	DAY OF WEEK	FEBRUARY HIGH a.m.	Ht.	HIGH p.m.	Ht.	LOW a.m.	LOW p.m.
1	T	2 02	9.0	2 10	9.6	8 02	8 29	1	S	2 50	9.8	3 15	9.1	9 09	9 26
2	F	2 42	9.1	2 54	9.3	8 47	9 11	2	M	3 38	9.9	4 09	8.8	10 02	10 17
3	S	3 25	9.2	3 43	9.1	9 37	9 58	3	T	4 31	9.9	5 09	8.5	11 01	11 15
4	S	4 12	9.4	4 37	8.8	10 30	10 48	4	W	5 30	10.0	6 13	8.5	...	12 04
5	M	5 03	9.7	5 35	8.7	11 28	11 44	5	T	6 33	10.2	7 20	8.6	12 17	1 09
6	T	5 58	10.0	6 36	8.7	...	12 28	6	F	7 38	10.6	8 25	9.0	1 21	2 12
7	W	6 56	10.4	7 39	8.8	12 42	1 29	7	S	8 41	11.0	9 25	9.5	2 23	3 11
8	T	7 55	10.8	8 40	9.1	1 41	2 29	8	S	9 40	11.4	10 20	10.0	3 23	4 05
9	F	8 55	11.3	9 40	9.5	2 40	3 27	9	M	10 36	11.6	11 12	10.5	4 18	4 56
10	S	9 52	11.7	10 35	9.9	3 37	4 22	10	T	11 27	11.6	...	...	5 11	5 45
11	S	10 47	11.9	11 29	10.3	4 32	5 14	11	W	12 01	10.8	12 17	11.4	6 02	6 31
12	M	11 41	12.0	...	...	5 26	6 05	12	T	12 46	10.9	1 06	11.0	6 52	7 17
13	T	12 21	10.5	12 34	11.8	6 19	6 55	13	F	1 32	10.7	1 54	10.4	7 41	8 02
14	W	1 11	10.6	1 27	11.3	7 12	7 44	14	S	2 18	10.4	2 44	9.7	8 31	8 48
15	T	2 02	10.5	2 19	10.7	8 05	8 33	15	S	3 05	10.0	3 35	9.0	9 22	9 37
16	F	2 52	10.3	3 13	9.9	8 59	9 23	16	M	3 54	9.6	4 29	8.3	10 16	10 28
17	S	3 43	10.0	4 09	9.2	9 55	10 15	17	T	4 47	9.2	5 28	7.9	11 14	11 24
18	S	4 36	9.7	5 08	8.6	10 54	11 09	18	W	5 45	8.9	6 30	7.7	...	12 15
19	M	5 31	9.4	6 09	8.1	11 55	...	19	T	6 45	8.8	7 31	7.7	12 23	1 17
20	T	6 28	9.2	7 10	7.9	12 05	12 57	20	F	7 43	8.9	8 27	7.9	1 21	2 13
21	W	7 24	9.2	8 09	7.9	1 02	1 56	21	S	8 36	9.2	9 15	8.2	2 15	3 01
22	T	8 18	9.3	9 01	8.0	1 56	2 49	22	S	9 23	9.5	9 57	8.6	3 04	3 43
23	F	9 07	9.5	9 47	8.2	2 46	3 35	23	M	10 05	9.8	10 35	9.0	3 48	4 21
24	S	9 51	9.7	10 29	8.5	3 32	4 15	24	T	10 45	10.0	11 11	9.4	4 30	4 58
25	S	10 32	9.9	11 07	8.7	4 15	4 53	25	W	11 23	10.2	11 46	9.7	5 10	5 34
26	M	11 11	10.0	11 43	9.0	4 56	5 29	26	T	...	...	12 01	10.2	5 50	6 11
27	T	11 48	10.1	...	...	5 35	6 05	27	F	12 21	10.0	12 40	10.1	6 30	6 49
28	W	12 18	9.2	12 26	10.1	6 15	6 41	28	S	12 58	10.2	1 21	9.9	7 12	7 29
29	T	12 53	9.4	1 04	10.0	6 55	7 19								
30	F	1 30	9.6	1 44	9.8	7 36	7 58								
31	S	2 08	9.7	2 27	9.5	8 20	8 39								

Dates when Ht. of **Low** Water is below Mean Lower Low with Ht. of lowest given for each period and Date of lowest in ( ):

8th 15th: -1.9' (11th - 12th)   6th - 14th: -1.6' (9th - 10th)
26th - 28th: -0.3' (28th)

Average Rise and Fall 9.5 ft.
When a high tide exceeds av. ht., the *following* low tide will be lower than av.
Since there is a high degree of correlation between the height of High Water and the velocities of the Flood and Ebb Currents for that same day, we offer a rough rule of thumb for estimating the current velocities, for ALL the Current Charts and Diagrams in this book. **Rule of Thumb:** Refer to Boston High Water. If the height of High Water is 11.0' or over, use the Current Chart velocities as shown. When the height is 10.5', subtract 10%; at 10.0', subtract 20%; at 9.0', 30%; at 8.0', 40%; below 7.5', 50%.

2009 HIGH & LOW WATER
BOSTON, MA
42°21.3'N, 71°03.1'W

Daylight Time starts March 8 at 2 a.m.   **Daylight Saving Time**

DAY OF MONTH	DAY OF WEEK	MARCH HIGH a.m.	Ht.	HIGH p.m.	Ht.	LOW a.m.	LOW p.m.	DAY OF MONTH	DAY OF WEEK	APRIL HIGH a.m.	Ht.	HIGH p.m.	Ht.	LOW a.m.	LOW p.m.
1	S	1 38	10.4	2 06	9.6	7 57	8 13	1	W	3 54	10.7	4 38	9.1	10 25	10 40
2	M	2 22	10.4	2 55	9.2	8 46	9 01	2	T	4 53	10.4	5 41	8.9	11 25	11 42
3	T	3 11	10.3	3 50	8.9	9 40	9 55	3	F	5 58	10.1	6 47	8.9	...	12 29
4	W	4 08	10.2	4 52	8.6	10 41	10 55	4	S	7 06	10.1	7 54	9.2	12 49	1 34
5	T	5 10	10.1	5 58	8.5	11 45	...	5	S	8 14	10.1	8 55	9.6	1 55	2 36
6	F	6 18	10.1	7 07	8.7	12 01	12 51	6	M	9 17	10.3	9 51	10.1	2 59	3 33
7	S	7 25	10.3	8 11	9.2	1 07	1 55	7	T	10 15	10.4	10 41	10.6	3 57	4 24
8	S	*9 30	10.7	*10 10	9.7	*3 11	*3 53	8	W	11 07	10.5	11 26	10.9	4 50	5 11
9	M	10 29	11.0	11 03	10.3	4 10	4 47	9	T	11 55	10.4	...	...	5 39	5 55
10	T	11 22	11.1	11 50	10.7	5 05	5 35	10	F	12 09	11.0	12 39	10.2	6 24	6 38
11	W	...	...	12 12	11.1	5 56	6 21	11	S	12 50	10.9	1 22	10.0	7 08	7 19
12	T	12 35	11.0	12 59	10.9	6 44	7 05	12	S	1 30	10.7	2 04	9.6	7 51	8 01
13	F	1 18	11.0	1 44	10.5	7 30	7 48	13	M	2 11	10.4	2 48	9.2	8 34	8 44
14	S	2 00	10.8	2 29	10.0	8 16	8 31	14	T	2 54	10.0	3 33	8.8	9 18	9 28
15	S	2 43	10.4	3 14	9.4	9 01	9 15	15	W	3 40	9.6	4 21	8.4	10 05	10 16
16	M	3 27	10.0	4 02	8.8	9 49	10 01	16	T	4 29	9.2	5 13	8.1	10 56	11 09
17	T	4 14	9.5	4 53	8.3	10 39	10 50	17	F	5 23	8.9	6 07	8.1	11 49	...
18	W	5 06	9.1	5 49	7.9	11 33	11 45	18	S	6 20	8.8	7 02	8.2	12 04	12 44
19	T	6 02	8.8	6 48	7.7	...	12 31	19	S	7 17	8.8	7 55	8.4	1 02	1 37
20	F	7 02	8.7	7 48	7.8	12 43	1 31	20	M	8 12	8.9	8 44	8.9	1 58	2 28
21	S	8 02	8.8	8 43	8.0	1 42	2 27	21	T	9 04	9.2	9 29	9.4	2 50	3 15
22	S	8 57	9.0	9 33	8.5	2 38	3 17	22	W	9 52	9.5	10 12	9.9	3 40	3 59
23	M	9 46	9.3	10 16	9.0	3 29	4 01	23	T	10 39	9.7	10 54	10.5	4 27	4 43
24	T	10 31	9.7	10 56	9.5	4 16	4 42	24	F	11 24	10.0	11 36	10.9	5 12	5 26
25	W	11 14	9.9	11 33	9.9	4 59	5 22	25	S	...	...	12 10	10.1	5 58	6 10
26	T	11 54	10.1	...	...	5 42	6 01	26	S	12 19	11.3	12 56	10.1	6 44	6 56
27	F	12 11	10.4	12 35	10.2	6 24	6 40	27	M	1 04	11.5	1 44	10.1	7 32	7 44
28	S	12 49	10.7	1 17	10.2	7 06	7 21	28	T	1 53	11.5	2 35	9.9	8 22	8 34
29	S	1 29	10.9	2 01	10.0	7 51	8 05	29	W	2 45	11.3	3 30	9.7	9 15	9 29
30	M	2 13	11.0	2 49	9.7	8 38	8 51	30	T	3 41	11.0	4 29	9.5	10 11	10 28
31	T	3 00	10.9	3 41	9.4	9 29	9 43								

*Daylight Saving Time starts

Dates when Ht. of **Low** Water is below Mean Lower Low with Ht. of lowest given for each period and Date of lowest in ( ):

1st - 2nd: -0.3' (1st)	1st: -0.2'
8th - 15th: -1.1' (12th)	7th - 12th: -0.8' (9th - 10th)
27th - 31st: -0.8' (29th - 30th)	24th - 30th: -1.2' (27th)

Average Rise and Fall 9.5 ft.
When a high tide exceeds av. ht., the *following* low tide will be lower than av.
Since there is a high degree of correlation between the height of High Water and the velocities of the Flood and Ebb Currents for that same day, we offer a rough rule of thumb for estimating the current velocities, for ALL the Current Charts and Diagrams in this book. **Rule of Thumb:** Refer to Boston High Water. If the height of High Water is 11.0' or over, use the Current Chart velocities as shown. When the height is 10.5', subtract 10%; at 10.0', subtract 20%; at 9.0', 30%; at 8.0', 40%; below 7.5', 50%.

2009 HIGH & LOW WATER
BOSTON, MA
42°21.3'N, 71°03.1'W

		Daylight Saving Time						Daylight Saving Time			

D A Y O F M O N T H	D A Y O F W E E K	MAY				D A Y O F M O N T H	D A Y O F W E E K	JUNE			
		HIGH		LOW				HIGH		LOW	
		a.m.	Ht.	p.m.	Ht.	a.m.	p.m.			a.m.	Ht.	p.m.	Ht.	a.m.	p.m.

D A Y O F M O N T H	D A Y O F W E E K	a.m.	Ht.	p.m.	Ht.	a.m.	p.m.	D A Y O F M O N T H	D A Y O F W E E K	a.m.	Ht.	p.m.	Ht.	a.m.	p.m.
1	F	4 42	10.6	5 30	9.4	11 10	11 31	1	M	6 36	9.8	7 11	10.1	12 20	12 47
2	S	5 47	10.3	6 34	9.5	...	12 12	2	T	7 39	9.5	8 08	10.2	1 24	1 45
3	S	6 54	10.0	7 36	9.7	12 36	1 13	3	W	8 40	9.3	9 01	10.3	2 25	2 39
4	M	7 59	9.9	8 34	10.1	1 42	2 12	4	T	9 37	9.2	9 50	10.4	3 22	3 31
5	T	9 01	9.8	9 28	10.4	2 44	3 08	5	F	10 29	9.1	10 36	10.4	4 14	4 19
6	W	9 58	9.8	10 16	10.6	3 41	3 58	6	S	11 16	9.1	11 19	10.4	5 01	5 04
7	T	10 49	9.8	11 01	10.7	4 33	4 45	7	S	11 59	9.0	...	...	5 45	5 47
8	F	11 36	9.7	11 43	10.7	5 20	5 29	8	M	12 01	10.3	12 41	9.0	6 26	6 28
9	S	...	...	12 20	9.6	6 04	6 11	9	T	12 42	10.3	1 22	8.9	7 06	7 10
10	S	12 23	10.6	1 01	9.4	6 46	6 53	10	W	1 21	10.1	2 01	8.9	7 45	7 51
11	M	1 03	10.5	1 42	9.2	7 27	7 34	11	T	2 02	10.0	2 41	8.8	8 25	8 33
12	T	1 44	10.2	2 24	9.0	8 09	8 16	12	F	2 44	9.8	3 22	8.8	9 06	9 17
13	W	2 26	10.0	3 06	8.8	8 51	9 00	13	S	3 27	9.6	4 04	8.9	9 48	10 04
14	T	3 10	9.7	3 51	8.6	9 35	9 46	14	S	4 13	9.3	4 48	9.0	10 31	10 52
15	F	3 57	9.4	4 39	8.5	10 21	10 35	15	M	5 00	9.1	5 33	9.2	11 16	11 43
16	S	4 47	9.2	5 27	8.5	11 09	11 27	16	T	5 51	8.9	6 21	9.4	...	12 04
17	S	5 39	9.0	6 17	8.7	11 58	...	17	W	6 44	8.9	7 09	9.8	12 37	12 54
18	M	6 32	8.9	7 07	8.9	12 22	12 49	18	T	7 40	8.9	8 00	10.2	1 32	1 46
19	T	7 26	8.9	7 55	9.4	1 16	1 39	19	F	8 36	9.0	8 52	10.6	2 28	2 40
20	W	8 20	9.0	8 43	9.8	2 10	2 28	20	S	9 32	9.2	9 45	11.1	3 23	3 33
21	T	9 13	9.2	9 29	10.4	3 03	3 17	21	S	10 28	9.5	10 39	11.6	4 18	4 27
22	F	10 04	9.5	10 17	10.9	3 53	4 06	22	M	11 24	9.8	11 33	11.9	5 11	5 21
23	S	10 55	9.7	11 04	11.4	4 43	4 55	23	T	...	...	12 18	10.1	6 05	6 15
24	S	11 45	9.9	11 54	11.7	5 33	5 44	24	W	12 28	12.0	1 12	10.3	6 57	7 09
25	M	...	...	12 36	10.1	6 23	6 34	25	T	1 23	12.0	2 06	10.5	7 49	8 04
26	T	12 44	11.9	1 29	10.1	7 14	7 26	26	F	2 19	11.8	3 00	10.5	8 42	9 00
27	W	1 37	11.8	2 22	10.1	8 06	8 19	27	S	3 15	11.3	3 55	10.5	9 34	9 58
28	T	2 32	11.6	3 18	10.1	9 00	9 16	28	S	4 13	10.7	4 50	10.5	10 28	10 57
29	F	3 30	11.2	4 15	10.0	9 55	10 15	29	M	5 12	10.1	5 45	10.4	11 22	11 58
30	S	4 30	10.8	5 14	10.0	10 52	11 16	30	T	6 13	9.5	6 41	10.2	...	12 18
31	S	5 32	10.3	6 13	10.1	11 49	...								

Dates when Ht. of **Low** Water is below Mean Lower Low with Ht. of lowest given for each period and Date of lowest in ( ):

7th - 10th: -0.4' (8th - 9th) 21st - 28th: -1.6' (24th - 25th)
23rd - 30th: -1.4' (26th - 27th)

Average Rise and Fall 9.5 ft.
When a high tide exceeds av. ht., the *following* low tide will be lower than av.
Since there is a high degree of correlation between the height of High Water and the velocities of the Flood and Ebb Currents for that same day, we offer a rough rule of thumb for estimating the current velocities, for ALL the Current Charts and Diagrams in this book. **Rule of Thumb:** Refer to Boston High Water. If the height of High Water is 11.0' or over, use the Current Chart velocities as shown. When the height is 10.5', subtract 10%; at 10.0', subtract 20%; at 9.0', 30%; at 8.0', 40%; below 7.5', 50%.

2009 HIGH & LOW WATER
BOSTON, MA
42°21.3'N, 71°03.1'W
Daylight Saving Time   **Daylight Saving Time**

Day of Month	Day of Week	JULY HIGH a.m.	Ht.	JULY HIGH p.m.	Ht.	JULY LOW a.m.	JULY LOW p.m.	Day of Month	Day of Week	AUGUST HIGH a.m.	Ht.	AUGUST HIGH p.m.	Ht.	AUGUST LOW a.m.	AUGUST LOW p.m.
1	W	7 14	9.1	7 37	10.1	1 00	1 14	1	S	8 46	8.2	8 56	9.6	2 32	2 34
2	T	8 15	8.8	8 32	10.0	2 01	2 09	2	S	9 41	8.3	9 47	9.7	3 28	3 27
3	F	9 14	8.6	9 24	10.0	3 00	3 03	3	M	10 29	8.4	10 34	9.8	4 17	4 15
4	S	10 07	8.6	10 12	10.0	3 53	3 53	4	T	11 13	8.6	11 17	10.0	4 59	4 59
5	S	10 55	8.6	10 57	10.0	4 41	4 40	5	W	11 52	8.9	11 57	10.1	5 38	5 41
6	M	11 39	8.7	11 40	10.1	5 25	5 23	6	T	...	...	12 29	9.1	6 15	6 21
7	T	...	...	12 19	8.8	6 05	6 05	7	F	12 35	10.1	1 04	9.3	6 51	7 01
8	W	12 20	10.1	12 57	8.9	6 43	6 46	8	S	1 13	10.0	1 39	9.5	7 26	7 41
9	T	1 00	10.1	1 36	9.0	7 20	7 27	9	S	1 52	9.9	2 16	9.6	8 03	8 21
10	F	1 38	10.0	2 12	9.1	7 57	8 07	10	M	2 30	9.7	2 52	9.8	8 41	9 04
11	S	2 18	9.9	2 50	9.2	8 35	8 49	11	T	3 11	9.5	3 32	9.9	9 21	9 49
12	S	2 58	9.7	3 28	9.3	9 14	9 33	12	W	3 56	9.2	4 16	10.0	10 04	10 39
13	M	3 41	9.4	4 09	9.4	9 55	10 19	13	T	4 46	8.9	5 05	10.0	10 53	11 34
14	T	4 26	9.2	4 52	9.6	10 38	11 09	14	F	5 42	8.7	6 01	10.2	11 47	...
15	W	5 15	9.0	5 39	9.8	11 25	...	15	S	6 42	8.6	7 00	10.3	12 33	12 46
16	T	6 09	8.8	6 31	10.0	12 02	12 16	16	S	7 45	8.7	8 03	10.6	1 35	1 48
17	F	7 07	8.7	7 26	10.3	12 59	1 12	17	M	8 49	9.1	9 05	11.0	2 37	2 50
18	S	8 07	8.8	8 23	10.7	1 58	2 09	18	T	9 49	9.6	10 06	11.4	3 37	3 50
19	S	9 07	9.0	9 22	11.1	2 58	3 08	19	W	10 46	10.1	11 03	11.7	4 33	4 47
20	M	10 07	9.4	10 20	11.6	3 56	4 06	20	T	11 40	10.6	11 58	11.8	5 26	5 42
21	T	11 04	9.8	11 17	11.9	4 52	5 03	21	F	...	...	12 30	11.0	6 16	6 36
22	W	11 59	10.3	...	...	5 46	5 58	22	S	12 51	11.7	1 19	11.2	7 04	7 27
23	T	12 13	12.0	12 53	10.6	6 38	6 53	23	S	1 42	11.3	2 08	11.2	7 52	8 19
24	F	1 08	12.0	1 45	10.9	7 29	7 47	24	M	2 33	10.8	2 56	11.0	8 40	9 10
25	S	2 02	11.7	2 36	11.0	8 19	8 41	25	T	3 25	10.1	3 45	10.6	9 28	10 03
26	S	2 56	11.2	3 27	10.9	9 09	9 36	26	W	4 18	9.5	4 36	10.1	10 18	10 58
27	M	3 50	10.5	4 19	10.7	9 59	10 32	27	T	5 14	8.8	5 30	9.7	11 11	11 56
28	T	4 46	9.8	5 12	10.4	10 51	11 30	28	F	6 13	8.4	6 28	9.4	...	12 07
29	W	5 45	9.2	6 07	10.0	11 45	...	29	S	7 14	8.1	7 27	9.2	12 57	1 05
30	T	6 45	8.7	7 04	9.8	12 30	12 41	30	S	8 14	8.1	8 25	9.3	1 58	2 03
31	F	7 47	8.4	8 01	9.6	1 32	1 38	31	M	9 08	8.2	9 18	9.4	2 55	2 58

Dates when Ht. of **Low** Water is below Mean Lower Low with Ht. of lowest given for each period and Date of lowest in ( ):

20th - 27th: -1.6' (23rd - 24th)   18th - 24th: -1.4' (21st)

Average Rise and Fall 9.5 ft.
When a high tide exceeds av. ht., the *following* low tide will be lower than av.
Since there is a high degree of correlation between the height of High Water and the velocities of the Flood and Ebb Currents for that same day, we offer a rough rule of thumb for estimating the current velocities, for ALL the Current Charts and Diagrams in this book. **Rule of Thumb:** Refer to Boston High Water. If the height of High Water is 11.0' or over, use the Current Chart velocities as shown. When the height is 10.5', subtract 10%; at 10.0', subtract 20%; at 9.0', 30%; at 8.0', 40%; below 7.5', 50%.

2009 HIGH & LOW WATER
BOSTON, MA
42°21.3'N, 71°03.1'W

		Daylight Saving Time						**Daylight Saving Time**				
D A Y O F M O N T H	D A Y O F W E E K	SEPTEMBER				D A Y O F M O N T H	D A Y O F W E E K	OCTOBER				
		HIGH		LOW				HIGH		LOW		
		a.m.	Ht.	**p.m.**	Ht.	a.m.	**p.m.**		a.m.	Ht.	**p.m.**	Ht.	a.m.	**p.m.**

D.M.	D.W.	a.m.	Ht.	p.m.	Ht.	a.m.	p.m.	D.M.	D.W.	a.m.	Ht.	p.m.	Ht.	a.m.	p.m.
1	T	9 57	8.5	10 06	9.6	3 44	3 47	1	T	9 59	9.1	10 15	9.6	3 44	4 00
2	W	10 39	8.8	10 49	9.8	4 26	4 32	2	F	10 38	9.5	10 57	9.8	4 25	4 43
3	T	11 18	9.2	11 29	10.0	5 04	5 14	3	S	11 15	9.9	11 37	9.9	5 03	5 25
4	F	11 54	9.5	...	...	5 41	5 54	4	S	11 52	10.3	...	...	5 41	6 05
5	S	12 07	10.0	12 29	9.8	6 17	6 34	5	M	12 17	9.9	12 29	10.5	6 20	6 47
6	S	12 45	10.0	1 04	10.0	6 53	7 13	6	T	12 58	9.9	1 08	10.7	7 00	7 30
7	M	1 24	9.9	1 40	10.2	7 30	7 54	7	W	1 40	9.7	1 49	10.8	7 42	8 15
8	T	2 03	9.7	2 18	10.3	8 09	8 37	8	T	2 25	9.5	2 35	10.7	8 26	9 04
9	W	2 47	9.5	3 01	10.3	8 51	9 24	9	F	3 16	9.3	3 27	10.6	9 16	9 57
10	T	3 33	9.2	3 47	10.3	9 37	10 15	10	S	4 10	9.0	4 22	10.4	10 10	10 55
11	F	4 25	8.9	4 40	10.2	10 28	11 12	11	S	5 10	8.9	5 25	10.2	11 11	11 56
12	S	5 23	8.7	5 39	10.2	11 26	...	12	M	6 13	9.0	6 31	10.2	...	12 15
13	S	6 26	8.7	6 43	10.3	12 13	12 28	13	T	7 18	9.3	7 37	10.2	12 59	1 21
14	M	7 31	8.9	7 49	10.5	1 17	1 33	14	W	8 19	9.7	8 41	10.4	2 00	2 25
15	T	8 34	9.3	8 53	10.8	2 19	2 37	15	T	9 16	10.3	9 40	10.5	2 58	3 24
16	W	9 33	9.9	9 53	11.1	3 18	3 37	16	F	10 08	10.8	10 35	10.6	3 51	4 20
17	T	10 28	10.5	10 49	11.3	4 13	4 33	17	S	10 56	11.1	11 26	10.6	4 41	5 11
18	F	11 18	11.0	11 42	11.3	5 04	5 27	18	S	11 42	11.3	...	...	5 28	6 00
19	S	...	...	12 06	11.3	5 52	6 17	19	M	12 14	10.4	12 26	11.3	6 13	6 46
20	S	12 32	11.1	12 52	11.4	6 39	7 07	20	T	1 00	10.1	1 09	11.0	6 57	7 31
21	M	1 21	10.8	1 37	11.2	7 24	7 55	21	W	1 46	9.7	1 53	10.7	7 41	8 17
22	T	2 09	10.3	2 23	10.9	8 10	8 43	22	T	2 32	9.3	2 38	10.2	8 26	9 03
23	W	2 58	9.7	3 10	10.4	8 56	9 33	23	F	3 19	8.9	3 25	9.8	9 13	9 52
24	T	3 48	9.1	3 59	9.9	9 45	10 25	24	S	4 08	8.5	4 16	9.4	10 02	10 42
25	F	4 41	8.6	4 52	9.5	10 36	11 20	25	S	5 01	8.3	5 10	9.1	10 55	11 36
26	S	5 38	8.2	5 50	9.1	11 31	...	26	M	5 55	8.2	6 06	8.9	11 51	...
27	S	6 37	8.1	6 49	9.0	12 18	12 29	27	T	6 50	8.3	7 03	8.8	12 30	12 48
28	M	7 35	8.1	7 47	9.0	1 17	1 28	28	W	7 42	8.5	7 58	8.9	1 22	1 44
29	T	8 28	8.4	8 41	9.2	2 12	2 24	29	T	8 30	8.9	8 49	9.0	2 12	2 36
30	W	9 16	8.7	9 30	9.4	3 01	3 14	30	F	9 14	9.4	9 37	9.2	2 58	3 24
								31	S	9 56	9.8	10 22	9.4	3 42	4 10

Dates when Ht. of **Low** Water is below Mean Lower Low with Ht. of lowest given for each period and Date of lowest in ( ):

16th - 22nd: -1.2' (19th)

5th - 8th: -0.4' (6th - 7th)
16th - 21st: -1.0' (18th)

Average Rise and Fall 9.5 ft.
When a high tide exceeds av. ht., the *following* low tide will be lower than av.
Since there is a high degree of correlation between the height of High Water and the velocities of the Flood and Ebb Currents for that same day, we offer a rough rule of thumb for estimating the current velocities, for ALL the Current Charts and Diagrams in this book. **Rule of Thumb:** Refer to Boston High Water. If the height of High Water is 11.0' or over, use the Current Chart velocities as shown. When the height is 10.5', subtract 10%; at 10.0', subtract 20%; at 9.0', 30%; at 8.0', 40%; below 7.5', 50%.

2009 HIGH & LOW WATER
BOSTON, MA

42°21.3'N, 71°03.1'W

Standard Time starts Nov. 1 at 2 a.m.　　　　**Standard Time**

DAY OF MONTH	DAY OF WEEK	NOVEMBER HIGH a.m.	Ht.	HIGH p.m.	Ht.	LOW a.m.	LOW p.m.	DAY OF MONTH	DAY OF WEEK	DECEMBER HIGH a.m.	Ht.	HIGH p.m.	Ht.	LOW a.m.	LOW p.m.
1	S	*9 36	10.3	*10 06	9.6	*3 24	*3 54	1	T	9 45	10.9	10 24	9.5	3 34	4 12
2	M	10 16	10.7	10 49	9.7	4 06	4 38	2	W	10 31	11.3	11 12	9.7	4 21	5 01
3	T	10 57	11.0	11 34	9.8	4 49	5 22	3	T	11 20	11.6	...	...	5 09	5 49
4	W	11 41	11.2	...	...	5 32	6 08	4	F	12 02	9.8	12 10	11.6	5 59	6 39
5	T	12 20	9.7	12 27	11.3	6 18	6 56	5	S	12 53	9.9	1 02	11.5	6 51	7 31
6	F	1 08	9.6	1 16	11.2	7 07	7 47	6	S	1 46	9.9	1 57	11.2	7 45	8 24
7	S	2 01	9.5	2 10	10.9	7 59	8 41	7	M	2 41	9.9	2 55	10.8	8 42	9 18
8	S	2 57	9.4	3 09	10.6	8 56	9 38	8	T	3 38	9.9	3 56	10.3	9 42	10 15
9	M	3 58	9.3	4 13	10.3	9 57	10 38	9	W	4 38	9.9	5 01	9.8	10 45	11 13
10	T	4 59	9.5	5 17	10.0	11 02	11 38	10	T	5 37	10.0	6 05	9.4	11 50	...
11	W	6 00	9.7	6 23	9.9	...	12 07	11	F	6 36	10.2	7 09	9.2	12 12	12 54
12	T	7 00	10.1	7 27	9.8	12 38	1 11	12	S	7 33	10.3	8 11	9.1	1 10	1 55
13	F	7 55	10.5	8 26	9.8	1 35	2 11	13	S	8 26	10.4	9 07	9.0	2 05	2 51
14	S	8 47	10.8	9 21	9.8	2 28	3 06	14	M	9 16	10.5	9 57	9.0	2 57	3 42
15	S	9 35	10.9	10 11	9.8	3 18	3 56	15	T	10 02	10.5	10 43	9.0	3 45	4 28
16	M	10 20	11.0	10 58	9.6	4 05	4 43	16	W	10 45	10.4	11 25	9.0	4 30	5 10
17	T	11 03	10.9	11 42	9.5	4 50	5 27	17	T	11 26	10.4	...	...	5 12	5 51
18	W	11 45	10.7	...	...	5 33	6 10	18	F	12 05	8.9	12 07	10.2	5 54	6 30
19	T	12 25	9.3	12 27	10.4	6 16	6 53	19	S	12 44	8.9	12 47	10.1	6 35	7 09
20	F	1 08	9.0	1 10	10.1	6 59	7 36	20	S	1 24	8.8	1 28	9.8	7 17	7 49
21	S	1 51	8.8	1 55	9.8	7 44	8 20	21	M	2 04	8.8	2 10	9.5	8 00	8 30
22	S	2 36	8.6	2 41	9.4	8 30	9 06	22	T	2 45	8.8	2 54	9.2	8 45	9 12
23	M	3 23	8.5	3 31	9.1	9 19	9 53	23	W	3 28	8.8	3 41	8.9	9 33	9 56
24	T	4 12	8.4	4 23	8.8	10 11	10 42	24	T	4 13	8.9	4 31	8.6	10 23	10 43
25	W	5 02	8.5	5 17	8.7	11 05	11 32	25	F	5 01	9.0	5 25	8.4	11 17	11 33
26	T	5 51	8.8	6 11	8.6	11 59	...	26	S	5 50	9.2	6 20	8.3	...	12 12
27	F	6 40	9.1	7 04	8.7	12 21	12 54	27	S	6 41	9.6	7 17	8.4	12 25	1 08
28	S	7 27	9.5	7 56	8.8	1 11	1 46	28	M	7 33	10.0	8 13	8.6	1 19	2 03
29	S	8 13	10.0	8 46	9.0	1 59	2 36	29	T	8 26	10.5	9 08	9.0	2 12	2 57
30	M	8 59	10.5	9 35	9.3	2 46	3 24	30	W	9 18	11.0	10 01	9.4	3 05	3 49
								31	T	10 11	11.5	10 53	9.7	3 58	4 41

*Standard Time starts

Dates when Ht. of **Low** Water is below Mean Lower Low with Ht. of lowest given for each period and Date of lowest in ( ):

2nd - 8th: -0.9' (4th - 5th)　　　　1st - 8th: -1.4' (4th)
14th - 18th: -0.7' (16th)　　　　14th - 16th: -0.3' (15th)
30th: -0.2'　　　　24th - 31st: -1.3' (31st)

Average Rise and Fall 9.5 ft.

When a high tide exceeds av. ht., the *following* low tide will be lower than av.
Since there is a high degree of correlation between the height of High Water and the velocities of the Flood and Ebb Currents for that same day, we offer a rough rule of thumb for estimating the current velocities, for ALL the Current Charts and Diagrams in this book. **Rule of Thumb:** Refer to Boston High Water. If the height of High Water is 11.0' or over, use the Current Chart velocities as shown. When the height is 10.5', subtract 10%; at 10.0', subtract 20%; at 9.0', 30%; at 8.0', 40%; below 7.5', 50%.

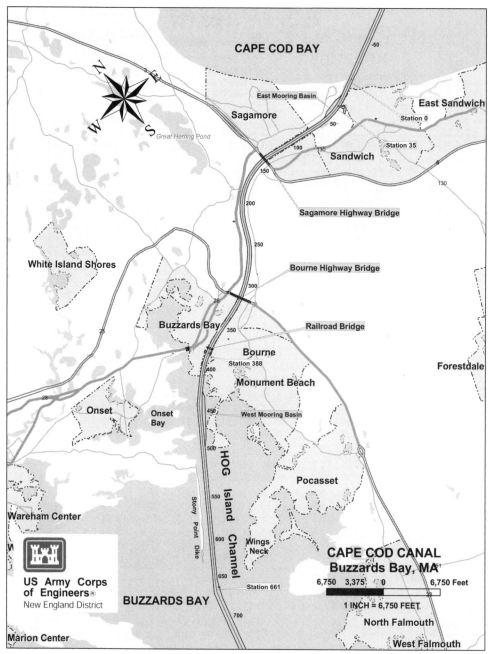

SMALL BOAT BASINS ON EITHER END OF THE CANAL: On E. end, 13-ft. mean low water, on S. side of Sandwich, available for mooring small boat traffic; On W. end, channel 13-ft. at mean low water, 100 ft. wide leads from NE side of Hog Is. Ch. abreast of Hog Is. to harbor in Onset Bay. Fuel, supplies and phone services at both locations.

See Cape Cod Canal Currents pp. 46-51

CAPE COD CANAL NAVIGATION REGULATIONS

For complete regulations see 33 USC, Part 207 and 36 CFR, Part 327

Speed Limit 10 m.p.h. – No excessive wake
All vessels going with the current have right of way over those going against it.

Clearance under all bridges: 135 ft. at mean high water. Buzz. Bay R.R. Bridge maintained in up, or open position, except when lowered for trains or maintenance.

Obtaining Clearance
Vessels over 65' shall not enter the Canal until clearance has been given by radio from the Marine Traffic Controller.

Vessels of any kind unable to make a through transit of the Canal against a head current of 6 kts. within a time limit of 2-1/2 hrs. are required to obtain helper tug assistance or wait for a fair current prior to receiving clearance from the Controller.

Two-way traffic through the Canal for all vessels is allowed when Controller on duty considers conditions suitable.

Communications
Direct communications are available at all hours by VHF radio or by phoning 508-759-4431. Call on Channel 13 to establish contact. Transmissions will then be switched to Channels 12 or 14 as the working channels. Channel 16 is also available but should be limited to emergency situations. Vessels shall maintain a radio guard on Channel 13 during the entire passage.

Traffic Lights
Traffic Lights are at Eastern End at Sandwich (Cape Cod Bay entrance) and at Western End near Wings Neck (Buzzards Bay entrance).

Entering From EASTERN END: (Lights S. side of ent. to Canal.)
RED LIGHT: Any type of vessel 65 feet in length and over must stop clear of the Cape Cod Bay entrance channel.
YELLOW LIGHT: Vessels 65 feet in length and over and drawing less than 25 feet may proceed as far as the East Mooring Basin where they must stop.
GREEN LIGHT: Vessels may proceed westward through the Canal.

When traffic lights are extinguished: all vessels over 65 feet cautioned not to enter Canal until clearance given, as above.

Entering From WESTERN END: (Lights near Wings Neck at W. Entr. to Hog Is. Ch.)
RED LIGHT: Vessels 65 feet and over in length and drawing less than 25 feet must keep southerly of Hog Island Channel Entrance Buoys Nos. 1 and 2 and utilize the general anchorage areas adjacent to the improved channel. Vessel traffic drawing 25 feet and over are directed not to enter the Canal channel at the Cleveland Ledge Light entrance and shall lay to or anchor in Buzzards Bay until clearance is granted by the Marine Traffic Controller or a green traffic light at Wings Neck is displayed.
YELLOW LIGHT: Vessels may proceed through Hog Island Channel as far as the West Mooring Basin where they must stop.
GREEN LIGHT: Vessels may proceed eastward through the Canal.

When traffic lights are extinguished: all vessels over 65 feet cautioned not to enter Canal until clearance given, as above.

Minimum Time Limits of Passage
Station 35: East Mooring Basin; Station 388: Admin. Office Buzzards Bay;
Station 661: Hog Island Channel westerly entrance Buoy No. 1.

	Station 35 to 388	Station 388 to 661
Against Head Current	60 minutes	46 minutes
With Fair Current	30 minutes	23 minutes
Slack Water	45 minutes	35 minutes

2009 CURRENT TABLE
CAPE COD CANAL
41°44.5'N, 70°36.8'W at R.R. Bridge

Standard Time						Standard Time				
JANUARY						**FEBRUARY**				
DAY OF MONTH	DAY OF WEEK	CURRENT TURNS TO				DAY OF MONTH	DAY OF WEEK	CURRENT TURNS TO		
		EAST Flood Starts		WEST Ebb Starts				EAST Flood Starts		WEST Ebb Starts

Day	Wk	a.m. / p.m. / Kts. (EAST)	a.m. / p.m. / Kts. (WEST)	Day	Wk	a.m. / p.m. / Kts. (EAST)	a.m. / p.m. / Kts. (WEST)
1	T	6 11 6 34 p4.1	12 30 12 21 p4.3	1	S	7 07 7 32 4.0	1 10 1 25 4.2
2	F	6 53 7 17 p4.0	1 09 1 04 p4.2	2	M	7 57 8 23 3.9	1 54 2 22 a4.1
3	S	7 40 8 04 p3.9	1 51 1 53 p4.1	3	T	8 55 9 23 a3.9	2 46 3 29 a4.1
4	S	8 32 8 56 p3.8	2 37 2 51 p4.0	4	W	10 01 10 31 a3.8	3 46 4 43 a4.1
5	M	9 29 9 54 p3.8	3 28 3 55 a4.0	5	T	11 10 11 41 a4.0	4 53 5 55 a4.2
6	T	10 31 10 57 3.8	4 24 5 04 a4.1	6	F	... 12 18 4.2	5 59 7 00 a4.4
7	W	11 34 ... 4.0	5 22 6 11 a4.2	7	S	12 47 1 20 p4.5	7 01 7 58 a4.6
8	T	12 01 12 35 p4.2	6 21 7 14 a4.4	8	S	1 46 2 16 p4.7	7 58 8 50 a4.9
9	F	1 02 1 35 p4.5	7 17 8 11 a4.7	9	M	2 41 3 08 p4.9	8 51 9 39 a5.0
10	S	1 59 2 29 p4.7	8 11 9 05 a4.9	10	T	3 30 3 56 p4.9	9 42 10 25 a5.1
11	S	2 53 3 21 p4.9	9 03 9 56 a5.1	11	W	4 17 4 42 p4.9	10 30 11 09 a5.1
12	M	3 45 4 12 p4.9	9 54 10 46 a5.2	12	T	5 02 5 27 4.7	11 18 11 52 a5.0
13	T	4 35 5 01 p4.9	10 45 11 34 a5.2	13	F	5 47 6 12 4.5	... 12 06 4.8
14	W	5 24 5 50 p4.8	11 36 ... 5.1	14	S	6 33 6 57 a4.3	12 36 12 56 a4.6
15	T	6 14 6 39 p4.6	12 22 12 28 p4.9	15	S	7 21 7 44 a4.1	1 21 1 48 a4.4
16	F	7 04 7 29 4.3	1 11 1 22 p4.6	16	M	8 12 8 36 a3.8	2 09 2 46 a4.2
17	S	7 57 8 22 a4.1	2 00 2 19 a4.4	17	T	9 09 9 34 a3.6	3 02 3 49 a3.9
18	S	8 53 9 17 3.8	2 52 3 20 a4.2	18	W	10 12 10 39 a3.4	4 00 4 55 a3.8
19	M	9 53 10 17 a3.7	3 47 4 24 a4.0	19	T	11 17 11 44 a3.4	5 02 5 58 a3.7
20	T	10 55 11 18 a3.6	4 44 5 28 a3.9	20	F	... 12 17 3.6	6 02 6 55 a3.8
21	W	11 55 ... 3.6	5 40 6 29 a3.9	21	S	12 42 1 10 p3.8	6 56 7 44 a4.0
22	T	12 17 12 51 p3.8	6 34 7 24 a4.0	22	S	1 31 1 56 p4.0	7 43 8 27 a4.2
23	F	1 11 1 40 p3.9	7 24 8 12 a4.1	23	M	2 14 2 36 p4.2	8 26 9 06 a4.4
24	S	1 58 2 24 p4.0	8 10 8 56 a4.2	24	T	2 53 3 14 p4.3	9 06 9 42 a4.5
25	S	2 41 3 04 p4.1	8 51 9 36 a4.4	25	W	3 29 3 50 p4.4	9 43 10 16 a4.6
26	M	3 19 3 41 p4.2	9 30 10 13 a4.5	26	T	4 03 4 25 p4.4	10 19 10 49 a4.7
27	T	3 56 4 17 p4.3	10 07 10 48 a4.6	27	F	4 39 5 01 4.4	10 56 11 22 a4.7
28	W	4 31 4 52 p4.3	10 43 11 22 a4.6	28	S	5 16 5 39 a4.4	11 35 11 57 a4.6
29	T	5 06 5 28 p4.3	11 19 11 57 a4.6				
30	F	5 43 6 06 p4.3	11 56 ... 4.5				
31	S	6 23 6 46 p4.2	12 32 12 38 p4.4				

The Kts. (knots) columns show the **maximum** predicted velocities of the stronger one of the Flood Currents and the stronger one of the Ebb Currents for each day.

The letter "a" means the velocity shown should occur **after** the **a.m.** Current Change.

The letter "p" means the velocity shown should occur **after** the **p.m.** Current Change (even if next morning).

No "a" or "p" means a.m. and p.m. velocities are the same for that day.

Av. Max. Vel.: Fl. 4.0 Kts., Ebb 4.5 Kts.

Max. Flood & Max. Ebb 3 hrs. after Flood Starts & Ebb Starts, ±20 min.

Average rise and fall: canal east end, 8.7 ft. (time of high water same as Boston); west end, at Monument Beach, 4.0 ft. (time of high water 20 min. after Newport).

See pp. 22-29 for Current Change at other points.

CAPE COD CANAL
41°44.5'N, 70°36.8'W at R.R. Bridge
Daylight Time starts Mar. 8 at 2 a.m.   **Daylight Saving Time**

MARCH

D A Y O F M O N T H	D A Y O F W E E K	EAST Flood Starts a.m.	EAST Flood Starts p.m.	EAST Flood Starts Kts.	WEST Ebb Starts a.m.	WEST Ebb Starts p.m.	WEST Ebb Starts Kts.
1	S	5 56	6 20	a4.3	...	12 18	4.4
2	M	6 41	7 07	a4.3	12 36	1 07	a4.4
3	T	7 33	8 00	a4.1	1 21	2 06	a4.3
4	W	8 32	9 03	a4.0	2 16	3 15	a4.2
5	T	9 41	10 15	a3.9	3 21	4 30	a4.1
6	F	10 53	11 28	a4.0	4 34	5 42	a4.2
7	S	...	12 03	4.2	5 44	6 45	a4.4
8	S	12 35	*2 05	p4.4	*7 48	*8 41	a4.6
9	M	2 34	3 01	p4.6	8 45	9 31	a4.8
10	T	3 24	3 50	p4.8	9 38	10 16	a4.9
11	W	4 11	4 36	p4.8	10 27	10 59	a5.0
12	T	4 55	5 20	4.7	11 14	11 40	a5.0
13	F	5 38	6 02	a4.7	11 59	...	4.9
14	S	6 20	6 43	a4.5	12 21	12 44	a4.7
15	S	7 02	7 24	a4.3	1 01	1 31	a4.6
16	M	7 46	8 09	a4.1	1 42	2 20	a4.4
17	T	8 34	8 58	a3.8	2 27	3 14	a4.1
18	W	9 28	9 55	a3.6	3 18	4 14	a3.9
19	T	10 29	10 59	a3.4	4 17	5 19	a3.7
20	F	11 34	...	3.4	5 22	6 21	a3.7
21	S	12 06	12 36	p3.5	6 25	7 18	a3.7
22	S	1 05	1 31	p3.7	7 21	8 07	a3.9
23	M	1 56	2 19	p3.9	8 11	8 50	a4.1
24	T	2 40	3 01	p4.1	8 56	9 29	a4.3
25	W	3 20	3 41	p4.3	9 37	10 06	a4.5
26	T	3 57	4 18	p4.4	10 16	10 40	4.6
27	F	4 34	4 56	4.4	10 54	11 15	a4.7
28	S	5 11	5 34	a4.5	11 34	11 50	4.7
29	S	5 51	6 15	a4.6	...	12 17	4.6
30	M	6 34	6 59	a4.5	12 28	1 04	a4.7
31	T	7 22	7 49	a4.4	1 10	1 57	a4.6

APRIL

D A Y O F M O N T H	D A Y O F W E E K	EAST Flood Starts a.m.	EAST Flood Starts p.m.	EAST Flood Starts Kts.	WEST Ebb Starts a.m.	WEST Ebb Starts p.m.	WEST Ebb Starts Kts.
1	W	8 16	8 46	a4.3	1 59	2 58	a4.5
2	T	9 18	9 52	a4.1	2 58	4 06	a4.3
3	F	10 27	11 05	a4.0	4 06	5 17	a4.2
4	S	11 38	...	4.1	5 20	6 25	a4.3
5	S	12 16	12 45	p4.2	6 30	7 25	a4.4
6	M	1 20	1 46	p4.4	7 33	8 18	a4.5
7	T	2 15	2 40	p4.5	8 30	9 06	a4.7
8	W	3 05	3 29	p4.6	9 22	9 50	a4.8
9	T	3 51	4 14	4.6	10 10	10 32	4.8
10	F	4 33	4 55	a4.6	10 56	11 11	4.7
11	S	5 14	5 35	a4.5	11 40	11 49	p4.7
12	S	5 54	6 15	a4.4	...	12 24	4.4
13	M	6 34	6 55	a4.3	12 28	1 08	a4.5
14	T	7 17	7 37	a4.1	1 07	1 55	a4.3
15	W	8 02	8 25	a3.8	1 50	2 45	a4.1
16	T	8 53	9 19	a3.6	2 39	3 41	a3.9
17	F	9 49	10 20	a3.5	3 36	4 40	a3.8
18	S	10 49	11 23	a3.5	4 39	5 39	a3.7
19	S	11 49	...	3.5	5 42	6 34	a3.8
20	M	12 22	12 45	p3.7	6 40	7 23	a3.9
21	T	1 14	1 35	p3.9	7 32	8 07	4.1
22	W	2 00	2 21	p4.1	8 20	8 48	4.3
23	T	2 43	3 04	p4.2	9 04	9 26	p4.5
24	F	3 24	3 45	4.3	9 48	10 04	p4.7
25	S	4 05	4 27	a4.5	10 31	10 42	p4.8
26	S	4 47	5 10	a4.6	11 16	11 22	p4.8
27	M	5 31	5 55	a4.7	...	12 03	4.6
28	T	6 18	6 44	a4.6	12 05	12 54	a4.8
29	W	7 09	7 37	a4.5	12 52	1 49	a4.7
30	T	8 05	8 37	a4.4	1 46	2 50	a4.6

*Daylight Saving Time starts

The Kts. (knots) columns show the **maximum** predicted velocities of the stronger one of the Flood Currents and the stronger one of the Ebb Currents for each day.

The letter "a" means the velocity shown should occur **after** the **a.m.** Current Change.

The letter "p" means the velocity shown should occur **after** the **p.m.** Current Change (even if next morning).

No "a" or "p" means a.m. and p.m. velocities are the same for that day.

Av. Max. Vel.: Fl. 4.0 Kts., Ebb 4.5 Kts.

Max. Flood & Max. Ebb 3 hrs. after Flood Starts & Ebb Starts, ±20 min.

Average rise and fall: canal east end, 8.7 ft. (time of high water same as Boston); west end, at Monument Beach, 4.0 ft. (time of high water 20 min. after Newport).

See pp. 22-29 for Current Change at other points.

Daylight Saving Time   Daylight Saving Time

		MAY										JUNE					
D A Y O F M O N T H	D A Y O F W E E K	CURRENT TURNS TO						D A Y O F M O N T H	D A Y O F W E E K	CURRENT TURNS TO							
		EAST Flood Starts			WEST Ebb Starts					EAST Flood Starts			WEST Ebb Starts				
		a.m.	p.m.	Kts.	a.m.	p.m.	Kts.			a.m.	p.m.	Kts.	a.m.	p.m.	Kts.		
1	F	9 06	9 42	a4.3	2 47	3 54	a4.5	1	M	10 55	11 32	a4.2	4 45	5 32	a4.4		
2	S	10 12	10 51	a4.2	3 54	4 59	a4.4	2	T	11 57	...	4.1	5 51	6 28	4.3		
3	S	11 19	11 58	a4.2	5 05	6 01	a4.3	3	W	12 33	12 56	p4.1	6 54	7 20	p4.3		
4	M	...	12 23	4.2	6 13	6 59	a4.4	4	T	1 29	1 50	a4.1	7 52	8 09	p4.4		
5	T	12 59	1 23	p4.3	7 15	7 51	4.4	5	F	2 20	2 40	a4.2	8 46	8 55	p4.4		
6	W	1 54	2 16	p4.3	8 12	8 38	4.5	6	S	3 07	3 26	a4.2	9 36	9 38	p4.5		
7	T	2 43	3 05	4.3	9 05	9 22	p4.6	7	S	3 51	4 09	a4.2	10 22	10 19	p4.5		
8	F	3 29	3 50	a4.4	9 53	10 04	p4.6	8	M	4 32	4 49	a4.2	11 05	10 58	p4.5		
9	S	4 12	4 32	a4.4	10 39	10 43	p4.6	9	T	5 12	5 28	a4.2	11 47	11 37	p4.4		
10	S	4 52	5 11	a4.4	11 23	11 21	p4.6	10	W	5 50	6 06	a4.1	...	12 27	4.0		
11	M	5 31	5 49	a4.3	...	12 05	4.2	11	T	6 29	6 46	a4.1	12 16	1 08	a4.4		
12	T	6 11	6 29	a4.2	12 01	12 48	a4.5	12	F	7 09	7 28	a4.0	12 56	1 49	a4.3		
13	W	6 51	7 11	a4.0	12 39	1 32	a4.3	13	S	7 51	8 13	a3.9	1 38	2 32	a4.2		
14	T	7 35	7 56	a3.9	1 21	2 18	a4.2	14	S	8 35	9 00	a3.9	2 24	3 17	a4.1		
15	F	8 21	8 46	a3.8	2 07	3 07	a4.0	15	M	9 23	9 52	a3.8	3 14	4 04	a4.0		
16	S	9 11	9 41	a3.7	2 59	3 59	a3.9	16	T	10 14	10 46	a3.8	4 09	4 53	a3.9		
17	S	10 05	10 37	a3.6	3 55	4 52	a3.8	17	W	11 08	11 41	a3.8	5 07	5 43	3.9		
18	M	11 00	11 34	a3.6	4 55	5 44	a3.8	18	T	...	12 03	3.8	6 08	6 33	p4.1		
19	T	11 55	...	3.7	5 54	6 34	3.9	19	F	12 37	12 59	3.9	7 08	7 23	p4.3		
20	W	12 27	12 48	p3.9	6 50	7 20	p4.1	20	S	1 32	1 54	a4.1	8 05	8 13	p4.5		
21	T	1 18	1 38	p4.0	7 42	8 04	p4.3	21	S	2 25	2 48	a4.3	9 00	9 02	p4.7		
22	F	2 05	2 27	4.1	8 33	8 47	p4.5	22	M	3 17	3 41	a4.6	9 54	9 52	p4.9		
23	S	2 52	3 14	4.3	9 22	9 30	p4.7	23	T	4 09	4 32	a4.7	10 46	10 42	p5.0		
24	S	3 39	4 01	a4.5	10 11	10 14	p4.9	24	W	5 00	5 24	a4.8	11 37	11 33	p5.1		
25	M	4 26	4 49	a4.7	11 00	10 59	p4.9	25	T	5 51	6 16	a4.9	...	12 29	4.7		
26	T	5 14	5 39	a4.8	11 51	11 47	p4.9	26	F	6 43	7 09	a4.8	12 26	1 21	a5.0		
27	W	6 05	6 31	a4.8	...	12 44	4.5	27	S	7 36	8 04	a4.7	1 20	2 13	a4.9		
28	T	6 57	7 26	a4.7	12 39	1 38	a4.9	28	S	8 31	9 01	a4.5	2 18	3 07	a4.8		
29	F	7 53	8 24	a4.6	1 34	2 36	a4.8	29	M	9 27	10 00	a4.3	3 18	4 02	a4.6		
30	S	8 51	9 26	a4.4	2 34	3 34	a4.7	30	T	10 26	11 01	a4.1	4 21	4 58	a4.3		
31	S	9 52	10 29	a4.3	3 39	4 34	a4.5										

The Kts. (knots) columns show the **maximum** predicted velocities of the stronger one of the Flood Currents and the stronger one of the Ebb Currents for each day.

The letter "a" means the velocity shown should occur **after** the **a.m.** Current Change.

The letter "p" means the velocity shown should occur **after** the **p.m.** Current Change (even if next morning).

No "a" or "p" means a.m. and p.m. velocities are the same for that day.

Av. Max. Vel.: Fl. 4.0 Kts., Ebb 4.5 Kts.

Max. Flood & Max. Ebb 3 hrs. after Flood Starts & Ebb Starts, ±20 min.

Average rise and fall: canal east end, 8.7 ft. (time of high water same as Boston); west end, at Monument Beach, 4.0 ft. (time of high water 20 min. after Newport).

See pp. 22-29 for Current Change at other points.

2009 CURRENT TABLE
CAPE COD CANAL
41°44.5'N, 70°36.8'W at R.R. Bridge

Daylight Saving Time								Daylight Saving Time						
JULY								**AUGUST**						
DAY OF MONTH	DAY OF WEEK	CURRENT TURNS TO						DAY OF MONTH	DAY OF WEEK	CURRENT TURNS TO				
		EAST Flood Starts			WEST Ebb Starts					EAST Flood Starts			WEST Ebb Starts	
		a.m.	p.m.	Kts.	a.m.	p.m.	Kts.			a.m.	p.m.	Kts.	a.m.	p.m.	Kts.
1	W	11 26	...	3.9	5 26	5 54	p4.2	1	S	12 32	12 56	a3.7	7 07	7 13	p4.0
2	T	12 03	12 26	a3.9	6 30	6 49	p4.2	2	S	1 30	1 53	a3.8	8 04	8 05	p4.1
3	F	1 01	1 23	a3.9	7 31	7 41	p4.2	3	M	2 22	2 42	a3.9	8 55	8 52	p4.2
4	S	1 56	2 16	a4.0	8 26	8 29	p4.2	4	T	3 08	3 26	a4.0	9 39	9 36	p4.3
5	S	2 46	3 04	a4.0	9 17	9 15	p4.3	5	W	3 49	4 05	a4.1	10 20	10 16	p4.4
6	M	3 31	3 48	a4.1	10 03	9 58	p4.4	6	T	4 27	4 41	a4.2	10 58	10 54	p4.5
7	T	4 12	4 28	a4.1	10 45	10 38	p4.4	7	F	5 03	5 17	a4.2	11 33	11 30	p4.5
8	W	4 51	5 06	a4.2	11 25	11 16	p4.4	8	S	5 38	5 51	a4.3	...	12 07	4.3
9	T	5 29	5 44	a4.2	11 59	11 54	p4.4	9	S	6 14	6 28	a4.2	12 06	12 40	a4.5
10	F	6 05	6 20	a4.2	...	12 40	4.1	10	M	6 49	7 05	a4.2	12 42	1 14	a4.4
11	S	6 42	6 58	a4.1	12 31	1 16	a4.4	11	T	7 27	7 46	a4.1	1 21	1 50	a4.3
12	S	7 20	7 38	a4.1	1 10	1 54	a4.3	12	W	8 10	8 33	a4.0	2 06	2 31	a4.2
13	M	8 00	8 22	a4.0	1 51	2 33	a4.2	13	T	8 58	9 28	3.8	2 58	3 18	4.0
14	T	8 44	9 09	a3.9	2 36	3 15	a4.1	14	F	9 54	10 30	p3.8	4 00	4 15	p4.0
15	W	9 33	10 03	a3.8	3 28	4 02	a4.0	15	S	10 58	11 37	p3.9	5 10	5 19	p4.1
16	T	10 27	11 01	3.7	4 28	4 55	p4.0	16	S	...	12 07	3.7	6 22	6 26	p4.3
17	F	11 26	...	3.7	5 34	5 51	p4.1	17	M	12 44	1 13	a4.1	7 28	7 29	p4.5
18	S	12 03	12 28	a3.9	6 40	6 50	p4.3	18	T	1 47	2 14	a4.3	8 27	8 28	p4.8
19	S	1 05	1 30	a4.1	7 44	7 47	p4.5	19	W	2 45	3 10	a4.6	9 21	9 22	p5.0
20	M	2 04	2 29	a4.3	8 43	8 43	p4.8	20	T	3 38	4 01	a4.8	10 10	10 14	p5.1
21	T	3 00	3 24	a4.6	9 38	9 36	p5.0	21	F	4 28	4 49	a4.9	10 57	11 04	p5.2
22	W	3 54	4 17	a4.8	10 30	10 28	p5.1	22	S	5 16	5 36	a4.9	11 42	11 53	p5.1
23	T	4 45	5 08	a4.9	11 19	11 19	p5.2	23	S	6 02	6 23	a4.8	...	12 27	4.9
24	F	5 35	5 58	a4.9	...	12 08	4.8	24	M	6 49	7 10	a4.6	12 43	1 12	a4.9
25	S	6 25	6 48	a4.9	12 11	12 56	a5.1	25	T	7 36	7 59	4.3	1 34	1 58	a4.7
26	S	7 14	7 39	a4.7	1 03	1 44	a5.0	26	W	8 25	8 52	4.0	2 28	2 47	4.3
27	M	8 05	8 32	a4.5	1 57	2 34	a4.8	27	T	9 18	9 50	3.7	3 27	3 41	p4.1
28	T	8 58	9 27	a4.2	2 54	3 26	a4.5	28	F	10 18	10 53	p3.6	4 30	4 40	p3.9
29	W	9 53	10 27	a3.9	3 54	4 21	4.2	29	S	11 23	11 58	p3.6	5 36	5 43	p3.8
30	T	10 53	11 29	3.7	4 59	5 18	p4.0	30	S	...	12 27	3.3	6 40	6 43	p3.8
31	F	11 55	...	3.5	6 04	6 16	p4.0	31	M	12 59	1 25	a3.7	7 36	7 38	p4.0

The Kts. (knots) columns show the **maximum** predicted velocities of the stronger one of the Flood Currents and the stronger one of the Ebb Currents for each day.

The letter "a" means the velocity shown should occur **after** the **a.m.** Current Change.

The letter "p" means the velocity shown should occur **after** the **p.m.** Current Change (even if next morning).

No "a" or "p" means a.m. and p.m. velocities are the same for that day.

Av. Max. Vel.: Fl. 4.0 Kts., Ebb 4.5 Kts.

Max. Flood & Max. Ebb 3 hrs. after Flood Starts & Ebb Starts, ±20 min.

Average rise and fall: canal east end, 8.7 ft. (time of high water same as Boston); west end, at Monument Beach, 4.0 ft. (time of high water 20 min. after Newport).

See pp. 22-29 for Current Change at other points.

2009 CURRENT TABLE
CAPE COD CANAL

41°44.5'N, 70°36.8'W at R.R. Bridge

Daylight Saving Time   Daylight Saving Time

D A Y O F M O N T H	D A Y O F W E E K	EAST Flood Starts a.m.	p.m.	Kts.	WEST Ebb Starts a.m.	p.m.	Kts.	D A Y O F M O N T H	D A Y O F W E E K	EAST Flood Starts a.m.	p.m.	Kts.	WEST Ebb Starts a.m.	p.m.	Kts.
		SEPTEMBER — CURRENT TURNS TO								**OCTOBER** — CURRENT TURNS TO					
1	T	1 53	2 15	a3.8	8 26	8 27	p4.2	1	T	2 02	2 25	a3.9	8 32	8 40	p4.3
2	W	2 39	2 58	a4.0	9 09	9 10	p4.3	2	F	2 45	3 04	a4.1	9 11	9 21	p4.4
3	T	3 20	3 37	a4.1	9 48	9 51	p4.5	3	S	3 24	3 41	4.2	9 48	10 00	p4.5
4	F	3 58	4 13	a4.2	10 24	10 28	p4.6	4	S	4 00	4 16	4.3	10 22	10 38	p4.6
5	S	4 33	4 47	a4.3	10 59	11 04	p4.6	5	M	4 37	4 52	p4.4	10 55	11 16	4.6
6	S	5 08	5 21	4.3	11 31	11 40	p4.6	6	T	5 14	5 30	p4.5	11 29	11 57	a4.6
7	M	5 43	5 57	4.3	...	12 03	4.4	7	W	5 52	6 11	p4.4	...	12 04	4.6
8	T	6 19	6 35	p4.3	12 17	12 37	a4.5	8	T	6 34	6 56	p4.3	12 41	12 44	p4.5
9	W	6 59	7 18	p4.2	12 58	1 13	a4.4	9	F	7 23	7 48	p4.2	1 30	1 29	p4.4
10	T	7 42	8 06	p4.1	1 44	1 55	4.2	10	S	8 16	8 46	p4.1	2 28	2 24	p4.3
11	F	8 33	9 02	p3.9	2 38	2 45	p4.1	11	S	9 19	9 52	p4.0	3 33	3 30	p4.2
12	S	9 32	10 08	p3.9	3 44	3 47	p4.1	12	M	10 29	11 02	p4.0	4 43	4 44	p4.2
13	S	10 41	11 18	p3.9	4 56	4 58	p4.1	13	T	11 40	...	3.7	5 51	5 55	p4.3
14	M	11 53	...	3.6	6 08	6 10	p4.3	14	W	12 10	12 46	a4.2	6 52	7 01	p4.5
15	T	12 28	1 01	a4.1	7 12	7 16	p4.5	15	T	1 13	1 44	a4.4	7 47	8 00	p4.7
16	W	1 31	2 01	a4.4	8 09	8 15	p4.8	16	F	2 09	2 36	4.5	8 36	8 54	p4.8
17	T	2 28	2 54	a4.6	9 00	9 09	p4.9	17	S	3 00	3 23	4.6	9 22	9 44	4.8
18	F	3 20	3 43	a4.8	9 47	9 59	p5.0	18	S	3 47	4 08	p4.7	10 05	10 32	a4.9
19	S	4 08	4 29	4.8	10 31	10 48	p5.0	19	M	4 31	4 52	p4.7	10 47	11 19	a4.9
20	S	4 54	5 13	4.8	11 14	11 35	4.9	20	T	5 14	5 34	p4.6	11 27	...	4.8
21	M	5 38	5 57	p4.7	11 56	...	4.9	21	W	5 56	6 17	p4.4	12 05	12 08	p4.6
22	T	6 22	6 42	p4.5	12 23	12 38	4.7	22	T	6 38	7 00	p4.2	12 51	12 49	p4.5
23	W	7 06	7 28	4.2	1 12	1 22	4.5	23	F	7 22	7 47	p3.9	1 39	1 34	p4.2
24	T	7 52	8 18	p3.9	2 03	2 09	p4.2	24	S	8 11	8 38	p3.7	2 31	2 24	p4.0
25	F	8 44	9 13	p3.7	2 59	3 01	p4.0	25	S	9 05	9 34	p3.5	3 26	3 21	p3.8
26	S	9 42	10 14	p3.5	4 00	4 01	p3.8	26	M	10 06	10 34	p3.5	4 25	4 23	p3.7
27	S	10 47	11 19	p3.5	5 03	5 06	p3.7	27	T	11 09	11 33	p3.6	5 23	5 26	p3.7
28	M	11 52	...	3.2	6 05	6 08	p3.8	28	W	...	12 07	3.3	6 17	6 25	p3.8
29	T	12 20	12 51	a3.6	7 01	7 05	p3.9	29	T	12 29	12 59	a3.7	7 06	7 17	p4.0
30	W	1 15	1 41	a3.8	7 49	7 55	p4.1	30	F	1 19	1 45	a3.9	7 50	8 05	p4.2
								31	S	2 04	2 27	4.0	8 31	8 49	p4.3

The Kts. (knots) columns show the **maximum** predicted velocities of the stronger one of the Flood Currents and the stronger one of the Ebb Currents for each day.

The letter "a" means the velocity shown should occur **after** the **a.m.** Current Change.

The letter "p" means the velocity shown should occur **after** the **p.m.** Current Change (even if next morning).

No "a" or "p" means a.m. and p.m. velocities are the same for that day.

Av. Max. Vel.: Fl. 4.0 Kts., Ebb 4.5 Kts.

Max. Flood & Max. Ebb 3 hrs. after Flood Starts & Ebb Starts, ±20 min.

Average rise and fall: canal east end, 8.7 ft. (time of high water same as Boston); west end, at Monument Beach, 4.0 ft. (time of high water 20 min. after Newport).

See pp. 22-29 for Current Change at other points.

2009 CURRENT TABLE
CAPE COD CANAL
41°44.5'N, 70°36.8'W at R.R. Bridge

Standard Time starts Nov. 1 at 2 a.m.   Standard Time

NOVEMBER

DAY OF MONTH	DAY OF WEEK	EAST Flood Starts a.m.	p.m.	Kts.	WEST Ebb Starts a.m.	p.m.	Kts.
1	S	1 46	*2 07	p4.2	*8 09	*8 32	4.4
2	M	2 27	2 47	p4.4	8 45	9 13	a4.6
3	T	3 07	3 27	p4.5	9 22	9 56	a4.7
4	W	3 48	4 08	p4.6	9 59	10 41	a4.7
5	T	4 31	4 53	p4.6	10 39	11 28	a4.7
6	F	5 17	5 41	p4.5	11 24	...	4.7
7	S	6 07	6 34	p4.4	12 21	12 14	p4.6
8	S	7 04	7 33	p4.3	1 18	1 12	p4.5
9	M	8 07	8 37	p4.2	2 20	2 17	p4.4
10	T	9 14	9 43	p4.2	3 24	3 28	p4.3
11	W	10 22	10 49	p4.2	4 27	4 38	p4.3
12	T	11 26	11 50	p4.3	5 26	5 43	p4.4
13	F	...	12 24	4.2	6 21	6 44	p4.5
14	S	12 47	1 16	p4.4	7 11	7 39	a4.6
15	S	1 39	2 05	p4.5	7 57	8 30	a4.7
16	M	2 26	2 50	p4.5	8 41	9 18	a4.7
17	T	3 11	3 33	p4.5	9 22	10 04	a4.7
18	W	3 53	4 14	p4.4	10 03	10 48	a4.7
19	T	4 33	4 55	p4.3	10 43	11 32	a4.6
20	F	5 14	5 37	p4.1	11 23	...	4.4
21	S	5 56	6 20	p4.0	12 16	12 05	p4.3
22	S	6 40	7 05	p3.8	1 02	12 51	p4.1
23	M	7 29	7 54	p3.7	1 50	1 42	p4.0
24	T	8 23	8 46	p3.6	2 41	2 38	p3.8
25	W	9 19	9 41	p3.6	3 34	3 37	p3.8
26	T	10 16	10 37	p3.6	4 26	4 37	p3.8
27	F	11 11	11 30	p3.7	5 16	5 34	p3.9
28	S	...	12 02	3.7	6 03	6 27	4.0
29	S	12 21	12 49	p4.0	6 48	7 17	4.2
30	M	1 09	1 35	p4.2	7 30	8 05	a4.4

DECEMBER

DAY OF MONTH	DAY OF WEEK	EAST Flood Starts a.m.	p.m.	Kts.	WEST Ebb Starts a.m.	p.m.	Kts.
1	T	1 55	2 20	p4.4	8 12	8 52	a4.6
2	W	2 41	3 05	p4.6	8 54	9 39	a4.8
3	T	3 27	3 51	p4.7	9 37	10 27	a4.9
4	F	4 14	4 39	p4.7	10 22	11 17	a4.9
5	S	5 03	5 29	p4.7	11 10	...	4.9
6	S	5 55	6 22	p4.6	12 09	12 03	p4.8
7	M	6 50	7 18	p4.5	1 03	1 00	p4.7
8	T	7 49	8 17	p4.3	2 00	2 02	p4.6
9	W	8 53	9 20	p4.2	2 58	3 09	p4.4
10	T	9 56	10 23	p4.1	3 58	4 17	p4.3
11	F	11 00	11 25	p4.1	4 56	5 23	a4.3
12	S	11 59	...	4.1	5 52	6 26	a4.3
13	S	12 24	12 56	p4.2	6 44	7 23	a4.4
14	M	1 18	1 47	p4.3	7 33	8 16	a4.5
15	T	2 08	2 34	p4.3	8 19	9 05	a4.5
16	W	2 53	3 17	p4.3	9 02	9 50	a4.6
17	T	3 35	3 57	p4.3	9 43	10 32	a4.6
18	F	4 14	4 37	p4.3	10 23	11 12	a4.5
19	S	4 52	5 15	p4.2	11 02	11 52	a4.5
20	S	5 31	5 54	p4.1	11 41	...	4.4
21	M	6 11	6 34	p4.0	12 32	12 22	p4.3
22	T	6 54	7 16	p3.9	1 13	1 06	p4.1
23	W	7 40	8 02	p3.8	1 55	1 54	p4.0
24	T	8 29	8 51	p3.7	2 41	2 47	p3.9
25	F	9 23	9 44	p3.6	3 30	3 46	3.8
26	S	10 19	10 41	p3.6	4 21	4 47	3.8
27	S	11 16	11 38	3.7	5 13	5 49	a3.9
28	M	...	12 12	3.9	6 04	6 47	a4.1
29	T	12 34	1 06	p4.1	6 54	7 42	a4.4
30	W	1 28	1 57	p4.4	7 43	8 34	a4.6
31	T	2 19	2 47	p4.6	8 31	9 24	a4.8

*Standard Time starts

The Kts. (knots) columns show the **maximum** predicted velocities of the stronger one of the Flood Currents and the stronger one of the Ebb Currents for each day.

The letter "a" means the velocity shown should occur **after** the **a.m.** Current Change.

The letter "p" means the velocity shown should occur **after** the **p.m.** Current Change (even if next morning).

No "a" or "p" means a.m. and p.m. velocities are the same for that day.

Av. Max. Vel.: Fl. 4.0 Kts., Ebb 4.5 Kts.

Max. Flood & Max. Ebb 3 hrs. after Flood Starts & Ebb Starts, ±20 min.

Average rise and fall: canal east end, 8.7 ft. (time of high water same as Boston); west end, at Monument Beach, 4.0 ft. (time of high water 20 min. after Newport).

See pp. 22-29 for Current Change at other points.

2009 CURRENT TABLE
WOODS HOLE, MA

41°31.2'N, 70°41.1'W 0.1 mile SW of Devil's Ft. Is. (for South End, +10 Flood, +20 Ebb)

		Standard Time						Standard Time		

JANUARY   FEBRUARY

DAY OF MONTH	DAY OF WEEK	CURRENT TURNS TO				DAY OF MONTH	DAY OF WEEK	CURRENT TURNS TO			
		SOUTHEAST Flood Starts		NORTHWEST Ebb Starts				SOUTHEAST Flood Starts		NORTHWEST Ebb Starts	
		a.m.	p.m.	a.m.	p.m.			a.m.	p.m.	a.m.	p.m.
1	T	6 31	6 54	1 25	1 16	1	S	7 27	7 52	2 05	2 20
2	F	7 13	7 37	2 04	1 59	2	M	8 17	8 43	2 49	3 17
3	S	8 00	8 24	2 46	2 48	3	T	9 15	9 43	3 41	4 24
4	S	8 52	9 16	3 32	3 46	4	W	10 21	10 51	4 41	5 38
5	M	9 49	10 14	4 23	4 50	5	T	11 30	...	5 48	6 50
6	T	10 51	11 17	5 19	5 59	6	F	12 01	12 38	6 54	7 55
7	W	11 54	...	6 17	7 06	7	S	1 07	1 40	7 56	8 53
8	T	12 20	12 55	7 16	8 09	8	S	2 06	2 36	8 53	9 45
9	F	1 22	1 55	8 12	9 06	9	M	3 01	3 28	9 46	10 34
10	S	2 19	2 49	9 06	10 00	10	T	3 50	4 16	10 37	11 20
11	S	3 13	3 41	9 58	10 51	11	W	4 37	5 02	11 25	...
12	M	4 05	4 32	10 49	11 41	12	T	5 22	5 47	12 04	12 13
13	T	4 55	5 21	11 40	...	13	F	6 07	6 32	12 47	1 01
14	W	5 44	6 10	12 29	12 31	14	S	6 53	7 17	1 31	1 51
15	T	6 34	6 59	1 17	1 23	15	S	7 41	8 04	2 16	2 43
16	F	7 24	7 49	2 06	2 17	16	M	8 32	8 56	3 04	3 41
17	S	8 17	8 42	2 55	3 14	17	T	9 29	9 54	3 57	4 44
18	S	9 13	9 37	3 47	4 15	18	W	10 32	10 59	4 55	5 50
19	M	10 13	10 37	4 42	5 19	19	T	11 37	...	5 57	6 53
20	T	11 15	11 38	5 39	6 23	20	F	12 04	12 37	6 57	7 50
21	W	...	12 15	6 35	7 24	21	S	1 02	1 30	7 51	8 39
22	T	12 37	1 11	7 29	8 19	22	S	1 51	2 16	8 38	9 22
23	F	1 31	2 00	8 19	9 07	23	M	2 34	2 56	9 21	10 01
24	S	2 18	2 44	9 05	9 51	24	T	3 13	3 34	10 01	10 37
25	S	3 01	3 24	9 46	10 31	25	W	3 49	4 10	10 38	11 11
26	M	3 39	4 01	10 25	11 08	26	T	4 23	4 45	11 14	11 44
27	T	4 16	4 37	11 02	11 43	27	F	4 59	5 21	11 51	...
28	W	4 51	5 12	11 38	...	28	S	5 36	5 59	12 17	12 30
29	T	5 26	5 48	12 17	12 14						
30	F	6 03	6 26	12 52	12 51						
31	S	6 43	7 06	1 27	1 33						

Av. Max. Vel.: Fl. 4.5 Kts., Ebb 3.6 Kts.
(There are times when velocities can exceed 7 Kts.)
The best guide for determining daily velocities at Woods Hole is to refer to velocities at Cape Cod Canal (pp. 46-51). If Flood or Ebb Current on a certain day is stronger or weaker than the Av. Max. Vel. at Cape Cod Canal, the Current for Woods Hole for that day will be similarly stronger or weaker than the Av. Max. Vel. at Woods Hole.

To hold longest fair current from Buzzards Bay headed East through Vd. & Nant. Sds. go through Woods Hole 2 1/2 hrs. after flood starts SE in Woods Hole. (Any earlier means adverse currents in the Sounds.)

See pp. 22-29 for Current Change at other points.

2009 CURRENT TABLE
WOODS HOLE, MA

41°31.2'N, 70°41.1'W   0.1 mile SW of Devil's Ft. Is. (for South End, +10 Flood, +20 Ebb)

Daylight Time starts Mar. 8 at 2 a.m.   Daylight Saving Time

		MARCH							APRIL			
D A Y O F M O N T H	D A Y O F W E E K	CURRENT TURNS TO				D A Y O F M O N T H	D A Y O F W E E K	CURRENT TURNS TO			
		SOUTHEAST Flood Starts		NORTHWEST Ebb Starts					SOUTHEAST Flood Starts		NORTHWEST Ebb Starts	
		a.m.	p.m.	a.m.	p.m.			a.m.	p.m.	a.m.	p.m.
1	S	6 16	6 40	12 52	1 13	1	W	8 36	9 06	2 54	3 53
2	M	7 01	7 27	1 31	2 02	2	T	9 38	10 12	3 53	5 01
3	T	7 53	8 20	2 16	3 01	3	F	10 47	11 25	5 01	6 12
4	W	8 52	9 23	3 11	4 10	4	S	11 58	...	6 15	7 20
5	T	10 01	10 35	4 16	5 25	5	S	12 36	1 05	7 25	8 20
6	F	11 13	11 48	5 29	6 37	6	M	1 40	2 06	8 28	9 13
7	S	...	12 23	6 39	7 40	7	T	2 35	3 00	9 25	10 01
8	S	12 55	*2 25	*8 43	*9 36	8	W	3 25	3 49	10 17	10 45
9	M	2 54	3 21	9 40	10 26	9	T	4 11	4 34	11 05	11 27
10	T	3 44	4 10	10 33	11 11	10	F	4 53	5 15	11 51	...
11	W	4 31	4 56	11 22	11 54	11	S	5 34	5 55	12 06	12 35
12	T	5 15	5 40	...	12 09	12	S	6 14	6 35	12 44	1 19
13	F	5 58	6 22	12 35	12 54	13	M	6 54	7 15	1 23	2 03
14	S	6 40	7 03	1 16	1 39	14	T	7 37	7 57	2 02	2 50
15	S	7 22	7 44	1 56	2 26	15	W	8 22	8 45	2 45	3 40
16	M	8 06	8 29	2 37	3 15	16	T	9 13	9 39	3 34	4 36
17	T	8 54	9 18	3 22	4 09	17	F	10 09	10 40	4 31	5 35
18	W	9 48	10 15	4 13	5 09	18	S	11 09	11 43	5 34	6 34
19	T	10 49	11 19	5 12	6 14	19	S	...	12 09	6 37	7 29
20	F	11 54	...	6 17	7 16	20	M	12 42	1 05	7 35	8 18
21	S	12 26	12 56	7 20	8 13	21	T	1 34	1 55	8 27	9 02
22	S	1 25	1 51	8 16	9 02	22	W	2 20	2 41	9 15	9 43
23	M	2 16	2 39	9 06	9 45	23	T	3 03	3 24	9 59	10 21
24	T	3 00	3 21	9 51	10 24	24	F	3 44	4 05	10 43	10 59
25	W	3 40	4 01	10 32	11 01	25	S	4 25	4 47	11 26	11 37
26	T	4 17	4 38	11 11	11 35	26	S	5 07	5 30	...	12 11
27	F	4 54	5 16	11 49	...	27	M	5 51	6 15	12 17	12 58
28	S	5 31	5 54	12 10	12 29	28	T	6 38	7 04	1 00	1 49
29	S	6 11	6 35	12 45	1 12	29	W	7 29	7 57	1 47	2 44
30	M	6 54	7 19	1 23	1 59	30	T	8 25	8 57	2 41	3 45
31	T	7 42	8 09	2 05	2 52						

*Daylight Saving Time starts

Av. Max. Vel.: Fl. 4.5 Kts., Ebb 3.6 Kts.

(There are times when velocities can exceed 7 Kts.)

The best guide for determining daily velocities at Woods Hole is to refer to velocities at Cape Cod Canal (pp. 46-51). If Flood or Ebb Current on a certain day is stronger or weaker than the Av. Max. Vel. at Cape Cod Canal, the Current for Woods Hole for that day will be similarly stronger or weaker than the Av. Max. Vel. at Woods Hole.

CAUTION: Going *from* Buzzards Bay *into* Vineyard Sound, whether through Woods Hole, or Robinsons Hole or Quicks Hole, *Red* Buoys must be kept on the LEFT or PORT hand, *Green* Buoys kept on the RIGHT or STARBOARD hand. You are considered to be proceeding seaward and should thus follow the rules for LEAVING a harbor.

41°31.2'N, 70°41.1'W   0.1 mile SW of Devil's Ft. Is. (for South End, +10 Flood, +20 Ebb)

Daylight Saving Time	Daylight Saving Time

		MAY						JUNE			
D A Y O F M O N T H	D A Y O F W E E K	CURRENT TURNS TO				D A Y O F M O N T H	D A Y O F W E E K	CURRENT TURNS TO			
		SOUTHEAST Flood Starts		NORTHWEST Ebb Starts				SOUTHEAST Flood Starts		NORTHWEST Ebb Starts	
		a.m.	**p.m.**	a.m.	**p.m.**			a.m.	**p.m.**	a.m.	**p.m.**
1	F	9 26	**10 02**	3 42	**4 49**	1	M	11 15	**11 52**	5 40	**6 27**
2	S	10 32	**11 11**	4 49	**5 54**	2	T	...	**12 17**	6 46	**7 23**
3	S	11 39	**...**	6 00	**6 56**	3	W	12 53	**1 16**	7 49	**8 15**
4	M	12 18	**12 43**	7 08	**7 54**	4	T	1 49	**2 10**	8 47	**9 04**
5	T	1 19	**1 43**	8 10	**8 46**	5	F	2 40	**3 00**	9 41	**9 50**
6	W	2 14	**2 36**	9 07	**9 33**	6	S	3 27	**3 46**	10 31	**10 33**
7	T	3 03	**3 25**	10 00	**10 17**	7	S	4 11	**4 29**	11 17	**11 14**
8	F	3 49	**4 10**	10 48	**10 59**	8	M	4 52	**5 09**	11 59	**11 53**
9	S	4 32	**4 52**	11 34	**11 38**	9	T	5 32	**5 48**	...	**12 42**
10	S	5 12	**5 31**	...	**12 18**	10	W	6 10	**6 26**	12 32	**1 22**
11	M	5 51	**6 09**	12 16	**1 00**	11	T	6 49	**7 06**	1 11	**2 03**
12	T	6 31	**6 49**	12 55	**1 43**	12	F	7 29	**7 48**	1 51	**2 44**
13	W	7 11	**7 31**	1 34	**2 27**	13	S	8 11	**8 33**	2 33	**3 27**
14	T	7 55	**8 16**	2 16	**3 13**	14	S	8 55	**9 20**	3 19	**4 12**
15	F	8 41	**9 06**	3 02	**4 02**	15	M	9 43	**10 12**	4 09	**4 59**
16	S	9 31	**10 01**	3 54	**4 54**	16	T	10 34	**11 06**	5 04	**5 48**
17	S	10 25	**10 57**	4 50	**5 47**	17	W	11 28	**...**	6 02	**6 38**
18	M	11 20	**11 54**	5 50	**6 39**	18	T	12 01	**12 23**	7 03	**7 28**
19	T	...	**12 15**	6 49	**7 29**	19	F	12 57	**1 19**	8 03	**8 18**
20	W	12 47	**1 08**	7 45	**8 15**	20	S	1 52	**2 14**	9 00	**9 08**
21	T	1 38	**1 58**	8 37	**8 59**	21	S	2 45	**3 08**	9 55	**9 57**
22	F	2 25	**2 47**	9 28	**9 42**	22	M	3 37	**4 01**	10 49	**10 47**
23	S	3 12	**3 34**	10 17	**10 25**	23	T	4 29	**4 52**	11 41	**11 37**
24	S	3 59	**4 21**	11 06	**11 09**	24	W	5 20	**5 44**	...	**12 32**
25	M	4 46	**5 09**	11 55	**11 54**	25	T	6 11	**6 36**	12 28	**1 24**
26	T	5 34	**5 59**	...	**12 46**	26	F	7 03	**7 29**	1 21	**2 16**
27	W	6 25	**6 51**	12 42	**1 39**	27	S	7 56	**8 24**	2 15	**3 08**
28	T	7 17	**7 46**	1 34	**2 33**	28	S	8 51	**9 21**	3 13	**4 02**
29	F	8 13	**8 44**	2 29	**3 31**	29	M	9 47	**10 20**	4 13	**4 57**
30	S	9 11	**9 46**	3 29	**4 29**	30	T	10 46	**11 21**	5 16	**5 53**
31	S	10 12	**10 49**	4 34	**5 29**						

Av. Max. Vel.: Fl. 4.5 Kts., Ebb 3.6 Kts.
(There are times when velocities can exceed 7 Kts.)

The best guide for determining daily velocities at Woods Hole is to refer to velocities at Cape Cod Canal (pp. 46-51). If Flood or Ebb Current on a certain day is stronger or weaker than the Av. Max. Vel. at Cape Cod Canal, the Current for Woods Hole for that day will be similarly stronger or weaker than the Av. Max. Vel. at Woods Hole.

To hold longest fair current from Buzzards Bay headed East through Vd. & Nant. Sds. go through Woods Hole 2 1/2 hrs. after flood starts SE in Woods Hole. (Any earlier means adverse currents in the Sounds.)

See pp. 22-29 for Current Change at other points.

2009 CURRENT TABLE
WOODS HOLE, MA

41°31.2'N, 70°41.1'W 0.1 mile SW of Devil's Ft. Is. (for South End, +10 Flood, +20 Ebb)

Daylight Saving Time   Daylight Saving Time

JULY						AUGUST			
DAY OF MONTH	DAY OF WEEK	CURRENT TURNS TO			DAY OF MONTH	DAY OF WEEK	CURRENT TURNS TO		
		SOUTHEAST Flood Starts		NORTHWEST Ebb Starts				SOUTHEAST Flood Starts		NORTHWEST Ebb Starts	
		a.m.	p.m.	a.m.	p.m.			a.m.	p.m.	a.m.	p.m.
1	W	11 46	...	6 21	6 49	1	S	12 52	1 16	8 02	8 08
2	T	12 23	12 46	7 25	7 44	2	S	1 50	2 13	8 59	9 00
3	F	1 21	1 43	8 26	8 36	3	M	2 42	3 02	9 50	9 47
4	S	2 16	2 36	9 21	9 24	4	T	3 28	3 46	10 34	10 31
5	S	3 06	3 24	10 12	10 10	5	W	4 09	4 25	11 15	11 11
6	M	3 51	4 08	10 58	10 53	6	T	4 47	5 01	11 53	11 49
7	T	4 32	4 48	11 40	11 33	7	F	5 23	5 37	...	12 28
8	W	5 11	5 26	...	12 20	8	S	5 58	6 11	12 25	1 02
9	T	5 49	6 04	12 11	12 58	9	S	6 34	6 48	1 01	1 35
10	F	6 25	6 40	12 49	1 35	10	M	7 09	7 25	1 37	2 09
11	S	7 02	7 18	1 26	2 11	11	T	7 47	8 06	2 16	2 45
12	S	7 40	7 58	2 05	2 49	12	W	8 30	8 53	3 01	3 26
13	M	8 20	8 42	2 46	3 28	13	T	9 18	9 48	3 53	4 13
14	T	9 04	9 29	3 31	4 10	14	F	10 14	10 50	4 55	5 10
15	W	9 53	10 23	4 23	4 57	15	S	11 18	11 57	6 05	6 14
16	T	10 47	11 21	5 23	5 50	16	S	...	12 27	7 17	7 21
17	F	11 46	...	6 29	6 46	17	M	1 04	1 33	8 23	8 24
18	S	12 23	12 48	7 35	7 45	18	T	2 07	2 34	9 22	9 23
19	S	1 25	1 50	8 39	8 42	19	W	3 05	3 30	10 16	10 17
20	M	2 24	2 49	9 38	9 38	20	T	3 58	4 21	11 05	11 09
21	T	3 20	3 44	10 33	10 31	21	F	4 48	5 09	11 52	11 59
22	W	4 14	4 37	11 25	11 23	22	S	5 36	5 56	...	12 37
23	T	5 05	5 28	...	12 14	23	S	6 22	6 43	12 48	1 22
24	F	5 55	6 18	12 14	1 03	24	M	7 09	7 30	1 38	2 07
25	S	6 45	7 08	1 06	1 51	25	T	7 56	8 19	2 29	2 53
26	S	7 34	7 59	1 58	2 39	26	W	8 45	9 12	3 23	3 42
27	M	8 25	8 52	2 52	3 29	27	T	9 38	10 10	4 22	4 36
28	T	9 18	9 47	3 49	4 21	28	F	10 38	11 13	5 25	5 35
29	W	10 13	10 47	4 49	5 16	29	S	11 43	...	6 31	6 38
30	T	11 13	11 49	5 54	6 13	30	S	12 18	12 47	7 35	7 38
31	F	...	12 15	6 59	7 11	31	M	1 19	1 45	8 31	8 33

Av. Max. Vel.: Fl. 4.5 Kts., Ebb 3.6 Kts.
(There are times when velocities can exceed 7 Kts.)

The best guide for determining daily velocities at Woods Hole is to refer to velocities at Cape Cod Canal (pp. 46-51). If Flood or Ebb Current on a certain day is stronger or weaker than the Av. Max. Vel. at Cape Cod Canal, the Current for Woods Hole for that day will be similarly stronger or weaker than the Av. Max. Vel. at Woods Hole.

CAUTION: Going *from* Buzzards Bay *into* Vineyard Sound, whether through Woods Hole, or Robinsons Hole or Quicks Hole, *Red* Buoys must be kept on the LEFT or PORT hand, *Green* Buoys kept on the RIGHT or STARBOARD hand. You are considered to be proceeding seaward and should thus follow the rules for LEAVING a harbor.

2009 CURRENT TABLE
WOODS HOLE, MA

41°31.2'N, 70°41.1'W 0.1 mile SW of Devil's Ft. Is. (for South End, +10 Flood, +20 Ebb)

Daylight Saving Time   Daylight Saving Time

SEPTEMBER						OCTOBER			
DAY OF MONTH	DAY OF WEEK	CURRENT TURNS TO				DAY OF MONTH	DAY OF WEEK	CURRENT TURNS TO			
		SOUTHEAST Flood Starts		NORTHWEST Ebb Starts				SOUTHEAST Flood Starts		NORTHWEST Ebb Starts	
		a.m.	p.m.	a.m.	p.m.			a.m.	p.m.	a.m.	p.m.
1	T	2 13	2 35	9 21	9 22	1	T	2 22	2 45	9 27	9 35
2	W	2 59	3 18	10 04	10 05	2	F	3 05	3 24	10 06	10 16
3	T	3 40	3 57	10 43	10 46	3	S	3 44	4 01	10 43	10 55
4	F	4 18	4 33	11 19	11 23	4	S	4 20	4 36	11 17	11 33
5	S	4 53	5 07	11 54	11 59	5	M	4 57	5 12	11 50	...
6	S	5 28	5 41	...	12 26	6	T	5 34	5 50	12 11	12 24
7	M	6 03	6 17	12 35	12 58	7	W	6 12	6 31	12 52	12 59
8	T	6 39	6 55	1 12	1 32	8	T	6 54	7 16	1 36	1 39
9	W	7 19	7 38	1 53	2 08	9	F	7 43	8 08	2 25	2 24
10	T	8 02	8 26	2 39	2 50	10	S	8 36	9 06	3 23	3 19
11	F	8 53	9 22	3 33	3 40	11	S	9 39	10 12	4 28	4 25
12	S	9 52	10 28	4 39	4 42	12	M	10 49	11 22	5 38	5 39
13	S	11 01	11 38	5 51	5 53	13	T	11 59	...	6 46	6 50
14	M	...	12 13	7 03	7 05	14	W	12 30	1 06	7 47	7 56
15	T	12 48	1 21	8 07	8 11	15	T	1 33	2 04	8 42	8 55
16	W	1 51	2 21	9 04	9 10	16	F	2 29	2 56	9 31	9 49
17	T	2 48	3 14	9 55	10 04	17	S	3 20	3 43	10 17	10 39
18	F	3 40	4 03	10 42	10 54	18	S	4 07	4 28	11 00	11 27
19	S	4 28	4 49	11 26	11 43	19	M	4 51	5 12	11 42	...
20	S	5 14	5 33	...	12 09	20	T	5 34	5 54	12 14	12 22
21	M	5 58	6 17	12 30	12 51	21	W	6 16	6 37	1 00	1 03
22	T	6 42	7 02	1 18	1 33	22	T	6 58	7 20	1 46	1 44
23	W	7 26	7 48	2 07	2 17	23	F	7 42	8 07	2 34	2 29
24	T	8 12	8 38	2 58	3 04	24	S	8 31	8 58	3 26	3 19
25	F	9 04	9 33	3 54	3 56	25	S	9 25	9 54	4 21	4 16
26	S	10 02	10 34	4 55	4 56	26	M	10 26	10 54	5 20	5 18
27	S	11 07	11 39	5 58	6 01	27	T	11 29	11 53	6 18	6 21
28	M	...	12 12	7 00	7 03	28	W	...	12 27	7 12	7 20
29	T	12 40	1 11	7 56	8 00	29	T	12 49	1 19	8 01	8 12
30	W	1 35	2 01	8 44	8 50	30	F	1 39	2 05	8 45	9 00
						31	S	2 24	2 47	9 26	9 44

Av. Max. Vel.: Fl. 4.5 Kts., Ebb 3.6 Kts.
(There are times when velocities can exceed 7 Kts.)
The best guide for determining daily velocities at Woods Hole is to refer to velocities at Cape Cod Canal (pp. 46-51). If Flood or Ebb Current on a certain day is stronger or weaker than the Av. Max. Vel. at Cape Cod Canal, the Current for Woods Hole for that day will be similarly stronger or weaker than the Av. Max. Vel. at Woods Hole.

To hold longest fair current from Buzzards Bay headed East through Vd. & Nant. Sds. go through Woods Hole 2 1/2 hrs. after flood starts SE in Woods Hole. (Any earlier means adverse currents in the Sounds.)

See pp. 22-29 for Current Change at other points.

41°31.2'N, 70°41.1'W 0.1 mile SW of Devil's Ft. Is. (for South End, +10 Flood, +20 Ebb)

Standard Time starts Nov. 1 at 2 a.m.   Standard Time

D A Y O F M O N T H	D A Y O F W E E K	SOUTHEAST Flood Starts		NORTHWEST Ebb Starts		D A Y O F M O N T H	D A Y O F W E E K	SOUTHEAST Flood Starts		NORTHWEST Ebb Starts	
		a.m.	**p.m.**	a.m.	**p.m.**			a.m.	**p.m.**	a.m.	**p.m.**
1	S	*2 06	*2 27	*9 04	*9 27	1	T	2 15	2 40	9 07	9 47
2	M	2 47	3 07	9 40	10 08	2	W	3 01	3 25	9 49	10 34
3	T	3 27	3 47	10 17	10 51	3	T	3 47	4 11	10 32	11 22
4	W	4 08	4 28	10 54	11 36	4	F	4 34	4 59	11 17	...
5	T	4 51	5 13	11 34	...	5	S	5 23	5 49	12 12	12 05
6	F	5 37	6 01	12 23	12 19	6	S	6 15	6 42	1 04	12 58
7	S	6 27	6 54	1 16	1 09	7	M	7 10	7 38	1 58	1 55
8	S	7 24	7 53	2 13	2 07	8	T	8 09	8 37	2 55	2 57
9	M	8 27	8 57	3 15	3 12	9	W	9 13	9 40	3 53	4 04
10	T	9 34	10 03	4 19	4 23	10	T	10 16	10 43	4 53	5 12
11	W	10 42	11 09	5 22	5 33	11	F	11 20	11 45	5 51	6 18
12	T	11 46	...	6 21	6 38	12	S	· ...	12 20	6 47	7 21
13	F	12 10	12 44	7 16	7 39	13	S	12 44	1 16	7 39	8 18
14	S	1 07	1 36	8 06	8 34	14	M	1 38	2 07	8 28	9 11
15	S	1 59	2 25	8 52	9 25	15	T	2 28	2 54	9 14	10 00
16	M	2 46	3 10	9 36	10 13	16	W	3 13	3 37	9 57	10 45
17	T	3 31	3 53	10 17	10 59	17	T	3 55	4 17	10 38	11 27
18	W	4 13	4 34	10 58	11 43	18	F	4 34	4 57	11 18	...
19	T	4 53	5 15	11 38	...	19	S	5 12	5 35	12 07	12 01
20	F	5 34	5 57	12 27	12 18	20	S	5 51	6 14	12 47	12 36
21	S	6 16	6 40	1 11	1 00	21	M	6 31	6 54	1 27	1 17
22	S	7 00	7 25	1 57	1 46	22	T	7 14	7 36	2 08	2 01
23	M	7 49	8 14	2 45	2 37	23	W	8 00	8 22	2 50	2 49
24	T	8 43	9 06	3 36	3 33	24	T	8 49	9 11	3 36	3 42
25	W	9 39	10 01	4 29	4 32	25	F	9 43	10 04	4 25	4 41
26	T	10 36	10 57	5 21	5 32	26	S	10 39	11 01	5 16	5 42
27	F	11 31	11 50	6 11	6 29	27	S	11 36	11 58	6 08	6 44
28	S	...	12 22	6 58	7 22	28	M	...	12 32	6 59	7 42
29	S	12 41	1 09	7 43	8 12	29	T	12 54	1 26	7 49	8 37
30	M	1 29	1 55	8 25	9 00	30	W	1 48	2 17	8 38	9 29
						31	T	2 39	3 07	9 26	10 19

*Standard Time starts

Av. Max. Vel.: Fl. 4.5 Kts., Ebb 3.6 Kts.
(There are times when velocities can exceed 7 Kts.)

The best guide for determining daily velocities at Woods Hole is to refer to velocities at Cape Cod Canal (pp. 46-51). If Flood or Ebb Current on a certain day is stronger or weaker than the Av. Max. Vel. at Cape Cod Canal, the Current for Woods Hole for that day will be similarly stronger or weaker than the Av. Max. Vel. at Woods Hole.

CAUTION: Going *from* Buzzards Bay *into* Vineyard Sound, whether through Woods Hole, or Robinsons Hole or Quicks Hole, *Red* Buoys must be kept on the LEFT or PORT hand, *Green* Buoys kept on the RIGHT or STARBOARD hand. You are considered to be proceeding seaward and should thus follow the rules for LEAVING a harbor.

My dear Captain and M. Mate,

As I cannot talk with you, I will do the next best thing. I will write you a letter.

Do you know, Captain and M. Mate of a place on the Atlantic Coast that is called "The Graveyard"? I propose to tell you something about it, and do what I can to keep vessels out of it. "The Graveyard" so called, is that part of the coast which lies — between Sow and Pigs Rocks and Naushon Island. This place has been called "The Graveyard" for many years, — because many a good craft has laid her bones there, and many a captain has lost his reputation there also. If a vessel gets into this graveyard, there must be a cause for it. Did it ever occur to you that seldom does a vessel go ashore on Gay Head, or on the south side of the Sound? but that hundreds of them have been piled up in "The Graveyard", or on the north side of the Sound? I will explain why this is so. if you are bound into Vineyard Sound in thick weather, you will probably refer to the "Gay Head and Cross Rip" table in this book, to see when the tide turns in or out. You will notice at the — head of each table that it says, "This table shows the time that the current turns Easterly and — Westerly, off Gay Head in ship channel." That — means off Gay Head when it bears about South. Now, as a rule, captains figure on the current, after they leave the Lightship, as running Easterly into the Sound, when as a matter of fact the first of the flood between the Lightship and Gay Head runs nearly North; and the current does not begin to run to the eastward until you are well into the Sound, as shown by the chart on the opposite page. Vessels bound into — Vineyard Sound from the Westward will have the current of ebb on the starboard bow. (see arrows on the hulls in the chart on the opposite page)

I have explained this matter, and I leave the rest to your judgment and careful consideration; and thus you will undoubtedly keep your vessel out of "The Graveyard".

Yours for a fair tide,

Geo. W. Eldridge.

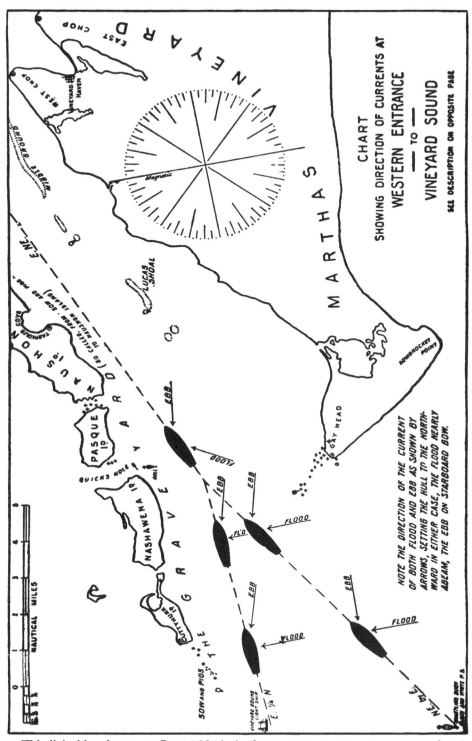

CHART
SHOWING DIRECTION OF CURRENTS AT
WESTERN ENTRANCE
— TO —
VINEYARD SOUND
SEE DESCRIPTION ON OPPOSITE PAGE

NOTE THE DIRECTION OF THE CURRENT OF BOTH FLOOD AND EBB AS SHOWN BY ARROWS, SETTING THE HULL TO THE NORTHWARD IN EITHER CASE, THE FLOOD NEARLY ABEAM, THE EBB ON STARBOARD BOW.

This lightship, shown on Capt. Eldridge's chart, on the western edge, was replaced many years ago by a buoy.

59

2009 CURRENT TABLE
POLLOCK RIP CHANNEL, MA
41°33'N, 69°59'W SE of Monomoy Pt. at Butler Hole

Standard Time   Standard Time

		JANUARY						FEBRUARY		
D A Y O F M O N T H	**D A Y O F W E E K**	**CURRENT TURNS TO**			**D A Y O F M O N T H**	**D A Y O F W E E K**	**CURRENT TURNS TO**		
		NORTHEAST Flood Starts	SOUTHWEST Ebb Starts				NORTHEAST Flood Starts	SOUTHWEST Ebb Starts	
		a.m. **p.m.** Kts.	a.m. **p.m.** Kts.				a.m. **p.m.** Kts.	a.m. **p.m.** Kts.	
1	T	6 06 **6 15** p2.1	12 22 **12 29** 1.8	1	S	6 53 **7 13** 1.9	1 08 **1 31** 1.8
2	F	6 48 **6 59** p2.0	1 04 **1 15** 1.8	2	M	7 41 **8 05** a1.9	1 55 **2 26** a1.8
3	S	7 33 **7 46** p1.9	1 48 **2 05** 1.7	3	T	8 35 **9 05** a1.8	2 47 **3 28** a1.7
4	S	8 21 **8 39** p1.8	2 35 **2 59** a1.7	4	W	9 37 **10 13** a1.7	3 46 **4 34** a1.6
5	M	9 13 **9 36** 1.7	3 25 **3 57** a1.7	5	T	10 43 **11 24** a1.7	4 49 **5 42** a1.5
6	T	10 09 **10 37** a1.8	4 18 **4 58** a1.6	6	F	11 51 ... 1.9	5 53 **6 46** a1.5
7	W	11 08 **11 40** a1.8	5 14 **6 00** a1.6	7	S	12 32 **12 55** p2.1	6 55 **7 46** a1.6
8	T	... **12 07** 1.9	6 11 **7 00** a1.7	8	S	1 34 **1 53** p2.2	7 53 **8 40** a1.8
9	F	12 44 **1 06** p2.1	7 07 **7 57** a1.7	9	M	2 31 **2 48** p2.3	8 47 **9 29** a1.9
10	S	1 42 **2 01** p2.2	8 02 **8 52** a1.8	10	T	3 21 **3 38** p2.4	9 38 **10 16** a2.0
11	S	2 39 **2 56** p2.3	8 56 **9 44** a1.9	11	W	4 08 **4 26** p2.4	10 27 **11 02** a2.1
12	M	3 33 **3 48** p2.4	9 49 **10 34** a2.0	12	T	4 54 **5 13** p2.3	11 16 **11 48** a2.0
13	T	4 24 **4 40** p2.4	10 41 **11 24** a2.0	13	F	5 40 **6 01** p2.2	... **12 06** 1.9
14	W	5 15 **5 32** p2.3	11 34 ... 2.0	14	S	6 27 **6 50** a2.1	12 34 **12 58** 1.8
15	T	6 06 **6 24** p2.2	12 14 **12 28** 1.9	15	S	7 16 **7 42** a1.9	1 22 **1 52** a1.7
16	F	6 59 **7 18** p2.1	1 05 **1 24** 1.8	16	M	8 09 **8 38** a1.9	2 13 **2 50** a1.6
17	S	7 53 **8 15** 1.9	1 57 **2 23** a1.7	17	T	9 06 **9 39** a1.8	3 08 **3 50** a1.5
18	S	8 49 **9 14** a1.9	2 51 **3 23** a1.6	18	W	10 05 **10 41** a1.8	4 06 **4 51** a1.4
19	M	9 48 **10 16** a1.9	3 47 **4 25** a1.6	19	T	11 04 **11 41** a1.9	5 04 **5 50** a1.4
20	T	10 46 **11 17** a1.9	4 43 **5 26** a1.5	20	F	11 59 ... 2.0	5 59 **6 43** a1.5
21	W	11 42 ... 2.0	5 39 **6 24** a1.5	21	S	12 35 **12 49** p2.1	6 50 **7 30** a1.6
22	T	12 14 **12 34** p2.0	6 31 **7 16** a1.6	22	S	1 22 **1 33** p2.1	7 36 **8 12** a1.7
23	F	1 07 **1 22** p2.1	7 20 **8 03** a1.6	23	M	2 03 **2 13** p2.2	8 18 **8 51** a1.8
24	S	1 53 **2 04** p2.2	8 05 **8 45** a1.7	24	T	2 40 **2 50** p2.2	8 57 **9 27** 1.8
25	S	2 34 **2 43** p2.2	8 46 **9 23** 1.7	25	W	3 15 **3 25** p2.2	9 35 **10 02** 1.9
26	M	3 12 **3 19** p2.2	9 25 **10 00** a1.8	26	T	3 48 **4 01** p2.2	10 11 **10 37** 2.0
27	T	3 47 **3 54** p2.2	10 02 **10 35** a1.9	27	F	4 22 **4 37** p2.2	10 49 **11 13** 2.0
28	W	4 21 **4 29** p2.2	10 39 **11 11** 1.9	28	S	4 57 **5 16** 2.1	11 30 **11 51** 2.0
29	T	4 55 **5 05** p2.2	11 18 **11 47** a2.0				
30	F	5 31 **5 44** p2.2	11 58 ... 1.9				
31	S	6 10 **6 26** p2.1	12 26 **12 42** 1.9				

The Kts. (knots) columns show the **maximum** predicted velocities of the stronger one of the Flood Currents and the stronger one of the Ebb Currents for each day.

The letter "a" means the velocity shown should occur **after** the **a.m.** Current Change. The letter "p" means the velocity shown should occur **after** the **p.m.** Current Change (even if next morning). No "a" or "p" means a.m. and p.m. velocities are the same for that day.

Av. Max. Vel.: Fl. 2.0 Kts., Ebb 1.8 Kts.

Max. Flood 3 hrs. 20 min. after Flood Starts, ±15 min.

Max. Ebb 2 hrs. 45 min. after Ebb Starts, ±15 min.

Gay Head (1 1/2 mi. NW of): av. max velocity, flood 2.0 kts., ebb 2.0 kts. Time of Fl. and Ebb 1 hr. 35 min. after Pollock Rip. Cross Rip: av. max. velocity, flood 1.3 kts., ebb 0.9 kts. Time of Fl. and Ebb 1 hr. 50 min. after Pollock Rip. Use POLLOCK RIP tables with current charts on pp. 66-77. See pp. 22-29 for Current Change at other points.

2009 CURRENT TABLE
POLLOCK RIP CHANNEL, MA
41°33'N, 69°59'W SE of Monomoy Pt. at Butler Hole

Daylight Time starts March 8 at 2 a.m.   Daylight Saving Time

MARCH							APRIL						
DAY OF MONTH	DAY OF WEEK	CURRENT TURNS TO					DAY OF MONTH	DAY OF WEEK	CURRENT TURNS TO				
		NORTHEAST Flood Starts			SOUTHWEST Ebb Starts				NORTHEAST Flood Starts			SOUTHWEST Ebb Starts	
		a.m.	p.m.	Kts.	a.m.	p.m.	Kts.		a.m.	p.m.	Kts.	a.m.	p.m.	Kts.
1	S	5 36	5 58	a2.1	...	12 14	1.9	1	W	7 49	8 29	a2.0	1 59	2 47	a1.8
2	M	6 20	6 47	a2.0	12 34	1 05	a1.9	2	T	8 51	9 38	a1.8	3 00	3 54	a1.6
3	T	7 10	7 42	a1.9	1 23	2 02	a1.8	3	F	10 02	10 53	a1.8	4 09	5 04	a1.5
4	W	8 08	8 47	a1.8	2 19	3 07	a1.6	4	S	11 16	...	1.8	5 20	6 12	1.4
5	T	9 15	10 00	a1.7	3 23	4 18	a1.5	5	S	12 06	12 27	p2.0	6 29	7 14	1.5
6	F	10 28	11 15	a1.8	4 33	5 28	a1.4	6	M	1 11	1 31	p2.1	7 32	8 10	p1.7
7	S	11 40	...	1.9	5 41	6 33	a1.5	7	T	2 07	2 28	p2.2	8 29	9 00	1.8
8	S	12 24	*1 45	p2.1	*7 45	*8 31	1.6	8	W	2 57	3 19	p2.3	9 21	9 46	1.9
9	M	2 25	2 44	p2.2	8 43	9 22	1.8	9	T	3 43	4 06	a2.3	10 09	10 29	1.9
10	T	3 16	3 35	p2.3	9 35	10 10	1.9	10	F	4 24	4 49	a2.3	10 54	11 10	1.9
11	W	4 03	4 23	p2.4	10 24	10 54	a2.0	11	S	5 05	5 30	a2.2	11 38	11 51	1.8
12	T	4 47	5 08	p2.3	11 11	11 37	a2.0	12	S	5 44	6 12	a2.2	...	12 22	1.7
13	F	5 30	5 52	2.2	11 57	...	1.9	13	M	6 25	6 55	a2.1	12 32	1 08	a1.7
14	S	6 12	6 36	a2.2	12 19	12 44	a1.9	14	T	7 08	7 42	a2.0	1 16	1 56	a1.7
15	S	6 55	7 21	a2.1	1 02	1 32	a1.8	15	W	7 55	8 32	a1.9	2 04	2 46	a1.6
16	M	7 41	8 10	a2.0	1 48	2 23	a1.7	16	T	8 46	9 27	a1.8	2 56	3 39	a1.5
17	T	8 30	9 04	a1.8	2 37	3 17	a1.6	17	F	9 41	10 24	a1.8	3 51	4 34	1.4
18	W	9 25	10 02	a1.8	3 31	4 15	a1.5	18	S	10 37	11 20	a1.8	4 48	5 28	1.4
19	T	10 23	11 03	a1.8	4 28	5 14	a1.4	19	S	11 33	...	1.9	5 44	6 19	p1.5
20	F	11 22	...	1.8	5 27	6 10	a1.4	20	M	12 13	12 25	p1.9	6 36	7 07	p1.6
21	S	12 03	12 18	p1.9	6 23	7 03	1.4	21	T	1 00	1 14	p2.0	7 25	7 51	p1.7
22	S	12 56	1 10	p2.0	7 16	7 51	p1.6	22	W	1 43	1 58	p2.0	8 11	8 32	p1.8
23	M	1 44	1 56	p2.1	8 03	8 34	p1.7	23	T	2 23	2 41	2.0	8 54	9 11	p1.9
24	T	2 26	2 38	p2.1	8 46	9 13	p1.8	24	F	3 01	3 22	a2.1	9 35	9 50	p2.0
25	W	3 04	3 17	p2.2	9 26	9 50	p1.9	25	S	3 39	4 03	a2.2	10 17	10 29	p2.0
26	T	3 39	3 54	p2.2	10 05	10 26	p2.0	26	S	4 18	4 46	a2.3	11 01	11 11	p2.0
27	F	4 13	4 31	2.2	10 43	11 02	2.0	27	M	5 01	5 32	a2.3	11 48	11 56	p2.0
28	S	4 48	5 10	a2.2	11 23	11 40	2.0	28	T	5 47	6 23	a2.2	...	12 39	1.8
29	S	5 26	5 51	a2.2	...	12 06	2.0	29	W	6 38	7 19	a2.2	12 47	1 34	a1.9
30	M	6 08	6 37	a2.2	12 21	12 54	a2.0	30	T	7 36	8 22	a2.0	1 44	2 35	a1.8
31	T	6 55	7 29	a2.1	1 07	1 47	a1.9								

*Daylight Saving Time starts

The Kts. (knots) columns show the **maximum** predicted velocities of the stronger one of the Flood Currents and the stronger one of the Ebb Currents for each day.

The letter "a" means the velocity shown should occur **after** the **a.m.** Current Change. The letter "p" means the velocity shown should occur **after** the **p.m.** Current Change (even if next morning). No "a" or "p" means a.m. and p.m. velocities are the same for that day.

Av. Max. Vel.: Fl. 2.0 Kts., Ebb 1.8 Kts.
Max. Flood 3 hrs. 20 min. after Flood Starts, ±15 min.
Max. Ebb 2 hrs. 45 min. after Ebb Starts, ±15 min.

Gay Head (1 1/2 mi. NW of): av. max velocity, flood 2.0 kts., ebb 2.0 kts. Time of Fl. and Ebb 1 hr. 35 min. after Pollock Rip. Cross Rip: av. max. velocity, flood 1.3 kts., ebb 0.9 kts. Time of Fl. and Ebb 1 hr. 50 min. after Pollock Rip. Use POLLOCK RIP tables with current charts on pp. 66-77. See pp. 22-29 for Current Change at other points.

41°33'N, 69°59'W SE of Monomoy Pt. at Butler Hole

Daylight Saving Time							Daylight Saving Time					

MAY — JUNE

Day of Month	Day of Week	NORTHEAST Flood Starts			SOUTHWEST Ebb Starts			Day of Month	Day of Week	NORTHEAST Flood Starts			SOUTHWEST Ebb Starts		
		a.m.	p.m.	Kts.	a.m.	p.m.	Kts.			a.m.	p.m.	Kts.	a.m.	p.m.	Kts.
1	F	8 40	9 31	a1.9	2 48	3 40	a1.6	1	M	10 41	11 24	a2.0	4 46	5 23	1.6
2	S	9 50	10 41	a1.9	3 56	4 46	1.5	2	T	11 47	...	2.0	5 52	6 21	p1.7
3	S	11 02	11 49	a2.0	5 06	5 50	1.5	3	W	12 24	12 49	2.0	6 54	7 16	p1.7
4	M	...	12 10	2.0	6 13	6 50	1.6	4	T	1 19	1 46	a2.1	7 51	8 07	p1.7
5	T	12 50	1 12	p2.1	7 15	7 44	1.7	5	F	2 10	2 38	a2.2	8 45	8 55	p1.7
6	W	1 45	2 09	2.1	8 12	8 34	p1.8	6	S	2 56	3 26	a2.2	9 34	9 39	p1.7
7	T	2 34	2 59	a2.2	9 04	9 20	1.8	7	S	3 39	4 09	a2.2	10 19	10 21	p1.7
8	F	3 20	3 46	a2.2	9 52	10 03	1.8	8	M	4 19	4 50	a2.2	11 01	11 01	p1.7
9	S	4 02	4 30	a2.2	10 37	10 44	p1.8	9	T	4 58	5 29	a2.1	11 41	11 41	p1.7
10	S	4 41	5 10	a2.2	11 20	11 25	1.7	10	W	5 35	6 07	a2.1	...	12 21	1.6
11	M	5 19	5 50	a2.1	...	12 02	1.6	11	T	6 13	6 47	a2.1	12 22	1 02	a1.7
12	T	5 59	6 31	a2.1	12 06	12 45	a1.7	12	F	6 54	7 30	a2.0	1 05	1 43	a1.7
13	W	6 40	7 15	a2.0	12 48	1 29	a1.7	13	S	7 36	8 14	a2.0	1 50	2 26	a1.7
14	T	7 23	8 02	a2.0	1 34	2 15	a1.6	14	S	8 22	9 00	a2.0	2 37	3 12	1.7
15	F	8 10	8 52	a1.9	2 23	3 03	a1.6	15	M	9 09	9 48	a1.9	3 27	3 58	p1.7
16	S	9 01	9 43	a1.9	3 14	3 53	1.5	16	T	10 00	10 37	a1.9	4 19	4 46	1.6
17	S	9 53	10 35	a1.9	4 08	4 43	1.5	17	W	10 52	11 26	1.8	5 12	5 35	p1.7
18	M	10 46	11 25	a1.9	5 02	5 32	p1.6	18	T	11 46	...	1.7	6 06	6 24	p1.7
19	T	11 38	...	1.9	5 54	6 20	p1.6	19	F	12 16	12 40	a1.8	7 00	7 13	p1.7
20	W	12 13	12 29	p1.9	6 45	7 06	p1.7	20	S	1 06	1 34	a1.9	7 53	8 02	p1.8
21	T	12 59	1 18	1.9	7 34	7 51	p1.8	21	S	1 56	2 28	a2.0	8 46	8 51	p1.8
22	F	1 43	2 05	a2.0	8 22	8 34	p1.8	22	M	2 47	3 21	a2.1	9 38	9 41	p1.9
23	S	2 26	2 52	a2.1	9 08	9 17	p1.9	23	T	3 38	4 14	a2.2	10 29	10 33	p2.0
24	S	3 09	3 39	a2.2	9 55	10 02	p2.0	24	W	4 29	5 07	a2.3	11 21	11 25	p2.0
25	M	3 55	4 27	a2.2	10 44	10 49	p2.0	25	T	5 23	6 01	a2.3	...	12 13	1.8
26	T	4 42	5 18	a2.3	11 34	11 39	p2.0	26	F	6 17	6 57	a2.3	12 20	1 07	a2.0
27	W	5 33	6 13	a2.3	...	12 27	1.8	27	S	7 14	7 54	a2.2	1 17	2 01	a1.9
28	T	6 27	7 10	a2.2	12 33	1 23	a1.9	28	S	8 12	8 53	a2.2	2 17	2 58	a1.8
29	F	7 26	8 12	a2.1	1 32	2 21	a1.8	29	M	9 13	9 53	a2.1	3 19	3 55	1.7
30	S	8 29	9 16	a2.1	2 34	3 22	a1.7	30	T	10 16	10 54	a2.0	4 22	4 53	p1.7
31	S	9 34	10 21	a2.0	3 40	4 23	1.6								

The Kts. (knots) columns show the **maximum** predicted velocities of the stronger one of the Flood Currents and the stronger one of the Ebb Currents for each day.

The letter "a" means the velocity shown should occur **after** the **a.m.** Current Change. The letter "p" means the velocity shown should occur **after** the **p.m.** Current Change (even if next morning). No "a" or "p" means a.m. and p.m. velocities are the same for that day.

Av. Max. Vel.: Fl. 2.0 Kts., Ebb 1.8 Kts.

Max. Flood 3 hrs. 20 min. after Flood Starts, ±15 min.

Max. Ebb 2 hrs. 45 min. after Ebb Starts, ±15 min.

Gay Head (1 1/2 mi. NW of): av. max velocity, flood 2.0 kts., ebb 2.0 kts. Time of Fl. and Ebb 1 hr. 35 min. after Pollock Rip. Cross Rip: av. max. velocity, flood 1.3 kts., ebb 0.9 kts. Time of Fl. and Ebb 1 hr. 50 min. after Pollock Rip. Use POLLOCK RIP tables with current charts on pp. 66-77. See pp. 22-29 for Current Change at other points.

2009 CURRENT TABLE
POLLOCK RIP CHANNEL, MA

41°33'N, 69°59'W SE of Monomoy Pt. at Butler Hole

Daylight Saving Time | Daylight Saving Time

D A Y O F M O N T H	D A Y O F W E E K	NORTHEAST Flood Starts a.m.	**p.m.**	Kts.	SOUTHWEST Ebb Starts a.m.	**p.m.**	Kts.	D A Y O F M O N T H	D A Y O F W E E K	NORTHEAST Flood Starts a.m.	**p.m.**	Kts.	SOUTHWEST Ebb Starts a.m.	**p.m.**	Kts.
		JULY — CURRENT TURNS TO								**AUGUST** — CURRENT TURNS TO					
1	W	11 20	11 54	p2.0	5 27	5 51	p1.6	1	S	12 20	12 55	a2.0	7 03	7 12	p1.6
2	T	...	12 22	1.9	6 29	6 47	p1.6	2	S	1 16	1 50	a2.1	7 58	8 04	p1.6
3	F	12 51	1 21	a2.1	7 29	7 40	p1.6	3	M	2 06	2 39	a2.1	8 47	8 50	p1.7
4	S	1 44	2 16	a2.1	8 23	8 29	p1.7	4	T	2 51	3 22	a2.2	9 30	9 33	p1.7
5	S	2 33	3 04	a2.2	9 13	9 15	p1.7	5	W	3 31	4 00	a2.2	10 10	10 12	p1.8
6	M	3 16	3 48	a2.2	9 58	9 58	p1.7	6	T	4 08	4 35	a2.2	10 46	10 50	p1.8
7	T	3 57	4 27	a2.2	10 38	10 38	p1.7	7	F	4 43	5 08	a2.2	11 21	11 27	p1.9
8	W	4 34	5 04	a2.2	11 17	11 17	p1.8	8	S	5 17	5 42	a2.2	11 56	...	1.8
9	T	5 11	5 41	a2.2	11 54	11 55	p1.8	9	S	5 53	6 18	a2.2	12 04	12 32	1.9
10	F	5 46	6 16	a2.1	...	12 31	1.7	10	M	6 29	6 54	a2.1	12 44	1 09	1.9
11	S	6 23	6 54	a2.1	12 35	1 09	1.8	11	T	7 09	7 34	a2.0	1 26	1 50	a1.9
12	S	7 02	7 34	a2.1	1 16	1 48	1.8	12	W	7 53	8 19	1.9	2 12	2 34	1.8
13	M	7 44	8 16	a2.0	2 00	2 30	1.8	13	T	8 42	9 10	1.8	3 03	3 23	p1.6
14	T	8 29	9 01	a1.9	2 47	3 15	1.7	14	F	9 38	10 07	p1.7	4 01	4 18	p1.6
15	W	9 18	9 50	1.8	3 38	4 02	p1.7	15	S	10 42	11 10	p1.7	5 04	5 18	p1.5
16	T	10 11	10 43	p1.8	4 33	4 53	p1.6	16	S	11 50	...	1.5	6 10	6 21	p1.5
17	F	11 09	11 39	p1.8	5 31	5 47	p1.6	17	M	12 16	12 57	a1.8	7 14	7 23	p1.6
18	S	...	12 09	1.6	6 31	6 42	p1.6	18	T	1 20	2 00	a2.0	8 13	8 21	p1.8
19	S	12 37	1 11	a1.9	7 31	7 38	p1.7	19	W	2 20	2 56	a2.1	9 08	9 16	p1.9
20	M	1 35	2 11	a2.0	8 28	8 33	p1.8	20	T	3 15	3 48	a2.3	9 58	10 08	p2.0
21	T	2 31	3 08	a2.1	9 23	9 27	p1.9	21	F	4 07	4 37	a2.4	10 47	10 59	p2.1
22	W	3 26	4 02	a2.2	10 15	10 20	p2.0	22	S	4 57	5 25	a2.4	11 34	11 49	2.0
23	T	4 19	4 54	a2.3	11 06	11 13	p2.0	23	S	5 46	6 12	a2.3	...	12 20	2.0
24	F	5 11	5 45	a2.4	11 56	...	1.9	24	M	6 36	7 01	a2.2	12 40	1 08	a2.0
25	S	6 04	6 37	a2.4	12 06	12 46	a2.0	25	T	7 26	7 51	2.0	1 33	1 58	1.8
26	S	6 56	7 29	a2.3	1 00	1 36	a2.0	26	W	8 20	8 46	p1.9	2 29	2 50	1.6
27	M	7 51	8 24	a2.1	1 56	2 29	a1.9	27	T	9 18	9 44	p1.8	3 28	3 47	1.5
28	T	8 48	9 21	a2.0	2 54	3 23	1.7	28	F	10 21	10 46	p1.8	4 30	4 46	1.4
29	W	9 48	10 20	p1.9	3 56	4 20	1.6	29	S	11 24	11 46	p1.9	5 32	5 46	p1.4
30	T	10 51	11 21	p1.9	4 59	5 19	p1.5	30	S	...	12 25	1.7	6 32	6 43	p1.5
31	F	11 55	...	1.7	6 03	6 17	p1.5	31	M	12 43	1 20	a2.0	7 26	7 35	p1.6

The Kts. (knots) columns show the **maximum** predicted velocities of the stronger one of the Flood Currents and the stronger one of the Ebb Currents for each day.

The letter "a" means the velocity shown should occur **after** the **a.m.** Current Change. The letter "p" means the velocity shown should occur **after** the **p.m.** Current Change (even if next morning). No "a" or "p" means a.m. and p.m. velocities are the same for that day.

Av. Max. Vel.: Fl. 2.0 Kts., Ebb 1.8 Kts.

Max. Flood 3 hrs. 20 min. after Flood Starts, ±15 min.

Max. Ebb 2 hrs. 45 min. after Ebb Starts, ±15 min.

Gay Head (1 1/2 mi. NW of): av. max velocity, flood 2.0 kts., ebb 2.0 kts. Time of Fl. and Ebb 1 hr. 35 min. after Pollock Rip. Cross Rip: av. max. velocity, flood 1.3 kts., ebb 0.9 kts. Time of Fl. and Ebb 1 hr. 50 min. after Pollock Rip. Use POLLOCK RIP tables with current charts on pp. 66-77. See pp. 22-29 for Current Change at other points.

POLLOCK RIP CHANNEL, MA
41°33'N, 69°59'W SE of Monomoy Pt. at Butler Hole

Daylight Saving Time	Daylight Saving Time

SEPTEMBER	OCTOBER

D A Y O F M O N T H	D A Y O F W E E K	CURRENT TURNS TO NORTHEAST Flood Starts a.m.	p.m.	Kts.	SOUTHWEST Ebb Starts a.m.	p.m.	Kts.	D A Y O F M O N T H	D A Y O F W E E K	CURRENT TURNS TO NORTHEAST Flood Starts a.m.	p.m.	Kts.	SOUTHWEST Ebb Starts a.m.	p.m.	Kts.
1	T	1 34	2 08	a2.1	8 14	8 22	p1.7	1	T	1 42	2 12	a2.1	8 17	8 32	1.7
2	W	2 20	2 50	a2.2	8 56	9 05	p1.8	2	F	2 24	2 50	a2.1	8 57	9 12	1.8
3	T	3 00	3 27	a2.2	9 35	9 44	p1.8	3	S	3 03	3 24	a2.2	9 34	9 51	1.9
4	F	3 37	4 01	a2.2	10 11	10 21	p1.9	4	S	3 39	3 58	2.1	10 09	10 28	1.9
5	S	4 12	4 34	a2.2	10 46	10 57	1.9	5	M	4 15	4 31	p2.2	10 44	11 07	a2.0
6	S	4 46	5 06	a2.2	11 20	11 34	p2.0	6	T	4 52	5 07	p2.2	11 20	11 47	a2.0
7	M	5 21	5 40	2.1	11 55	...	2.0	7	W	5 31	5 45	p2.2	11 59	...	2.0
8	T	5 58	6 17	2.1	12 13	12 32	1.9	8	T	6 14	6 29	p2.1	12 32	12 42	1.9
9	W	6 39	6 59	2.0	12 55	1 13	1.9	9	F	7 03	7 20	p2.0	1 22	1 31	p1.8
10	T	7 24	7 44	p1.9	1 43	1 58	1.8	10	S	7 59	8 17	p1.9	2 18	2 29	1.6
11	F	8 16	8 38	p1.8	2 37	2 51	1.6	11	S	9 03	9 24	p1.8	3 21	3 34	1.5
12	S	9 16	9 41	p1.7	3 38	3 52	1.5	12	M	10 15	10 36	p1.8	4 28	4 44	p1.5
13	S	10 25	10 51	p1.7	4 45	4 59	p1.5	13	T	11 27	11 48	p1.9	5 35	5 53	p1.5
14	M	11 38	...	1.5	5 53	6 07	p1.5	14	W	...	12 33	1.7	6 38	6 58	p1.6
15	T	12 02	12 47	a1.9	6 57	7 11	p1.6	15	T	12 55	1 32	a2.1	7 36	7 57	1.7
16	W	1 08	1 48	a2.0	7 56	8 10	p1.8	16	F	1 54	2 25	a2.2	8 28	8 51	p1.9
17	T	2 09	2 42	a2.2	8 49	9 04	p1.9	17	S	2 48	3 13	2.2	9 16	9 41	1.9
18	F	3 03	3 32	a2.3	9 38	9 55	p2.0	18	S	3 38	3 58	p2.3	10 02	10 29	1.9
19	S	3 53	4 18	a2.3	10 25	10 44	2.0	19	M	4 24	4 41	p2.3	10 45	11 16	a1.9
20	S	4 41	5 02	2.3	11 09	11 32	2.0	20	T	5 09	5 23	p2.2	11 28	...	1.8
21	M	5 27	5 47	2.2	11 54	...	1.9	21	W	5 53	6 06	p2.1	12 02	12 12	p1.8
22	T	6 14	6 32	p2.1	12 20	12 39	a1.9	22	T	6 39	6 51	p2.0	12 50	12 58	p1.7
23	W	7 02	7 20	p2.0	1 11	1 27	1.7	23	F	7 27	7 39	p1.9	1 39	1 47	p1.6
24	T	7 53	8 11	p1.9	2 03	2 18	1.6	24	S	8 19	8 31	p1.8	2 30	2 40	p1.5
25	F	8 49	9 07	p1.8	3 00	3 13	p1.5	25	S	9 14	9 26	p1.8	3 24	3 36	1.4
26	S	9 49	10 07	p1.8	3 58	4 12	p1.4	26	M	10 11	10 23	p1.8	4 18	4 33	1.4
27	S	10 50	11 07	p1.9	4 58	5 11	p1.4	27	T	11 07	11 19	p1.9	5 12	5 29	1.4
28	M	11 49	...	1.6	5 55	6 08	1.4	28	W	11 59	...	1.7	6 03	6 22	1.5
29	T	12 04	12 43	a1.9	6 47	7 01	1.5	29	T	12 11	12 46	a2.0	6 50	7 11	1.6
30	W	12 56	1 30	a2.0	7 34	7 49	p1.7	30	F	12 59	1 30	a2.0	7 34	7 56	1.7
								31	S	1 44	2 09	2.0	8 16	8 39	1.8

The Kts. (knots) columns show the **maximum** predicted velocities of the stronger one of the Flood Currents and the stronger one of the Ebb Currents for each day.

The letter "a" means the velocity shown should occur **after** the **a.m.** Current Change. The letter "p" means the velocity shown should occur **after** the **p.m.** Current Change (even if next morning). No "a" or "p" means a.m. and p.m. velocities are the same for that day.

Av. Max. Vel.: Fl. 2.0 Kts., Ebb 1.8 Kts.

Max. Flood 3 hrs. 20 min. after Flood Starts, ±15 min.

Max. Ebb 2 hrs. 45 min. after Ebb Starts, ±15 min.

Gay Head (1 1/2 mi. NW of): av. max velocity, flood 2.0 kts., ebb 2.0 kts. Time of Fl. and Ebb 1 hr. 35 min. after Pollock Rip. Cross Rip: av. max. velocity, flood 1.3 kts., ebb 0.9 kts. Time of Fl. and Ebb 1 hr. 50 min. after Pollock Rip. Use POLLOCK RIP tables with current charts on pp. 66-77. See pp. 22-29 for Current Change at other points.

2009 CURRENT TABLE
POLLOCK RIP CHANNEL, MA

41°33'N, 69°59'W SE of Monomoy Pt. at Butler Hole

Standard Time starts Nov. 1 at 2 a.m.   Standard Time

NOVEMBER

DAY OF MONTH	DAY OF WEEK	NORTHEAST Flood Starts			SOUTHWEST Ebb Starts		
		a.m.	p.m.	Kts.	a.m.	p.m.	Kts.
1	S	1 26	*1 47	p2.1	*7 55	*8 20	a1.9
2	M	2 06	2 23	p2.2	8 33	9 01	1.9
3	T	2 45	3 00	p2.2	9 10	9 43	a2.0
4	W	3 26	3 39	p2.2	9 50	10 27	a2.0
5	T	4 09	4 22	p2.2	10 32	11 14	a2.0
6	F	4 56	5 10	p2.2	11 19	...	1.9
7	S	5 49	6 04	p2.1	12 06	12 13	p1.8
8	S	6 47	7 04	p2.0	1 03	1 13	p1.7
9	M	7 53	8 12	p1.9	2 05	2 19	p1.6
10	T	9 01	9 21	p1.9	3 09	3 28	1.5
11	W	10 10	10 31	p2.0	4 14	4 37	1.5
12	T	11 14	11 37	p2.0	5 15	5 41	1.6
13	F	...	12 12	2.0	6 12	6 41	1.7
14	S	12 37	1 05	p2.2	7 05	7 36	a1.8
15	S	1 32	1 54	p2.3	7 54	8 28	1.8
16	M	2 22	2 39	p2.3	8 40	9 16	a1.8
17	T	3 09	3 21	p2.2	9 23	10 01	a1.8
18	W	3 52	4 02	p2.2	10 06	10 46	a1.8
19	T	4 35	4 43	p2.1	10 48	11 30	a1.7
20	F	5 17	5 25	p2.0	11 32	...	1.7
21	S	6 01	6 09	p2.0	12 14	12 18	p1.6
22	S	6 47	6 55	p1.9	1 00	1 06	p1.6
23	M	7 36	7 45	p1.9	1 47	1 58	1.5
24	T	8 27	8 37	p1.9	2 36	2 51	1.5
25	W	9 19	9 29	p1.9	3 26	3 45	1.5
26	T	10 09	10 22	p1.8	4 15	4 38	a1.6
27	F	10 58	11 13	1.8	5 03	5 30	a1.6
28	S	11 44	...	1.9	5 50	6 19	a1.7
29	S	12 02	12 28	p2.0	6 34	7 06	a1.7
30	M	12 49	1 10	p2.0	7 17	7 52	a1.8

DECEMBER

DAY OF MONTH	DAY OF WEEK	NORTHEAST Flood Starts			SOUTHWEST Ebb Starts		
		a.m.	p.m.	Kts.	a.m.	p.m.	Kts.
1	T	1 35	1 52	p2.1	8 00	8 37	a1.9
2	W	2 20	2 35	p2.2	8 42	9 23	a1.9
3	T	3 06	3 19	p2.3	9 26	10 10	a2.0
4	F	3 53	4 07	p2.3	10 13	11 00	a2.0
5	S	4 44	4 57	p2.3	11 04	11 52	a2.0
6	S	5 37	5 52	p2.2	11 59	...	1.9
7	M	6 35	6 52	p2.1	12 47	12 58	p1.8
8	T	7 36	7 55	p2.0	1 46	2 02	1.7
9	W	8 42	9 03	p1.9	2 46	3 09	1.6
10	T	9 46	10 10	p1.9	3 47	4 16	a1.6
11	F	10 49	11 16	1.9	4 48	5 22	a1.6
12	S	11 49	...	2.0	5 46	6 24	a1.7
13	S	12 18	12 45	p2.1	6 41	7 21	a1.7
14	M	1 15	1 35	p2.2	7 32	8 14	a1.7
15	T	2 07	2 22	p2.2	8 20	9 02	a1.8
16	W	2 54	3 04	p2.2	9 04	9 46	a1.7
17	T	3 36	3 44	p2.2	9 46	10 27	a1.7
18	F	4 16	4 23	p2.2	10 27	11 07	a1.7
19	S	4 55	5 01	p2.1	11 08	11 47	a1.7
20	S	5 34	5 40	p2.1	11 49	...	1.7
21	M	6 14	6 21	p2.0	12 27	12 33	p1.7
22	T	6 57	7 05	p2.0	1 09	1 19	1.7
23	W	7 42	7 51	p1.9	1 53	2 08	a1.7
24	T	8 29	8 41	p1.8	2 39	3 00	1.6
25	F	9 17	9 33	p1.8	3 26	3 53	a1.6
26	S	10 07	10 26	1.7	4 15	4 48	a1.6
27	S	10 58	11 21	a1.8	5 05	5 42	a1.6
28	M	11 48	...	1.9	5 54	6 35	a1.6
29	T	12 15	12 38	p2.0	6 43	7 27	a1.7
30	W	1 08	1 27	p2.1	7 32	8 17	a1.8
31	T	1 59	2 16	p2.2	8 20	9 07	a1.9

*Standard Time starts

The Kts. (knots) columns show the **maximum** predicted velocities of the stronger one of the Flood Currents and the stronger one of the Ebb Currents for each day.

The letter "a" means the velocity shown should occur **after** the **a.m.** Current Change. The letter "p" means the velocity shown should occur **after** the **p.m.** Current Change (even if next morning). No "a" or "p" means a.m. and p.m. velocities are the same for that day.

Av. Max. Vel.: Fl. 2.0 Kts., Ebb 1.8 Kts.

Max. Flood 3 hrs. 20 min. after Flood Starts, ±15 min.

Max. Ebb 2 hrs. 45 min. after Ebb Starts, ±15 min.

Gay Head (1 1/2 mi. NW of): av. max velocity, flood 2.0 kts., ebb 2.0 kts. Time of Fl. and Ebb 1 hr. 35 min. after Pollock Rip. Cross Rip: av. max. velocity, flood 1.3 kts., ebb 0.9 kts. Time of Fl. and Ebb 1 hr. 50 min. after Pollock Rip. Use POLLOCK RIP tables with current charts on pp. 66-77. See pp. 22-29 for Current Change at other points.

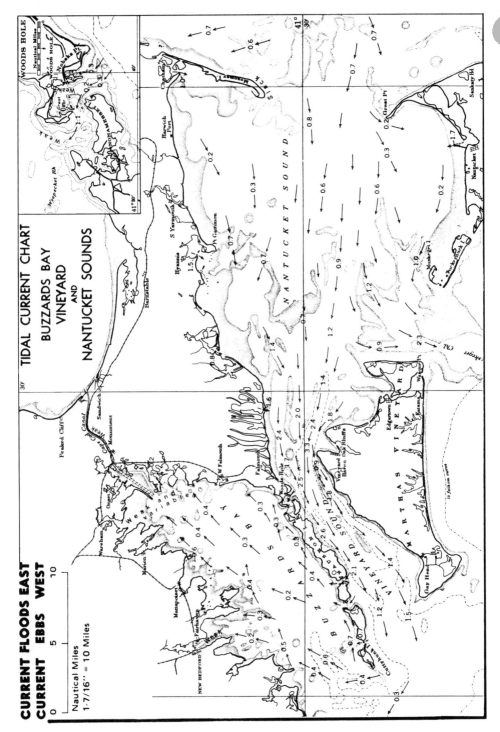

FLOOD STARTS AT POLLOCK RIP CHANNEL
OR: 4 HOURS **AFTER** HIGH WATER AT BOSTON

Velocities shown are at Spring Tides. **See Note at bottom of Boston Tables:
Rule-of-Thumb for Current Velocities.** *(Pollock Rip Ch. is SE of Monomoy Pt.)*

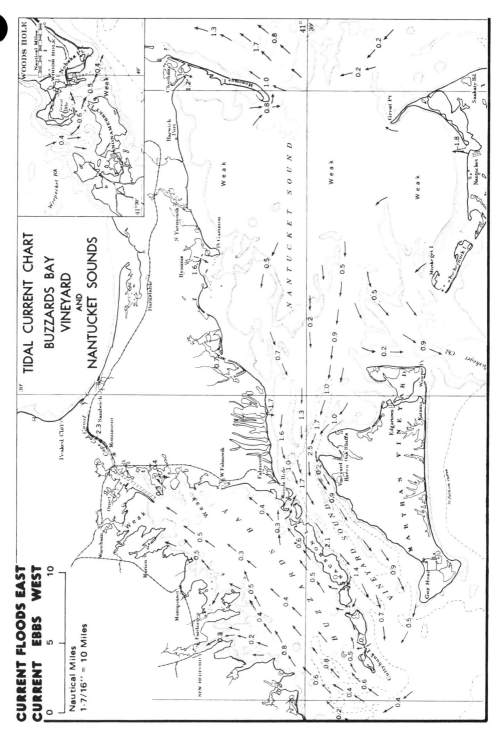

TIDAL CURRENT CHART

BUZZARDS BAY
VINEYARD
AND
NANTUCKET SOUNDS

CURRENT FLOODS EAST
CURRENT EBBS WEST

Nautical Miles
1-7/16'' = 10 Miles

1 HOUR **AFTER** FLOOD STARTS AT POLLOCK RIP CHANNEL
OR: 5 HOURS **AFTER** HIGH WATER AT BOSTON

Velocities shown are at Spring Tides. **See Note at bottom of Boston Tables:
Rule-of-Thumb for Current Velocities.**   *(Pollock Rip Ch. is SE of Monomoy Pt.)*

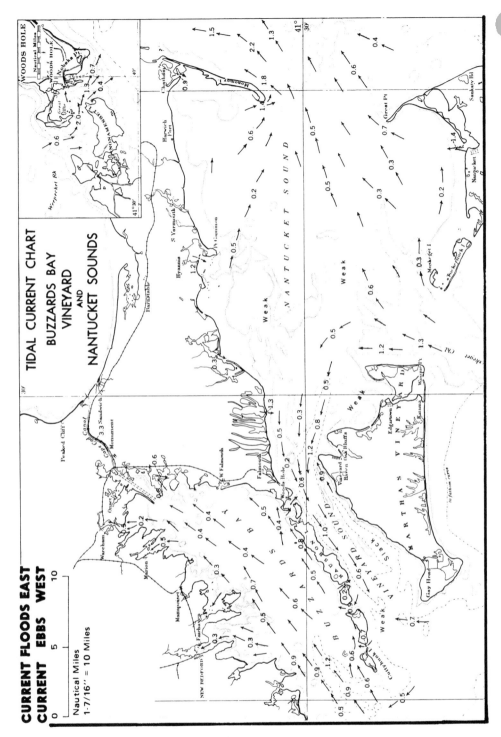

TIDAL CURRENT CHART

BUZZARDS BAY

VINEYARD

AND

NANTUCKET SOUNDS

CURRENT FLOODS EAST
CURRENT EBBS WEST

Nautical Miles
1-7/16" = 10 Miles

0   5   10

2 HOURS **AFTER** FLOOD STARTS AT POLLOCK RIP CHANNEL
OR: LOW WATER AT BOSTON

Velocities shown are at Spring Tides. **See Note at bottom of Boston Tables:**
Rule-of-Thumb for Current Velocities. *(Pollock Rip Ch. is SE of Monomoy Pt.)*

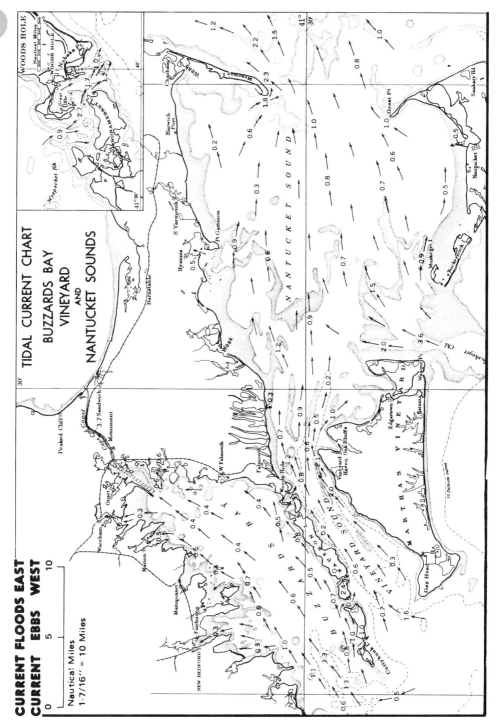

TIDAL CURRENT CHART

BUZZARDS BAY
VINEYARD
AND
NANTUCKET SOUNDS

CURRENT FLOODS EAST
CURRENT EBBS WEST

Nautical Miles
1-7/16" = 10 Miles

3 HOURS **AFTER** FLOOD STARTS AT POLLOCK RIP CHANNEL
OR: 1 HOUR **AFTER** LOW WATER AT BOSTON

Velocities shown are at Spring Tides. **See Note at bottom of Boston Tables:**
Rule-of-Thumb for Current Velocities. *(Pollock Rip Ch. is SE of Monomoy Pt.)*

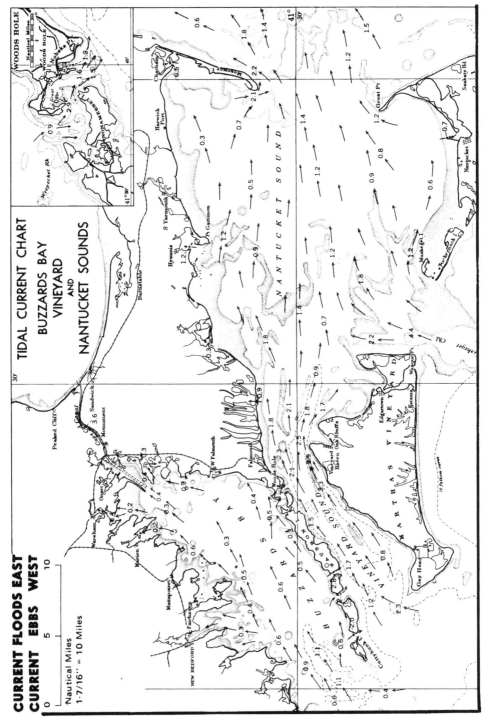

4 HOURS **AFTER** FLOOD STARTS AT POLLOCK RIP CHANNEL
OR: 2 HOURS **AFTER** LOW WATER AT BOSTON

Velocities shown are at Spring Tides. **See Note at bottom of Boston Tables:
Rule-of-Thumb for Current Velocities.** *(Pollock Rip Ch. is SE of Monomoy Pt.)*

70

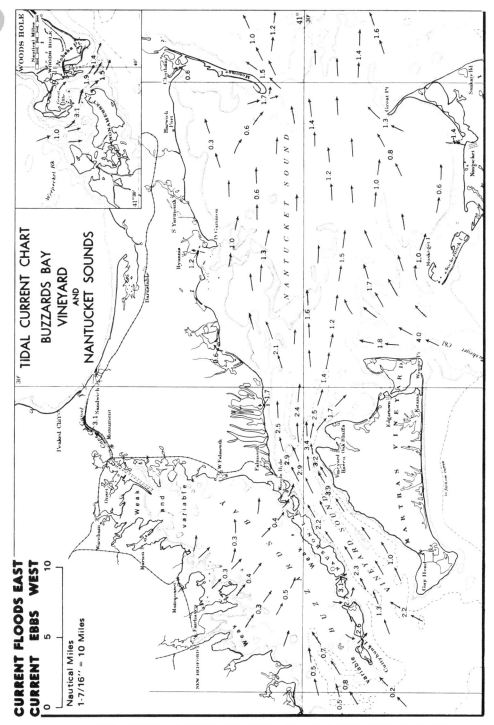

CURRENT FLOODS EAST
CURRENT EBBS WEST

TIDAL CURRENT CHART
BUZZARDS BAY
VINEYARD
AND
NANTUCKET SOUNDS

Nautical Miles
1-7/16" = 10 Miles

5 HOURS **AFTER** FLOOD STARTS AT POLLOCK RIP CHANNEL
OR: 3 HOURS **AFTER** LOW WATER AT BOSTON

Velocities shown are at Spring Tides. **See Note at bottom of Boston Tables:
Rule-of-Thumb for Current Velocities.** *(Pollock Rip Ch. is SE of Monomoy Pt.)*

71

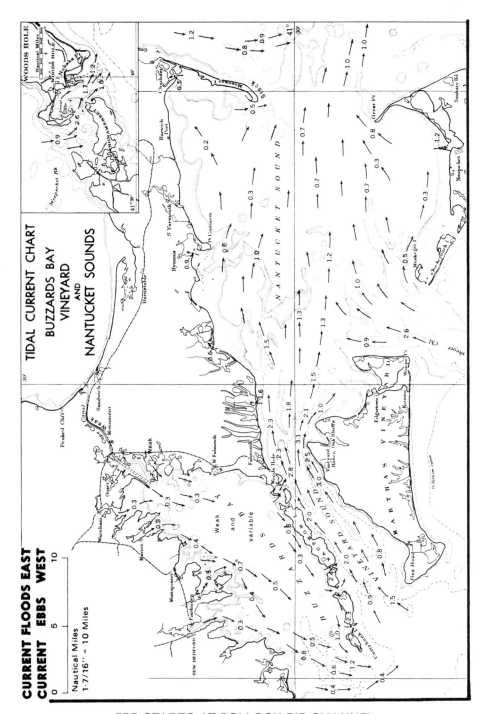

TIDAL CURRENT CHART
BUZZARDS BAY
VINEYARD
AND
NANTUCKET SOUNDS

CURRENT FLOODS EAST
CURRENT EBBS WEST

Nautical Miles
1-7/16" = 10 Miles

0   5   10

EBB STARTS AT POLLOCK RIP CHANNEL
OR: 4 HOURS **AFTER** LOW WATER AT BOSTON

Velocities shown are at Spring Tides. **See Note at bottom of Boston Tables:**
Rule-of-Thumb for Current Velocities. *(Pollock Rip Ch. is SE of Monomoy Pt.)*

72

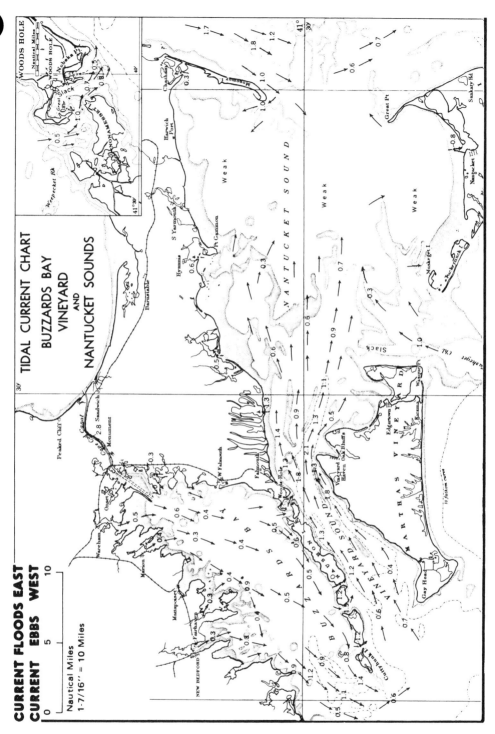

1 HOUR **AFTER** EBB STARTS AT POLLOCK RIP CHANNEL
OR: 5 HOURS **AFTER** LOW WATER AT BOSTON

Velocities shown are at Spring Tides. **See Note at bottom of Boston Tables:
Rule-of-Thumb for Current Velocities.** *(Pollock Rip Ch. is SE of Monomoy Pt.)*

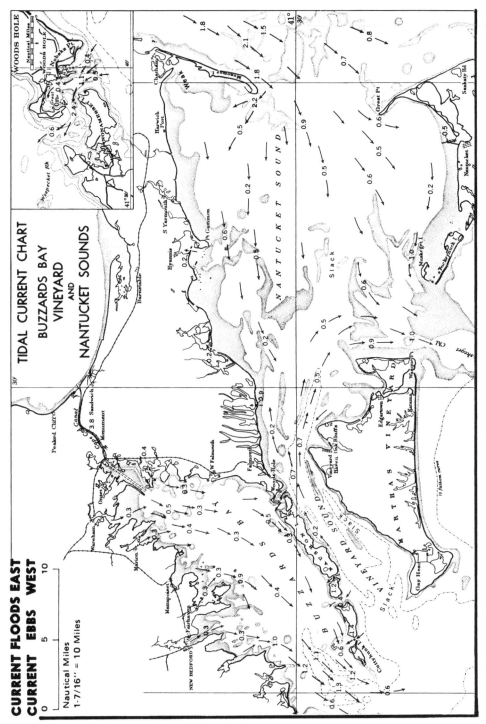

TIDAL CURRENT CHART
BUZZARDS BAY
VINEYARD
AND
NANTUCKET SOUNDS

CURRENT FLOODS EAST
CURRENT EBBS WEST

Nautical Miles
1-7/16'' = 10 Miles

2 HOURS **AFTER** EBB STARTS AT POLLOCK RIP CHANNEL
OR: HIGH WATER AT BOSTON

Velocities shown are at Spring Tides. **See Note at bottom of Boston Tables:**
Rule-of-Thumb for Current Velocities. *(Pollock Rip Ch. is SE of Monomoy Pt.)*

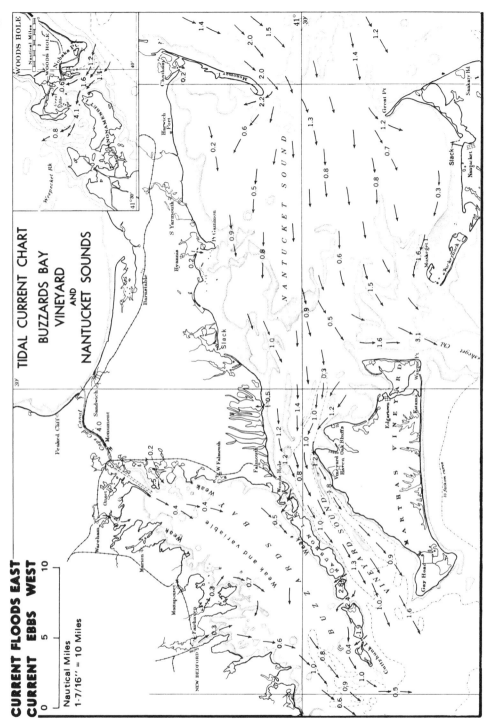

3 HOURS AFTER EBB STARTS AT POLLOCK RIP CHANNEL
OR: 1 HOUR AFTER HIGH WATER AT BOSTON

Velocities shown are at Spring Tides. **See Note at bottom of Boston Tables:**
Rule-of-Thumb for Current Velocities. *(Pollock Rip Ch. is SE of Monomoy Pt.)*

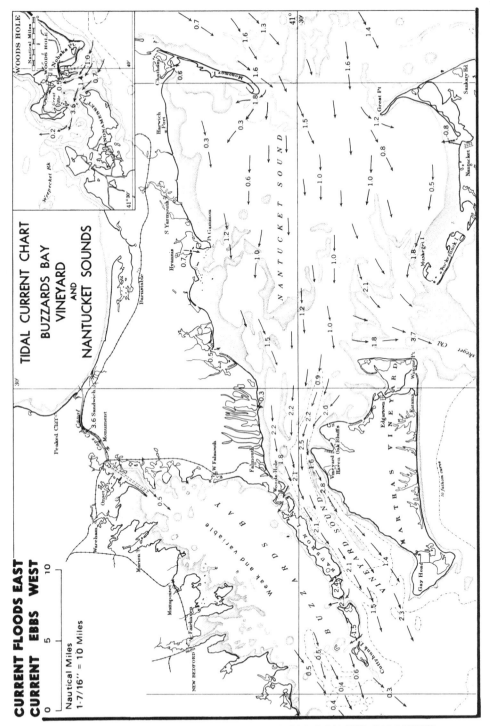

TIDAL CURRENT CHART
BUZZARDS BAY
VINEYARD
AND
NANTUCKET SOUNDS

CURRENT FLOODS EAST
CURRENT EBBS WEST

Nautical Miles
1-7/16" = 10 Miles

WOODS HOLE

4 HOURS **AFTER** EBB STARTS AT POLLOCK RIP CHANNEL
OR: 2 HOURS **AFTER** HIGH WATER AT BOSTON

Velocities shown are at Spring Tides. **See Note at bottom of Boston Tables:**
Rule-of-Thumb for Current Velocities. *(Pollock Rip Ch. is SE of Monomoy Pt.)*

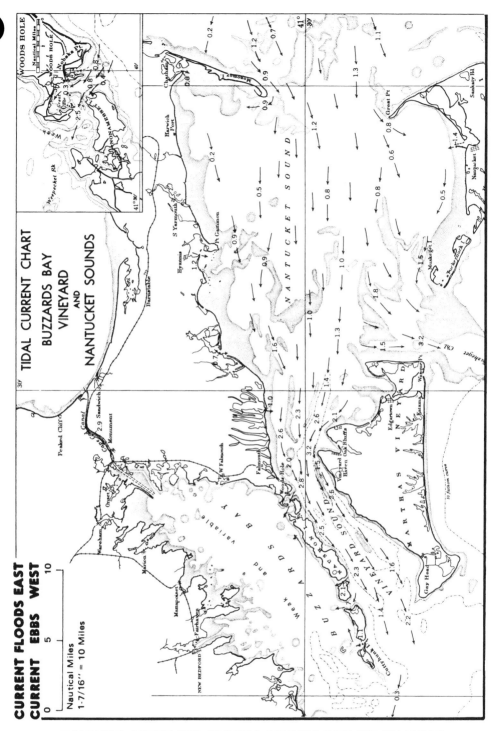

5 HOURS **AFTER** EBB STARTS AT POLLOCK RIP CHANNEL
OR: 3 HOURS **AFTER** HIGH WATER AT BOSTON

Velocities shown are at Spring Tides. **See Note at bottom of Boston Tables:
Rule-of-Thumb for Current Velocities.** *(Pollock Rip Ch. is SE of Monomoy Pt.)*

2009 HIGH & LOW WATER
NEWPORT, RI
41°30.3'N, 71°19.6'W

Standard Time   Standard Time

D A Y O F M O N T H	D A Y O F W E E K	JANUARY HIGH a.m.	Ht.	HIGH p.m.	Ht.	LOW a.m.	LOW p.m.	D A Y O F M O N T H	D A Y O F W E E K	FEBRUARY HIGH a.m.	Ht.	HIGH p.m.	Ht.	LOW a.m.	LOW p.m.
1	T	10 37	3.1	11 02	3.0	3 42	4 14	1	S	11 45	2.9	...	...	4 48	4 58
2	F	11 22	3.0	11 47	3.1	4 24	4 51	2	M	12 07	3.4	12 38	2.8	5 43	5 50
3	S	...	...	12 10	2.9	5 13	5 35	3	T	1 04	3.4	1 38	2.7	7 01	6 55
4	S	12 37	3.2	1 03	2.8	6 16	6 30	4	W	2 08	3.5	2 44	2.8	8 46	8 10
5	M	1 31	3.3	2 02	2.8	7 37	7 33	5	T	3 18	3.6	3 53	3.0	10 04	9 22
6	T	2 32	3.5	3 05	2.9	9 01	8 38	6	F	4 28	3.8	4 57	3.3	11 03	10 28
7	W	3 37	3.7	4 11	3.0	10 08	9 39	7	S	5 30	4.1	5 55	3.7	11 55	11 29
8	T	4 43	4.0	5 13	3.3	11 07	10 37	8	S	6 25	4.3	6 48	4.0	...	12 42
9	F	5 44	4.3	6 11	3.6	11 59	11 34	9	M	7 16	4.4	7 39	4.2	12 27	1 24
10	S	6 38	4.5	7 03	3.8	...	12 57	10	T	8 03	4.3	8 26	4.3	1 21	2 03
11	S	7 30	4.6	7 55	4.0	12 32	1 47	11	W	8 51	4.1	9 14	4.3	2 11	2 39
12	M	8 21	4.5	8 47	4.1	1 28	2 33	12	T	9 38	3.9	10 02	4.1	2 56	3 12
13	T	9 12	4.3	9 39	4.1	2 22	3 14	13	F	10 25	3.5	10 51	3.8	3 39	3 46
14	W	10 03	4.0	10 31	4.0	3 14	3 53	14	S	11 13	3.2	11 40	3.5	4 21	4 21
15	T	10 54	3.7	11 23	3.8	4 04	4 30	15	S	11 59	2.8	...	...	5 05	5 01
16	F	11 45	3.3	...	...	4 55	5 09	16	M	12 30	3.2	12 50	2.5	5 57	5 48
17	S	12 16	3.6	12 37	2.9	5 52	5 53	17	T	1 24	2.9	1 44	2.3	7 12	6 49
18	S	1 09	3.4	1 29	2.6	7 08	6 45	18	W	2 25	2.7	2 46	2.3	8 52	8 07
19	M	2 05	3.1	2 26	2.4	8 41	7 47	19	T	3 34	2.7	3 52	2.3	9 53	9 22
20	T	3 06	3.0	3 27	2.4	9 41	8 49	20	F	4 36	2.8	4 50	2.5	10 38	10 20
21	W	4 09	3.0	4 27	2.4	10 26	9 44	21	S	5 25	2.9	5 37	2.8	11 18	11 10
22	T	5 05	3.1	5 19	2.5	11 06	10 34	22	S	6 05	3.1	6 17	3.1	11 55	11 55
23	F	5 52	3.2	6 04	2.7	11 45	11 21	23	M	6 40	3.3	6 53	3.3	...	12 31
24	S	6 32	3.3	6 44	2.9	...	12 23	24	T	7 14	3.5	7 28	3.5	12 38	1 04
25	S	7 08	3.4	7 21	3.1	12 07	1 01	25	W	7 49	3.5	8 03	3.7	1 17	1 36
26	M	7 42	3.4	7 57	3.2	12 51	1 37	26	T	8 26	3.6	8 41	3.7	1 55	2 07
27	T	8 16	3.5	8 33	3.3	1 33	2 09	27	F	9 05	3.5	9 21	3.8	2 31	2 38
28	W	8 52	3.4	9 10	3.3	2 11	2 40	28	S	9 48	3.4	10 05	3.8	3 07	3 11
29	T	9 30	3.3	9 49	3.3	2 48	3 10								
30	F	10 11	3.2	10 31	3.4	3 24	3 41								
31	S	10 56	3.1	11 17	3.4	4 03	4 16								

Dates when Ht. of **Low** Water is below Mean Lower Low with Ht. of lowest given for each period and Date of lowest in ( ):

6th - 15th: -0.9' (11th - 12th)   1st: -0.2'
26th - 31st: -0.3' (28th - 29th)   5th - 14th: -0.9' (10th - 11th)
                                         24th - 28th: -0.4' (26th - 28th)

Average Rise and Fall 3.5 ft.

When a high tide exceeds av. ht., the *following* low tide will be lower than av.

2009 HIGH & LOW WATER
NEWPORT, RI
41°30.3'N, 71°19.6'W

Daylight Time starts March 8 at 2 a.m.   Daylight Saving Time

D A Y O F M O N T H	D A Y O F W E E K	MARCH HIGH a.m.	Ht.	MARCH HIGH p.m.	Ht.	MARCH LOW a.m.	MARCH LOW p.m.	D A Y O F M O N T H	D A Y O F W E E K	APRIL HIGH a.m.	Ht.	APRIL HIGH p.m.	Ht.	APRIL LOW a.m.	APRIL LOW p.m.
1	S	10 35	3.2	10 53	3.7	3 46	3 49	1	W	12 33	3.9	1 10	3.2	6 20	6 11
2	M	11 26	3.1	11 47	3.6	4 30	4 32	2	T	1 34	3.7	2 12	3.1	7 56	7 23
3	T	...	...	12 21	2.9	5 25	5 25	3	F	2 40	3.6	3 17	3.2	9 47	9 00
4	W	12 46	3.5	1 22	2.9	6 44	6 32	4	S	3 48	3.5	4 23	3.5	10 45	10 31
5	T	1 52	3.5	2 29	2.9	8 52	7 56	5	S	4 56	3.6	5 26	3.8	11 29	11 35
6	F	3 04	3.5	3 38	3.1	10 02	9 20	6	M	5 56	3.7	6 21	4.1	...	12 06
7	S	4 14	3.7	4 43	3.5	10 53	10 29	7	T	6 49	3.8	7 10	4.4	12 26	12 40
8	S	*6 15	3.9	*6 40	3.9	...	†12 36	8	W	7 36	3.8	7 56	4.5	1 12	1 12
9	M	7 10	4.1	7 31	4.2	12 29	1 15	9	T	8 21	3.8	8 40	4.5	1 53	1 45
10	T	7 57	4.1	8 18	4.4	1 22	1 51	10	F	9 03	3.7	9 22	4.4	2 31	2 20
11	W	8 43	4.1	9 03	4.5	2 10	2 26	11	S	9 45	3.5	10 04	4.1	3 08	2 56
12	T	9 27	3.9	9 48	4.4	2 53	2 59	12	S	10 27	3.3	10 46	3.8	3 44	3 33
13	F	10 11	3.7	10 32	4.2	3 33	3 32	13	M	11 10	3.1	11 30	3.5	4 21	4 12
14	S	10 55	3.4	11 17	3.8	4 11	4 07	14	T	11 54	2.9	...	...	5 00	4 52
15	S	11 40	3.1	...	...	4 49	4 43	15	W	12 15	3.2	12 41	2.7	5 43	5 38
16	M	12 03	3.5	12 26	2.8	5 28	5 22	16	T	1 02	3.0	1 30	2.7	6 35	6 33
17	T	12 51	3.2	1 13	2.6	6 13	6 08	17	F	1 52	2.8	2 23	2.7	7 45	7 49
18	W	1 42	2.9	2 05	2.4	7 12	7 06	18	S	2 47	2.7	3 20	2.7	9 05	9 23
19	T	2 39	2.7	3 03	2.4	8 44	8 27	19	S	3 46	2.7	4 17	2.9	10 03	10 31
20	F	3 44	2.6	4 08	2.5	10 07	9 58	20	M	4 44	2.8	5 10	3.2	10 46	11 21
21	S	4 49	2.7	5 10	2.7	10 57	11 02	21	T	5 37	3.0	5 57	3.6	11 24	...
22	S	5 43	2.8	5 59	3.0	11 37	11 51	22	W	6 23	3.3	·6 40	3.9	12 05	12 02
23	M	6 26	3.1	6 41	3.3	...	12 14	23	T	7 08	3.5	7 23	4.2	12 48	12 39
24	T	7 05	3.3	7 19	3.6	12 35	12 49	24	F	7 51	3.6	8 06	4.4	1 31	1 19
25	W	7 42	3.5	7 55	3.9	1 16	1 23	25	S	8 36	3.7	8 51	4.6	2 15	2 00
26	T	8 20	3.6	8 33	4.1	1 56	1 57	26	S	9 22	3.7	9 38	4.6	3 00	2 43
27	F	9 00	3.7	9 14	4.2	2 35	2 32	27	M	10 12	3.7	10 30	4.5	3 45	3 28
28	S	9 43	3.6	9 57	4.2	3 14	3 08	28	T	11 05	3.6	11 25	4.3	4 32	4 15
29	S	10 29	3.5	10 45	4.2	3 54	3 47	29	W	...	...	12 01	3.5	5 24	5 07
30	M	11 19	3.4	11 37	4.0	4 36	4 28	30	T	12 23	4.1	1 00	3.5	6 27	6 08
31	T	...	...	12 12	3.2	5 23	5 15								

*Daylight Saving Time starts

Dates when Ht. of **Low** Water is below Mean Lower Low with Ht. of lowest given for each period and Date of lowest in ( ):

1st - 2nd: -0.3' (1st)
7th - 14th: -0.7' (11th - 12th)
25th - 30th: -0.4' (27th - 29th)

7th - 11th: -0.4' (9th - 10th)
23rd - 28th: -0.4' (25th - 27th)

Average Rise and Fall 3.5 ft.

When a high tide exceeds av. ht., the *following* low tide will be lower than av.

2009 HIGH & LOW WATER
NEWPORT, RI
41°30.3'N, 71°19.6'W

Daylight Saving Time   Daylight Saving Time

DAY OF MONTH	DAY OF WEEK	MAY HIGH a.m.	Ht.	MAY HIGH p.m.	Ht.	MAY LOW a.m.	MAY LOW p.m.	DAY OF MONTH	DAY OF WEEK	JUNE HIGH a.m.	Ht.	JUNE HIGH p.m.	Ht.	JUNE LOW a.m.	JUNE LOW p.m.
1	F	1 23	3.9	2 00	3.6	7 59	7 27	1	M	3 04	3.4	3 40	4.0	9 29	10 26
2	S	2 25	3.7	3 02	3.7	9 20	9 19	2	T	4 04	3.3	4 39	4.0	10 11	11 18
3	S	3 29	3.5	4 04	3.8	10 14	10 37	3	W	5 04	3.2	5 35	4.1	10 46	11 59
4	M	4 32	3.4	5 04	4.0	10 55	11 32	4	T	5 59	3.2	6 26	4.1	11 19	...
5	T	5 32	3.4	5 59	4.2	11 28	...	5	F	6 49	3.2	7 13	4.1	12 38	12 01
6	W	6 25	3.5	6 48	4.4	12 17	12 01	6	S	7 34	3.2	7 56	4.1	1 13	12 35
7	T	7 13	3.5	7 34	4.4	12 56	12 31	7	S	8 16	3.2	8 37	4.0	1 49	1 17
8	F	7 57	3.5	8 16	4.4	1 33	1 07	8	M	8 57	3.2	9 16	3.9	2 26	2 01
9	S	8 40	3.5	8 59	4.2	2 08	1 45	9	T	9 38	3.2	9 56	3.7	3 04	2 45
10	S	9 20	3.4	9 38	4.0	2 45	2 24	10	W	10 18	3.2	10 33	3.6	3 42	3 29
11	M	10 01	3.2	10 19	3.8	3 22	3 05	11	T	11 00	3.2	11 12	3.4	4 20	4 12
12	T	10 43	3.1	11 00	3.5	4 00	3 47	12	F	11 42	3.1	11 52	3.3	4 57	4 55
13	W	11 27	3.0	11 42	3.3	4 39	4 30	13	S	...	...	12 24	3.2	5 34	5 40
14	T	...	...	12 12	2.9	5 20	5 15	14	S	12 34	3.1	1 06	3.2	6 14	6 32
15	F	12 26	3.1	12 58	2.9	6 05	6 06	15	M	1 19	3.0	1 51	3.3	7 00	7 37
16	S	1 10	3.0	1 45	2.9	6 57	7 10	16	T	2 08	3.0	2 39	3.4	7 52	8 53
17	S	1 58	2.9	2 33	3.0	7 58	8 32	17	W	3 02	3.0	3 32	3.6	8 47	10 01
18	M	2 50	2.8	3 24	3.2	8 57	9 47	18	T	4 01	3.0	4 31	3.9	9 42	10 59
19	T	3 46	2.9	4 17	3.4	9 47	10 43	19	F	5 04	3.1	5 31	4.2	10 34	11 54
20	W	4 45	3.0	5 11	3.8	10 32	11 31	20	S	6 04	3.3	6 29	4.5	11 26	...
21	T	5 41	3.2	6 03	4.1	11 15	...	21	S	7 00	3.6	7 24	4.7	12 48	12 19
22	F	6 34	3.4	6 53	4.4	12 18	12 01	22	M	7 54	3.8	8 17	4.9	1 44	1 14
23	S	7 24	3.6	7 42	4.7	1 07	12 45	23	T	8 47	4.0	9 10	4.9	2 38	2 10
24	S	8 13	3.8	8 32	4.8	1 57	1 33	24	W	9 40	4.2	10 03	4.8	3 30	3 06
25	M	9 04	3.8	9 24	4.8	2 49	2 23	25	T	10 35	4.2	10 57	4.6	4 18	4 03
26	T	9 56	3.9	10 17	4.7	3 40	3 15	26	F	11 30	4.3	11 51	4.3	5 04	5 00
27	W	10 51	3.9	11 13	4.5	4 31	4 08	27	S	...	...	12 25	4.3	5 50	6 00
28	T	11 48	3.9	...	...	5 23	5 04	28	S	12 46	3.9	1 20	4.2	6 37	7 13
29	F	12 10	4.2	12 46	3.9	6 19	6 08	29	M	1 40	3.6	2 15	4.1	7 29	8 48
30	S	1 07	4.0	1 43	3.9	7 25	7 31	30	T	2 35	3.3	3 12	4.0	8 24	10 03
31	S	2 05	3.7	2 41	4.0	8 34	9 16								

Dates when Ht. of **Low** Water is below Mean Lower Low with Ht. of lowest given for each period and Date of lowest in ( ):

22nd - 27th: -0.5' (25th)   20th - 26th: -0.5' (22nd - 23rd)

Average Rise and Fall 3.5 ft.

When a high tide exceeds av. ht., the *following* low tide will be lower than av.

2009 HIGH & LOW WATER
NEWPORT, RI
41°30.3'N, 71°19.6'W

		Daylight Saving Time						Daylight Saving Time			

DAY OF MONTH	DAY OF WEEK	JULY				DAY OF MONTH	DAY OF WEEK	AUGUST			
		HIGH		LOW				HIGH		LOW	
		a.m. / Ht.	p.m. / Ht.	a.m.	p.m.			a.m. / Ht.	p.m. / Ht.	a.m.	p.m.
1	W	3 33 3.1	4 11 3.9	9 17	10 58	1	S	5 04 2.8	5 43 3.5	10 20	11 59
2	T	4 33 2.9	5 10 3.8	10 03	11 43	2	S	6 01 2.9	6 34 3.6	11 12	...
3	F	5 32 2.9	6 05 3.8	10 46	...	3	M	6 49 3.1	7 17 3.7	12 34	12 01
4	S	6 25 3.0	6 54 3.8	12 21	-A-	4	T	7 31 3.3	7 54 3.7	1 08	12 47
5	S	7 12 3.1	7 37 3.9	12 55	12 13	5	W	8 09 3.4	8 29 3.8	1 43	1 33
6	M	7 55 3.2	8 17 3.9	1 31	12 58	6	T	8 46 3.6	9 02 3.8	2 18	2 16
7	T	8 35 3.3	8 54 3.8	2 08	1 45	7	F	9 21 3.6	9 36 3.7	2 52	2 57
8	W	9 13 3.3	9 30 3.7	2 45	2 30	8	S	9 56 3.7	10 12 3.7	3 24	3 35
9	T	9 53 3.4	10 07 3.6	3 21	3 14	9	S	10 34 3.7	10 52 3.5	3 54	4 11
10	F	10 30 3.4	10 42 3.5	3 56	3 54	10	M	11 12 3.7	11 33 3.4	4 25	4 48
11	S	11 08 3.4	11 21 3.4	4 28	4 34	11	T	11 55 3.7	... ...	4 57	5 28
12	S	11 48 3.4	... ...	5 00	5 14	12	W	12 20 3.3	12 42 3.8	5 35	6 18
13	M	12 02 3.3	12 29 3.5	5 34	5 58	13	T	1 10 3.1	1 35 3.8	6 21	7 24
14	T	12 46 3.2	1 13 3.5	6 13	6 51	14	F	2 06 3.1	2 34 3.8	7 20	9 03
15	W	1 35 3.1	2 02 3.6	7 00	8 01	15	S	3 09 3.1	3 41 3.9	8 30	10 31
16	T	2 29 3.0	2 58 3.8	7 57	9 24	16	S	4 16 3.2	4 50 4.1	9 44	11 31
17	F	3 30 3.0	4 01 3.9	9 00	10 37	17	M	5 24 3.5	5 55 4.4	10 53	...
18	S	4 35 3.1	5 07 4.2	10 03	11 39	18	T	6 25 3.9	6 53 4.6	12 22	12 01
19	S	5 41 3.3	6 10 4.5	11 04	...	19	W	7 20 4.3	7 45 4.8	1 09	12 55
20	M	6 41 3.7	7 08 4.7	12 35	12 03	20	T	8 11 4.6	8 35 4.8	1 54	1 52
21	T	7 37 4.0	8 02 4.9	1 29	1 02	21	F	9 01 4.8	9 24 4.7	2 35	2 46
22	W	8 30 4.3	8 54 4.9	2 20	2 01	22	S	9 50 4.9	10 12 4.4	3 14	3 36
23	T	9 22 4.5	9 45 4.8	3 08	2 58	23	S	10 40 4.7	11 01 4.1	3 52	4 23
24	F	10 14 4.6	10 36 4.5	3 51	3 53	24	M	11 30 4.5	11 51 3.7	4 28	5 08
25	S	11 07 4.6	11 28 4.2	4 31	4 46	25	T	... ...	12 21 4.2	5 05	5 55
26	S	11 59 4.5	... ...	5 10	5 39	26	W	12 42 3.4	1 14 3.9	5 45	6 51
27	M	12 20 3.9	12 52 4.3	5 49	6 37	27	T	1 34 3.1	2 09 3.6	6 31	8 27
28	T	1 12 3.5	1 46 4.0	6 31	7 54	28	F	2 29 2.8	3 09 3.4	7 30	10 02
29	W	2 05 3.2	2 41 3.8	7 20	9 29	29	S	3 29 2.7	4 14 3.3	8 46	10 52
30	T	3 01 2.9	3 41 3.6	8 19	10 34	30	S	4 33 2.8	5 16 3.3	10 03	11 30
31	F	4 02 2.8	4 44 3.5	9 23	11 22	31	M	5 33 2.9	6 07 3.4	11 02	11 59

A also at 11:28 a.m.

Dates when Ht. of **Low** Water is below Mean Lower Low with Ht. of lowest given for each period and Date of lowest in ( ):

20th - 26th: -0.5' (22nd - 24th)   18th - 24th: -0.5' (20th - 21st)

Average Rise and Fall 3.5 ft.

When a high tide exceeds av. ht., the *following* low tide will be lower than av.

2009 HIGH & LOW WATER
NEWPORT, RI
41°30.3'N, 71°19.6'W

Daylight Saving Time | Daylight Saving Time

D A Y O F M O N T H	D A Y O F W E E K	SEPTEMBER						D A Y O F M O N T H	D A Y O F W E E K	OCTOBER					
		HIGH				LOW				HIGH				LOW	
		a.m.	Ht.	p.m.	Ht.	a.m.	p.m.			a.m.	Ht.	p.m.	Ht.	a.m.	p.m.
1	T	6 22	3.2	6 49	3.5	11 50	...	1	T	6 27	3.5	6 48	3.5	...	12 15
2	W	7 03	3.4	7 25	3.7	12 36	12 35	2	F	7 03	3.8	7 24	3.7	12 29	12 55
3	T	7 39	3.7	7 58	3.8	1 09	1 17	3	S	7 39	4.0	8 00	3.8	1 02	1 34
4	F	8 13	3.8	8 32	3.8	1 43	1 58	4	S	8 14	4.2	8 37	3.8	1 36	2 13
5	S	8 47	3.9	9 06	3.8	2 15	2 36	5	M	8 52	4.3	9 18	3.8	2 10	2 51
6	S	9 22	4.0	9 44	3.7	2 47	3 13	6	T	9 33	4.3	10 02	3.6	2 45	3 30
7	M	10 00	4.0	10 24	3.6	3 18	3 48	7	W	10 18	4.3	10 50	3.5	3 22	4 10
8	T	10 42	4.0	11 09	3.5	3 51	4 25	8	T	11 09	4.2	11 43	3.4	4 02	4 55
9	W	11 29	4.0	...	...	4 26	5 07	9	F	...	...	12 05	4.1	4 47	5 49
10	T	12 01	3.3	12 20	3.9	5 06	5 57	10	S	12 40	3.3	1 04	3.9	5 40	7 06
11	F	12 53	3.2	1 17	3.9	5 55	7 06	11	S	1 41	3.3	2 07	3.9	6 45	9 12
12	S	1 52	3.1	2 19	3.9	6 56	9 12	12	M	2 44	3.4	3 13	3.8	8 13	10 14
13	S	2 56	3.2	3 27	3.9	8 14	10 30	13	T	3 49	3.6	4 19	3.9	9 49	10 59
14	M	4 03	3.4	4 36	4.0	9 40	11 20	14	W	4 52	4.0	5 21	4.0	10 59	11 36
15	T	5 09	3.7	5 40	4.2	10 53	11 59	15	T	5 50	4.3	6 16	4.1	11 53	...
16	W	6 08	4.2	6 36	4.4	11 54	...	16	F	6 42	4.6	7 06	4.1	12 11	12 41
17	T	7 01	4.6	7 27	4.5	12 42	12 49	17	S	7 30	4.8	7 53	4.1	12 46	1 26
18	F	7 51	4.8	8 14	4.5	1 20	1 40	18	S	8 16	4.8	8 37	4.0	1 21	2 08
19	S	8 38	5.0	9 00	4.4	1 58	2 28	19	M	9 00	4.7	9 22	3.8	1 58	2 48
20	S	9 25	4.9	9 47	4.2	2 35	3 13	20	T	9 45	4.5	10 06	3.6	2 36	3 28
21	M	10 12	4.7	10 33	3.9	3 12	3 55	21	W	10 31	4.2	10 52	3.3	3 15	4 07
22	T	11 00	4.4	11 21	3.5	3 49	4 36	22	T	11 18	3.8	11 40	3.1	3 55	4 47
23	W	11 49	4.0	...	...	4 27	5 18	23	F	...	...	12 07	3.5	4 37	5 32
24	T	12 11	3.2	12 41	3.7	5 07	6 06	24	S	12 30	2.9	12 58	3.2	5 23	6 25
25	F	1 02	3.0	1 35	3.4	5 53	7 10	25	S	1 22	2.8	1 50	3.0	6 18	7 37
26	S	1 56	2.8	2 33	3.2	6 50	9 03	26	M	2 16	2.8	2 44	2.9	7 33	8 59
27	S	2 55	2.8	3 35	3.1	8 11	10 08	27	T	3 11	2.9	3 39	2.9	9 11	9 53
28	M	3 57	2.8	4 35	3.1	9 46	10 48	28	W	4 06	3.0	4 32	3.0	10 19	10 33
29	T	4 56	3.0	5 27	3.2	10 48	11 22	29	T	4 57	3.3	5 21	3.1	11 06	11 10
30	W	5 45	3.3	6 10	3.3	11 34	11 55	30	F	5 42	3.5	6 05	3.3	11 48	11 45
								31	S	6 24	3.8	6 47	3.5	...	12 28

Dates when Ht. of **Low** Water is below Mean Lower Low with Ht. of lowest given for each period and Date of lowest in ( ):

17th - 21st: -0.4' (18th - 19th)   16th - 19th: -0.3' (18th)

Average Rise and Fall 3.5 ft.

When a high tide exceeds av. ht., the *following* low tide will be lower than av.

2009 HIGH & LOW WATER
NEWPORT, RI
41°30.3'N, 71°19.6'W

Standard Time starts Nov. 1 at 2 a.m.   Standard Time

DAY OF MONTH	DAY OF WEEK	NOVEMBER HIGH a.m.	Ht.	NOVEMBER HIGH p.m.	Ht.	LOW a.m.	LOW p.m.	DAY OF MONTH	DAY OF WEEK	DECEMBER HIGH a.m.	Ht.	DECEMBER HIGH p.m.	Ht.	LOW a.m.	LOW p.m.
1	S	*6 04	4.1	*6 29	3.6	12 21	*12 09	1	T	6 20	4.3	6 48	3.5	...	12 32
2	M	6 45	4.3	7 11	3.7	12 01	12 50	2	W	7 08	4.5	7 37	3.6	12 10	1 21
3	T	7 27	4.4	7 55	3.7	12 37	1 33	3	T	7 57	4.5	8 27	3.7	12 58	2 10
4	W	8 13	4.5	8 43	3.6	1 18	2 18	4	F	8 49	4.5	9 20	3.7	1 48	2 59
5	T	9 02	4.4	9 34	3.6	2 01	3 03	5	S	9 43	4.4	10 16	3.7	2 39	3 48
6	F	9 55	4.3	10 30	3.5	2 47	3 52	6	S	10 39	4.1	11 13	3.7	3 33	4 39
7	S	10 53	4.1	11 28	3.5	3 37	4 49	7	M	11 36	3.9	...	...	4 31	5 36
8	S	11 52	3.9	...	...	4 34	6 05	8	T	12 10	3.7	12 33	3.6	5 41	6 45
9	M	12 29	3.5	12 54	3.8	5 43	7 41	9	W	1 10	3.8	1 33	3.4	7 19	7 55
10	T	1 29	3.6	1 55	3.6	7 22	8 43	10	T	2 08	3.8	2 32	3.2	8 51	8 48
11	W	2 31	3.8	2 58	3.6	8 59	9 28	11	F	3 08	3.8	3 33	3.1	9 52	9 31
12	T	3 32	4.0	3 58	3.6	10 00	10 05	12	S	4 08	3.9	4 32	3.1	10 41	10 08
13	F	4 29	4.2	4 55	3.6	10 49	10 39	13	S	5 03	4.0	5 25	3.1	11 24	10 45
14	S	5 22	4.4	5 45	3.6	11 32	11 13	14	M	5 53	4.0	6 13	3.2	11 59	11 23
15	S	6 10	4.5	6 32	3.6	-A-	12 11	15	T	6 39	4.0	6 57	3.2	...	12 36
16	M	6 56	4.5	7 16	3.6	...	12 49	16	W	7 22	3.9	7 39	3.2	12 04	1 12
17	T	7 39	4.3	7 59	3.5	12 27	1 27	17	T	8 03	3.8	8 20	3.1	12 47	1 48
18	W	8 22	4.1	8 42	3.3	1 07	2 05	18	F	8 43	3.6	9 01	3.1	1 30	2 24
19	T	9 06	3.9	9 26	3.2	1 48	2 44	19	S	9 22	3.4	9 42	3.0	2 13	3 01
20	F	9 49	3.6	10 11	3.0	2 30	3 23	20	S	10 01	3.2	10 23	3.0	2 56	3 38
21	S	10 34	3.3	10 57	2.9	3 14	4 04	21	M	10 41	3.0	11 05	2.9	3 39	4 15
22	S	11 19	3.1	11 45	2.8	3 59	4 49	22	T	11 21	2.9	11 46	2.9	4 23	4 54
23	M	...	...	12 04	2.9	4 50	5 40	23	W	...	...	12 03	2.7	5 11	5 37
24	T	12 32	2.8	12 50	2.8	5 51	6 40	24	T	12 30	2.9	12 49	2.7	6 09	6 28
25	W	1 20	2.9	1 39	2.7	7 12	7 41	25	F	1 16	3.0	1 40	2.6	7 23	7 25
26	T	2 09	3.0	2 31	2.8	8 32	8 34	26	S	2 08	3.1	2 37	2.6	8 39	8 23
27	F	3 00	3.2	3 26	2.8	9 28	9 18	27	S	3 06	3.3	3 39	2.7	9 41	9 16
28	S	3 52	3.4	4 20	3.0	10 15	10 00	28	M	4 07	3.5	4 40	2.9	10 35	10 08
29	S	4 43	3.7	5 12	3.2	11 00	10 42	29	T	5 06	3.9	5 36	3.2	11 27	10 59
30	M	5 32	4.0	6 00	3.4	11 45	11 25	30	W	6 01	4.2	6 29	3.4	-B-	12 19
								31	T	6 53	4.4	7 20	3.7	...	1 10

*Standard Time starts   A also at 11:49 p.m.   B also at 11:52 p.m.

Dates when Ht. of **Low** Water is below Mean Lower Low with Ht. of lowest given for each period and Date of lowest in ( ):

3rd - 6th: -0.3' (3rd - 5th)   1st - 6th: -0.6' (3rd - 4th)
15th: -0.2'   28th - 31st: -0.7' (30th)
29th - 30th: -0.3' (30th)

Average Rise and Fall 3.5 ft.

When a high tide exceeds av. ht., the *following* low tide will be lower than av.

Narragansett Bay Currents

This Current Diagram shows average maximum currents with a normal range (3.5 ft.) of tides at Newport. See pages 78-83.

Maximum Ebb Currents, 3 hours after High Water at Newport, are shown by double-headed arrows and velocities are <u>underlined.</u>

Maximum flood Currents, 2 1/2 hours before High Water at Newport, are shown by single-headed arrows and velocities are <u>not</u> underlined.

When height shown is 3.0 ft., subtract 30% from all velocities shown. When height is 4.0 ft., add 20%; when 4.5 ft., add 40%; when 5.0 ft., add 60%.

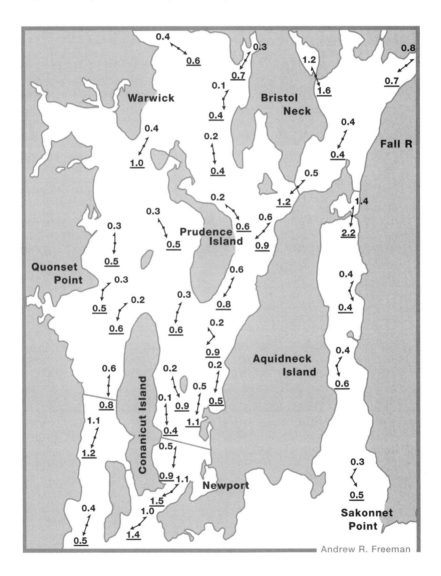

Andrew R. Freeman

Holding a Fair Current between Eastern Long Island and Nantucket

There is a curious phenomenon which can be used to advantage by every vessel, and particularly the slower cruiser or auxiliary, in making the passage *either* way between eastern Long Island Sound, on the west, and Buzzards Bay, Vineyard and Nantucket Sounds on the east.

Note in the very simplified diagram below, that in Long Island Sound, the Ebb Current flows to the *east*, and in Buzzards Bay, Vineyard and Nantucket Sounds the Ebb Current flows to the *west*. (Off Newport, these opposed Ebb Currents merge and flow *south*.) The reverse is also true: the Flood Current flows *west* through Long Island Sound and *east* through Buzzards Bay, Vineyard and Nantucket Sounds. (Half arrow indicates Ebb Current, whole arrow indicates Flood Current.)

In making a *complete* passage through the area of the diagram, simply ride the favoring Ebb Current toward Newport from either direction and, pick up the favoring Flood Current in leaving the Newport area.

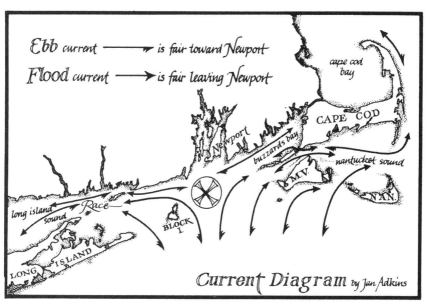

Arrive at "X" at the times shown for "Current Turns to Northwest at The Race," tables pp. 86-91.

The E-W currents between Pt. Judith and Cuttyhunk are only 1/2 to 1 kt., while those to the West of Pt. Judith and to the East of Cuttyhunk are much greater. Bearing this in mind, those making *only a partial trip* through the area may find it better even to buck a slight head current in the Pt. Judith-Cuttyhunk area so as to pick up the maximum hours of strong favoring currents beyond those points.

For example, if headed for the Cape Cod Canal, refer to the Vineyard & Nantucket Sound Current Charts, pp. 66-77 and arrive just N. of Cuttyhunk as Flood Starts at Pollock Rip, pp. 60-65 to ensure the most favorable currents. If headed for Nantucket, refer to the same Charts and arrive just S. of Cuttyhunk at 3 hours after Flood Starts at Pollock Rip, pp. 60-65. If headed into Long Island Sound, refer to the Long Island Sound Current Charts, pp. 92-97 and arrive at Pt. Judith when Flood Current turns West at The Race, p.95.

85

THE RACE, LONG ISLAND SOUND
41°14'N, 72°03.58'W S. of W. end of Fishers Is., 0.6 nm N.W. of Valiant Rock

	Standard Time						Standard Time				
		JANUARY						**FEBRUARY**			
DAY OF MONTH	DAY OF WEEK	CURRENT TURNS TO				DAY OF MONTH	DAY OF WEEK	CURRENT TURNS TO			
		NORTHWEST Flood Starts			SOUTHEAST Ebb Starts			NORTHWEST Flood Starts		SOUTHEAST Ebb Starts	
		a.m.	**p.m.**	Kts.	a.m.	**p.m.**	Kts.	a.m.	**p.m.** Kts.	a.m.	**p.m.** Kts.

Day	Wk	a.m.	p.m.	Kts.	a.m.	p.m.	Kts.	Day	Wk	a.m.	p.m.	Kts.	a.m.	p.m.	Kts.
1	T	8 18	8 39	p2.5	2 09	2 14	p2.7	1	S	9 29	9 34	p2.6	2 59	3 23	a2.9
2	F	9 07	9 21	p2.5	2 50	3 00	2.6	2	M	10 28	10 31	p2.5	3 51	4 23	a2.8
3	S	10 00	10 09	p2.5	3 36	3 53	a2.6	3	T	11 34	11 36	p2.5	4 52	5 32	a2.8
4	S	11 00	11 04	p2.5	4 27	4 53	a2.7	4	W	...	12 44	2.2	6 00	6 44	a2.9
5	M	...	12 03	2.1	5 25	5 59	a2.8	5	T	12 46	1 50	a2.6	7 09	7 53	a3.1
6	T	12 03	1 07	a2.6	6 26	7 06	a3.0	6	F	1 54	2 50	a2.8	8 14	8 55	a3.4
7	W	1 06	2 08	a2.8	7 28	8 10	a3.3	7	S	2 57	3 44	3.1	9 14	9 51	a3.7
8	T	2 07	3 06	a3.0	8 28	9 09	a3.6	8	S	3 55	4 34	3.4	10 08	10 42	a4.0
9	F	3 08	4 01	a3.3	9 25	10 05	a3.9	9	M	4 50	5 22	3.6	10 59	11 30	a4.1
10	S	4 04	4 51	a3.5	10 20	10 58	a4.1	10	T	5 40	6 06	p3.7	11 48	...	4.1
11	S	4 59	5 40	3.6	11 12	11 48	a4.2	11	W	6 29	6 50	p3.6	12 16	12 35	3.9
12	M	5 53	6 28	3.6	...	12 03	4.2	12	T	7 17	7 34	p3.4	1 02	1 21	a3.8
13	T	6 46	7 16	p3.6	12 38	12 54	p4.0	13	F	8 06	8 18	3.0	1 47	2 08	a3.6
14	W	7 39	8 03	p3.4	1 28	1 44	3.7	14	S	8 56	9 05	p2.7	2 33	2 57	a3.3
15	T	8 33	8 52	p3.1	2 18	2 36	a3.5	15	S	9 49	9 55	p2.3	3 21	3 49	a2.9
16	F	9 29	9 43	p2.8	3 09	3 30	a3.2	16	M	10 48	10 52	p2.0	4 14	4 47	a2.6
17	S	10 28	10 36	p2.5	4 02	4 27	a3.0	17	T	11 51	11 54	p1.8	5 12	5 51	a2.3
18	S	11 29	11 34	p2.2	4 57	5 28	a2.7	18	W	...	12 54	1.6	6 14	6 56	a2.2
19	M	...	12 31	1.8	5 55	6 31	a2.6	19	T	12 57	1 53	a1.8	7 14	7 55	a2.3
20	T	12 33	1 32	a2.1	6 53	7 32	a2.5	20	F	1 56	2 44	1.9	8 10	8 47	a2.5
21	W	1 31	2 27	a2.0	7 49	8 28	a2.5	21	S	2 48	3 28	2.1	8 59	9 32	a2.7
22	T	2 25	3 16	a2.1	8 40	9 17	a2.7	22	S	3 34	4 07	p2.4	9 42	10 12	a2.9
23	F	3 14	3 59	2.2	9 26	10 02	a2.8	23	M	4 16	4 43	p2.6	10 22	10 48	a3.1
24	S	3 59	4 39	a2.4	10 08	10 42	a3.0	24	T	4 54	5 16	p2.8	10 59	11 22	a3.2
25	S	4 39	5 15	2.5	10 47	11 19	a3.1	25	W	5 31	5 49	p3.0	11 34	11 55	3.3
26	M	5 18	5 49	2.6	11 23	11 54	a3.2	26	T	6 08	6 22	p3.1	...	12 10	3.2
27	T	5 55	6 21	p2.8	11 58	...	3.2	27	F	6 46	6 57	p3.1	12 29	12 47	a3.3
28	W	6 32	6 54	p2.8	12 27	12 33	p3.1	28	S	7 28	7 35	p3.0	1 05	1 26	a3.3
29	T	7 10	7 27	p2.8	1 00	1 09	p3.1								
30	F	7 51	8 04	p2.8	1 36	1 48	a3.0								
31	S	8 37	8 46	p2.7	2 14	2 32	a2.9								

The Kts. (knots) columns show the **maximum** predicted velocities of the stronger one of the Flood Currents and the stronger one of the Ebb Currents for each day. The letter "a" means the velocity shown should occur **after** the **a.m.** Current Change. The letter "p" means the velocity shown should occur **after** the **p.m.** Current Change (even if next morning). No "a" or "p" means a.m. and p.m. velocities are the same for that day.

Av. Max. Vel.: Fl. 2.9 Kts., Ebb 3.5 Kts.
Max. Flood 2 hrs. 45 min. after Flood Starts, ±15 min.
Max. Ebb 3 hrs. 25 min. after Ebb Starts, ±15 min.
Use THE RACE tables with current charts pp. 92-97

See pp. 22-29 for Current Change at other points.

2009 CURRENT TABLE
THE RACE, LONG ISLAND SOUND
41°14'N, 72°03.58'W S. of W. end of Fishers Is., 0.6 nm N.W. of Valiant Rock

Daylight Time starts March 8 at 2 a.m.   Daylight Saving Time

Day of Month	Day of Week	MARCH — NORTHWEST Flood Starts a.m.	p.m.	Kts.	SOUTHEAST Ebb Starts a.m.	p.m.	Kts.	Day of Month	Day of Week	APRIL — NORTHWEST Flood Starts a.m.	p.m.	Kts.	SOUTHEAST Ebb Starts a.m.	p.m.	Kts.
1	S	8 13	8 19	p2.9	1 45	2 11	a3.3	1	W	10 50	11 02	2.5	4 12	4 56	a3.1
2	M	9 05	9 10	p2.7	2 32	3 03	a3.1	2	T	11 58	...	2.3	5 19	6 07	a2.9
3	T	10 06	10 11	p2.5	3 27	4 05	a3.0	3	F	12 15	1 06	2.4	6 32	7 18	a2.9
4	W	11 14	11 21	p2.4	4 31	5 16	a2.9	4	S	1 30	2 11	p2.6	7 43	8 23	a3.0
5	T	...	12 25	2.2	5 44	6 31	a2.9	5	S	2 38	3 09	p2.8	8 48	9 21	p3.2
6	F	12 36	1 32	a2.5	6 56	7 39	a3.0	6	M	3 38	4 01	p3.1	9 46	10 12	p3.5
7	S	1 47	2 32	p2.8	8 02	8 40	a3.3	7	T	4 32	4 49	p3.3	10 38	10 59	p3.7
8	S	*3 49	*4 25	p3.1	*10 01	*10 34	3.5	8	W	5 20	5 33	p3.3	11 26	11 43	p3.8
9	M	4 46	5 14	p3.4	10 55	11 22	3.7	9	T	6 06	6 16	p3.3	...	12 10	3.4
10	T	5 36	5 58	p3.5	11 44	...	3.8	10	F	6 48	6 55	3.1	12 24	12 53	a3.7
11	W	6 24	6 41	3.5	12 08	12 30	a3.9	11	S	7 30	7 35	2.9	1 04	1 34	a3.6
12	T	7 09	7 23	3.4	12 51	1 14	a3.9	12	S	8 11	8 14	2.7	1 44	2 15	a3.4
13	F	7 53	8 04	3.2	1 33	1 57	a3.8	13	M	8 54	8 57	2.4	2 23	2 57	a3.1
14	S	8 38	8 45	2.9	2 14	2 40	a3.5	14	T	9 39	9 43	2.1	3 06	3 43	a2.8
15	S	9 23	9 29	2.5	2 57	3 25	a3.2	15	W	10 29	10 37	1.9	3 52	4 35	a2.5
16	M	10 12	10 17	2.2	3 41	4 13	a2.8	16	T	11 24	11 37	a1.8	4 46	5 32	a2.3
17	T	11 07	11 12	p1.9	4 31	5 08	a2.5	17	F	...	12 22	1.7	5 46	6 33	a2.2
18	W	...	12 07	1.6	5 28	6 11	a2.3	18	S	12 42	1 19	p1.8	6 48	7 30	a2.2
19	T	12 15	1 10	a1.7	6 31	7 16	a2.1	19	S	1 44	2 11	p2.0	7 48	8 22	2.3
20	F	1 21	2 10	a1.7	7 34	8 16	a2.2	20	M	2 39	2 58	p2.2	8 42	9 09	p2.6
21	S	2 23	3 02	p1.9	8 33	9 08	a2.3	21	T	3 28	3 41	p2.5	9 31	9 51	p3.0
22	S	3 17	3 47	p2.2	9 24	9 54	a2.6	22	W	4 13	4 22	p2.8	10 16	10 31	p3.3
23	M	4 04	4 28	p2.5	10 10	10 34	2.8	23	T	4 56	5 01	p3.1	10 59	11 10	p3.6
24	T	4 47	5 04	p2.7	10 51	11 11	p3.1	24	F	5 38	5 41	p3.2	11 41	11 51	p3.8
25	W	5 26	5 40	p3.0	11 30	11 46	p3.4	25	S	6 20	6 23	p3.4	...	12 24	3.2
26	T	6 05	6 14	p3.1	...	12 08	3.2	26	S	7 04	7 07	p3.4	12 33	1 08	a3.9
27	F	6 43	6 51	p3.2	12 22	12 46	a3.6	27	M	7 51	7 55	p3.3	1 18	1 56	a3.8
28	S	7 24	7 29	p3.3	12 59	1 26	a3.6	28	T	8 42	8 49	p3.1	2 07	2 48	a3.7
29	S	8 08	8 12	p3.2	1 38	2 09	a3.6	29	W	9 37	9 49	a2.9	3 02	3 45	a3.5
30	M	8 56	9 00	p3.0	2 23	2 57	a3.5	30	T	10 37	10 57	a2.7	4 02	4 48	a3.3
31	T	9 49	9 56	p2.8	3 13	3 52	a3.3								

*Daylight Saving Time starts

The Kts. (knots) columns show the **maximum** predicted velocities of the stronger one of the Flood Currents and the stronger one of the Ebb Currents for each day. The letter "a" means the velocity shown should occur **after** the a.m. Current Change. The letter "p" means the velocity shown should occur **after** the p.m. Current Change (even if next morning). No "a" or "p" means a.m. and p.m. velocities are the same for that day.

Av. Max. Vel.: Fl. 2.9 Kts., Ebb 3.5 Kts.

Max. Flood 2 hrs. 45 min. after Flood Starts, ±15 min.

Max. Ebb 3 hrs. 25 min. after Ebb Starts, ±15 min.

Use THE RACE tables with current charts pp. 92-97

See pp. 22-29 for Current Change at other points.

2009 CURRENT TABLE
THE RACE, LONG ISLAND SOUND

41°14'N, 72°03.58'W S. of W. end of Fishers Is., 0.6 nm N.W. of Valiant Rock

Daylight Saving Time   Daylight Saving Time

DAY OF MONTH	DAY OF WEEK	NORTHWEST Flood Starts		SOUTHEAST Ebb Starts			DAY OF MONTH	DAY OF WEEK	NORTHWEST Flood Starts		SOUTHEAST Ebb Starts		
		a.m.	**p.m.** Kts.	a.m.	**p.m.** Kts.				a.m.	**p.m.** Kts.	a.m.	**p.m.** Kts.	
MAY							**JUNE**						
1	F	11 40	... 2.6	5 08	5 55 a3.1		1	M	12 59	1 15 p2.7	7 02	7 35 p3.1	
2	S	12 08	12 43 p2.6	6 17	7 00 a3.0		2	T	2 03	2 12 p2.7	8 04	8 30 p3.1	
3	S	1 18	1 45 p2.7	7 25	8 01 p3.0		3	W	3 01	3 06 p2.7	9 03	9 22 p3.2	
4	M	2 23	2 42 p2.8	8 28	8 57 p3.2		4	T	3 54	3 56 p2.7	9 56	10 10 p3.2	
5	T	3 22	3 34 p3.0	9 26	9 48 p3.4		5	F	4 43	4 43 p2.7	10 45	10 55 p3.2	
6	W	4 14	4 22 p3.0	10 18	10 35 p3.5		6	S	5 27	5 26 p2.6	11 30	11 36 p3.2	
7	T	5 02	5 07 p3.0	11 06	11 18 p3.5		7	S	6 09	6 07 p2.6	...	12 12 2.6	
8	F	5 46	5 49 p2.9	11 50	11 59 p3.5		8	M	6 48	6 47 p2.5	12 16	12 53 a3.2	
9	S	6 29	6 30 2.8	...	12 32 2.8		9	T	7 28	7 27 2.4	12 54	1 31 a3.1	
10	S	7 08	7 09 p2.7	12 38	1 12 a3.4		10	W	8 04	8 07 a2.4	1 32	2 10 a3.0	
11	M	7 48	7 48 2.5	1 17	1 52 a3.2		11	T	8 42	8 49 a2.3	2 10	2 50 a2.9	
12	T	8 28	8 30 2.3	1 55	2 33 a3.0		12	F	9 22	9 35 a2.3	2 50	3 31 a2.7	
13	W	9 10	9 15 a2.2	2 36	3 17 a2.8		13	S	10 02	10 25 a2.3	3 33	4 15 a2.6	
14	T	9 55	10 05 a2.1	3 19	4 04 a2.6		14	S	10 46	11 18 a2.2	4 20	5 01 a2.4	
15	F	10 43	11 01 a2.0	4 08	4 54 a2.4		15	M	11 32	... 2.2	5 12	5 50 p2.4	
16	S	11 33	... 2.0	5 01	5 47 a2.3		16	T	12 15	12 22 p2.3	6 09	6 41 p2.6	
17	S	12 01	12 25 p2.0	5 59	6 40 2.2		17	W	1 13	1 14 p2.4	7 08	7 34 p2.8	
18	M	1 00	1 16 p2.1	6 58	7 32 p2.4		18	T	2 10	2 08 p2.6	8 08	8 27 p3.1	
19	T	1 56	2 05 p2.3	7 55	8 20 p2.7		19	F	3 06	3 02 p2.8	9 05	9 21 p3.4	
20	W	2 49	2 53 p2.6	8 49	9 07 p3.1		20	S	3 59	3 56 p3.1	10 00	10 13 p3.7	
21	T	3 38	3 39 p2.8	9 39	9 53 p3.4		21	S	4 51	4 50 p3.3	10 54	11 06 p3.9	
22	F	4 26	4 25 p3.1	10 28	10 38 p3.7		22	M	5 41	5 43 p3.5	11 46	11 58 p4.1	
23	S	5 12	5 12 p3.3	11 15	11 25 p3.9		23	T	6 31	6 37 p3.6	...	12 37 3.4	
24	S	5 59	6 00 p3.4	...	12 03 3.2		24	W	7 21	7 32 3.5	12 50	1 29 a4.1	
25	M	6 47	6 50 p3.5	12 13	12 52 a4.0		25	T	8 11	8 29 a3.5	1 43	2 22 a4.1	
26	T	7 37	7 43 p3.4	1 03	1 44 a4.0		26	F	9 02	9 27 a3.4	2 37	3 16 a3.9	
27	W	8 28	8 40 3.2	1 56	2 37 a3.9		27	S	9 55	10 28 a3.2	3 33	4 11 a3.6	
28	T	9 22	9 41 a3.1	2 51	3 34 a3.7		28	S	10 49	11 30 a3.0	4 31	5 08 a3.3	
29	F	10 18	10 46 a3.0	3 50	4 34 a3.5		29	M	11 45	... 2.8	5 31	6 06 p3.1	
30	S	11 17	11 53 a2.9	4 52	5 35 a3.2		30	T	12 34	12 43 p2.6	6 34	7 04 p3.0	
31	S	...	12 16 2.8	5 57	6 36 3.0								

The Kts. (knots) columns show the **maximum** predicted velocities of the stronger one of the Flood Currents and the stronger one of the Ebb Currents for each day. The letter "a" means the velocity shown should occur **after** the a.m. Current Change. The letter "p" means the velocity shown should occur **after** the p.m. Current Change (even if next morning). No "a" or "p" means a.m. and p.m. velocities are the same for that day.

Av. Max. Vel.: Fl. 2.9 Kts., Ebb 3.5 Kts.

Max. Flood 2 hrs. 45 min. after Flood Starts, ±15 min.

Max. Ebb 3 hrs. 25 min. after Ebb Starts, ±15 min.

Use THE RACE tables with current charts pp. 92-97

See pp. 22-29 for Current Change at other points.

2009 CURRENT TABLE
THE RACE, LONG ISLAND SOUND
41°14'N, 72°03.58'W S. of W. end of Fishers Is., 0.6 nm N.W. of Valiant Rock

		JULY						AUGUST			
		Daylight Saving Time						Daylight Saving Time			
DAY OF MONTH	DAY OF WEEK	CURRENT TURNS TO				DAY OF MONTH	DAY OF WEEK	CURRENT TURNS TO			
		NORTHWEST Flood Starts			SOUTHEAST Ebb Starts			NORTHWEST Flood Starts			SOUTHEAST Ebb Starts
		a.m.	**p.m.**	Kts.	a.m. **p.m.** Kts.			a.m.	**p.m.**	Kts.	a.m. **p.m.** Kts.
1	W	1 36	**1 41**	p2.5	7 37 **8 01** p2.9	1	S	3 06	**3 07**	p2.1	9 07 **9 22** p2.7
2	T	2 36	**2 38**	p2.4	8 37 **8 56** p2.9	2	S	3 57	**3 58**	p2.2	9 59 **10 10** p2.8
3	F	3 32	**3 31**	p2.4	9 33 **9 46** p2.9	3	M	4 43	**4 44**	p2.3	10 45 **10 54** p2.9
4	S	4 22	**4 20**	p2.4	10 24 **10 33** p3.0	4	T	5 23	**5 26**	p2.5	11 27 **11 33** p3.0
5	S	5 07	**5 05**	p2.4	11 10 **11 16** p3.0	5	W	6 00	**6 05**	p2.6	... **12 04** 2.7
6	M	5 49	**5 47**	p2.5	11 52 **11 55** p3.1	6	T	6 34	**6 42**	2.6	12 10 **12 39** a3.1
7	T	6 27	**6 27**	p2.5	... **12 31** 2.5	7	F	7 07	**7 19**	a2.7	12 46 **1 12** a3.1
8	W	7 03	**7 05**	2.5	12 33 **1 08** a3.1	8	S	7 38	**7 56**	a2.7	1 20 **1 45** a3.1
9	T	7 39	**7 45**	2.5	1 10 **1 44** a3.0	9	S	8 12	**8 36**	a2.8	1 55 **2 19** a3.0
10	F	8 12	**8 23**	a2.5	1 45 **2 19** a3.0	10	M	8 46	**9 17**	a2.7	2 31 **2 55** p2.9
11	S	8 47	**9 05**	a2.5	2 22 **2 56** a2.9	11	T	9 25	**10 05**	a2.6	3 12 **3 36** p2.8
12	S	9 23	**9 49**	a2.5	3 01 **3 34** a2.7	12	W	10 09	**11 00**	a2.5	3 59 **4 24** p2.7
13	M	10 03	**10 39**	a2.5	3 43 **4 16** 2.6	13	T	11 02	...	2.4	4 54 **5 21** p2.7
14	T	10 47	**11 34**	a2.4	4 31 **5 03** p2.6	14	F	12 03	**12 04**	p2.4	5 59 **6 26** p2.8
15	W	11 37	...	2.4	5 26 **5 56** p2.7	15	S	1 10	**1 12**	p2.5	7 10 **7 35** p3.0
16	T	12 34	**12 33**	p2.4	6 28 **6 55** p2.8	16	S	2 17	**2 21**	p2.7	8 19 **8 41** p3.2
17	F	1 36	**1 34**	p2.6	7 33 **7 56** p3.0	17	M	3 18	**3 25**	p3.0	9 22 **9 42** p3.6
18	S	2 38	**2 36**	p2.8	8 38 **8 57** p3.3	18	T	4 13	**4 25**	p3.3	10 19 **10 38** p3.9
19	S	3 37	**3 37**	p3.0	9 39 **9 55** p3.6	19	W	5 04	**5 20**	p3.6	11 11 **11 31** p4.1
20	M	4 32	**4 35**	p3.3	10 36 **10 51** p3.9	20	T	5 52	**6 12**	p3.7	... **12 01** 3.9
21	T	5 24	**5 31**	p3.5	11 29 **11 45** p4.1	21	F	6 38	**7 03**	a3.7	12 21 **12 48** a4.1
22	W	6 13	**6 25**	p3.6	... **12 21** 3.7	22	S	7 24	**7 52**	a3.7	1 09 **1 35** 4.0
23	T	7 02	**7 19**	3.6	12 37 **1 11** a4.2	23	S	8 09	**8 42**	a3.5	1 58 **2 22** 3.8
24	F	7 49	**8 13**	a3.7	1 28 **2 01** a4.1	24	M	8 56	**9 34**	a3.2	2 46 **3 10** p3.5
25	S	8 37	**9 07**	a3.6	2 19 **2 51** a3.9	25	T	9 44	**10 29**	a2.9	3 37 **4 00** p3.1
26	S	9 26	**10 03**	a3.3	3 11 **3 42** a3.6	26	W	10 37	**11 28**	a2.5	4 31 **4 54** p2.8
27	M	10 17	**11 01**	a3.0	4 05 **4 36** 3.2	27	T	11 35	...	2.1	5 30 **5 54** p2.5
28	T	11 11	...	2.7	5 02 **5 32** p2.9	28	F	12 31	**12 38**	p1.9	6 34 **6 56** p2.3
29	W	12 03	**12 09**	p2.4	6 03 **6 30** p2.7	29	S	1 35	**1 42**	p1.9	7 38 **7 58** p2.4
30	T	1 06	**1 10**	p2.2	7 06 **7 30** p2.6	30	S	2 34	**2 41**	p1.9	8 38 **8 53** p2.5
31	F	2 08	**2 10**	p2.1	8 09 **8 28** p2.6	31	M	3 26	**3 34**	p2.1	9 30 **9 43** p2.7

The Kts. (knots) columns show the **maximum** predicted velocities of the stronger one of the Flood Currents and the stronger one of the Ebb Currents for each day. The letter "a" means the velocity shown should occur **after** the **a.m.** Current Change. The letter "p" means the velocity shown should occur **after** the **p.m.** Current Change (even if next morning). No "a" or "p" means a.m. and p.m. velocities are the same for that day.
Av. Max. Vel.: Fl. 2.9 Kts., Ebb 3.5 Kts.
Max. Flood 2 hrs. 45 min. after Flood Starts, ±15 min.
Max. Ebb 3 hrs. 25 min. after Ebb Starts, ±15 min.
Use THE RACE tables with current charts pp. 92-97

See pp. 22-29 for Current Change at other points.

2009 CURRENT TABLE
THE RACE, LONG ISLAND SOUND
41°14'N, 72°03.58'W S. of W. end of Fishers Is., 0.6 nm N.W. of Valiant Rock

Daylight Saving Time	Daylight Saving Time

SEPTEMBER							OCTOBER						
DAY OF MONTH	DAY OF WEEK	CURRENT TURNS TO					DAY OF MONTH	DAY OF WEEK	CURRENT TURNS TO				
		NORTHWEST Flood Starts			SOUTHEAST Ebb Starts				NORTHWEST Flood Starts			SOUTHEAST Ebb Starts	
		a.m.	**p.m.**	Kts.	a.m.	**p.m.**	Kts.			a.m.	**p.m.**	Kts.	a.m.	**p.m.**	Kts.
1	T	4 11	4 20	p2.3	10 15	10 27	p2.8	1	T	4 10	4 32	p2.5	10 17	10 36	p2.9
2	W	4 50	5 01	p2.5	10 55	11 07	p3.0	2	F	4 47	5 11	2.7	10 54	11 14	a3.1
3	T	5 26	5 40	p2.7	11 31	11 44	p3.1	3	S	5 22	5 48	a2.9	11 29	11 51	a3.3
4	F	6 00	6 16	2.8	...	12 05	3.1	4	S	5 56	6 25	a3.0	...	12 02	3.4
5	S	6 32	6 52	a2.9	12 19	12 38	p3.2	5	M	6 30	7 04	a3.1	12 28	12 37	p3.5
6	S	7 04	7 28	a2.9	12 54	1 10	p3.2	6	T	7 07	7 44	a3.1	1 05	1 15	p3.5
7	M	7 37	8 07	a2.9	1 29	1 44	p3.2	7	W	7 47	8 30	a3.0	1 45	1 56	p3.4
8	T	8 13	8 50	a2.9	2 06	2 21	p3.1	8	T	8 32	9 20	a2.9	2 30	2 43	p3.2
9	W	8 55	9 40	a2.8	2 47	3 05	p3.0	9	F	9 26	10 19	a2.7	3 22	3 38	p3.0
10	T	9 42	10 35	a2.6	3 36	3 56	p2.9	10	S	10 28	11 23	a2.5	4 23	4 43	p2.9
11	F	10 40	11 40	a2.4	4 34	4 57	p2.8	11	S	11 40	...	2.4	5 32	5 54	p2.8
12	S	11 48	...	2.4	5 42	6 08	p2.8	12	M	12 30	12 54	2.4	6 42	7 07	p2.9
13	S	12 50	1 01	p2.4	6 55	7 20	p2.9	13	T	1 35	2 04	p2.6	7 48	8 14	p3.1
14	M	1 57	2 13	p2.6	8 05	8 28	p3.2	14	W	2 35	3 06	p2.9	8 48	9 14	p3.3
15	T	2 58	3 17	p2.9	9 06	9 29	p3.5	15	T	3 29	4 02	3.1	9 41	10 08	a3.6
16	W	3 52	4 14	p3.3	10 01	10 24	p3.7	16	F	4 18	4 53	3.3	10 30	10 58	a3.8
17	T	4 42	5 07	p3.5	10 51	11 15	p3.9	17	S	5 05	5 40	a3.5	11 16	11 45	a4.0
18	F	5 29	5 57	3.6	11 39	...	4.0	18	S	5 49	6 25	a3.4	11 59	...	3.9
19	S	6 13	6 44	a3.7	12 03	12 24	p4.1	19	M	6 32	7 09	a3.3	12 30	12 42	p3.8
20	S	6 57	7 30	a3.6	12 50	1 08	p4.0	20	T	7 14	7 53	a3.1	1 14	1 24	p3.5
21	M	7 41	8 17	a3.4	1 35	1 52	p3.7	21	W	7 57	8 37	a2.8	1 58	2 06	p3.2
22	T	8 25	9 05	a3.0	2 21	2 37	p3.4	22	T	8 41	9 24	a2.5	2 42	2 51	p2.9
23	W	9 11	9 56	a2.6	3 08	3 24	p3.0	23	F	9 30	10 15	a2.2	3 30	3 39	p2.6
24	T	10 02	10 51	a2.3	3 59	4 16	p2.6	24	S	10 25	11 10	a1.9	4 23	4 33	p2.3
25	F	10 59	11 52	a2.0	4 56	5 14	p2.3	25	S	11 26	...	1.7	5 20	5 32	p2.2
26	S	...	12 03	1.8	5 58	6 17	p2.2	26	M	12 07	12 30	1.7	6 20	6 34	p2.2
27	S	12 55	1 09	p1.7	7 02	7 20	p2.2	27	T	1 03	1 31	a1.8	7 16	7 33	p2.2
28	M	1 53	2 10	p1.8	8 01	8 18	p2.3	28	W	1 55	2 26	a2.0	8 08	8 27	p2.4
29	T	2 45	3 03	p2.0	8 53	9 09	p2.5	29	T	2 42	3 15	2.2	8 54	9 16	2.6
30	W	3 30	3 50	p2.3	9 38	9 55	p2.7	30	F	3 25	3 59	a2.5	9 35	10 01	a2.9
								31	S	4 05	4 40	2.7	10 15	10 42	a3.2

The Kts. (knots) columns show the **maximum** predicted velocities of the stronger one of the Flood Currents and the stronger one of the Ebb Currents for each day. The letter "a" means the velocity shown should occur **after** the **a.m.** Current Change. The letter "p" means the velocity shown should occur **after** the **p.m.** Current Change (even if next morning). No "a" or "p" means a.m. and p.m. velocities are the same for that day.

Av. Max. Vel.: Fl. 2.9 Kts., Ebb 3.5 Kts.
Max. Flood 2 hrs. 45 min. after Flood Starts, ±15 min.
Max. Ebb 3 hrs. 25 min. after Ebb Starts, ±15 min.
Use THE RACE tables with current charts pp. 92-97

See pp. 22-29 for Current Change at other points.

2009 CURRENT TABLE
THE RACE, LONG ISLAND SOUND

41°14'N, 72°03.58'W S. of W. end of Fishers Is., 0.6 nm N.W. of Valiant Rock

Standard Time starts Nov. 1 at 2 a.m. | **Standard Time**

NOVEMBER								DECEMBER							
DAY OF MONTH	DAY OF WEEK	CURRENT TURNS TO NORTHWEST Flood Starts			SOUTHEAST Ebb Starts			DAY OF MONTH	DAY OF WEEK	CURRENT TURNS TO NORTHWEST Flood Starts			SOUTHEAST Ebb Starts		
		a.m.	p.m.	Kts.	a.m.	p.m.	Kts.			a.m.	p.m.	Kts.	a.m.	p.m.	Kts.
1	S	*3 43	*4 20	a2.9	*9 52	*10 23	a3.4	1	T	3 51	4 39	a3.1	10 04	10 42	a3.7
2	M	4 21	5 01	a3.1	10 31	11 03	a3.6	2	W	4 37	5 25	a3.3	10 50	11 28	a3.8
3	T	5 01	5 42	a3.2	11 10	11 45	a3.7	3	T	5 25	6 11	a3.4	11 37	...	3.9
4	W	5 42	6 27	a3.2	11 53	...	3.7	4	F	6 15	7 00	a3.3	12 17	12 27	p3.9
5	T	6 28	7 14	a3.2	12 30	12 39	p3.6	5	S	7 09	7 51	3.2	1 08	1 20	p3.7
6	F	7 19	8 06	a3.0	1 19	1 30	p3.5	6	S	8 08	8 44	3.1	2 01	2 16	p3.5
7	S	8 16	9 03	a2.8	2 13	2 27	p3.3	7	M	9 10	9 41	p3.0	2 59	3 16	p3.3
8	S	9 21	10 04	2.6	3 14	3 31	p3.1	8	T	10 16	10 40	p2.9	3 59	4 20	3.1
9	M	10 32	11 08	p2.6	4 19	4 39	p3.0	9	W	11 25	11 41	p2.8	5 00	5 26	a3.1
10	T	11 43	...	2.5	5 24	5 49	p2.9	10	T	...	12 30	2.4	6 01	6 32	a3.1
11	W	12 09	12 50	a2.7	6 27	6 55	3.0	11	F	12 40	1 32	a2.8	7 00	7 34	a3.2
12	T	1 08	1 51	a2.9	7 26	7 55	a3.3	12	S	1 38	2 29	a2.8	7 56	8 32	a3.3
13	F	2 03	2 47	a3.0	8 19	8 51	a3.5	13	S	2 32	3 21	a2.8	8 48	9 24	a3.3
14	S	2 54	3 37	a3.1	9 08	9 41	a3.6	14	M	3 22	4 09	a2.8	9 36	10 12	a3.4
15	S	3 42	4 24	a3.1	9 55	10 28	a3.7	15	T	4 09	4 52	a2.8	10 20	10 56	a3.4
16	M	4 27	5 08	a3.1	10 38	11 13	a3.7	16	W	4 52	5 33	a2.7	11 02	11 38	a3.3
17	T	5 10	5 50	a3.0	11 20	11 55	a3.5	17	T	5 33	6 12	a2.6	11 41	...	3.2
18	W	5 52	6 32	a2.8	...	12 01	3.4	18	F	6 13	6 50	a2.6	12 17	12 19	p3.1
19	T	6 33	7 13	a2.6	12 37	12 41	p3.1	19	S	6 53	7 27	a2.5	12 56	12 57	p3.0
20	F	7 16	7 55	a2.4	1 19	1 22	p2.9	20	S	7 35	8 05	p2.4	1 35	1 36	p2.8
21	S	8 01	8 40	a2.2	2 03	2 05	p2.7	21	M	8 19	8 44	p2.3	2 14	2 17	p2.6
22	S	8 51	9 26	2.0	2 49	2 53	p2.5	22	T	9 06	9 26	p2.2	2 56	3 01	p2.5
23	M	9 46	10 16	p2.0	3 39	3 45	p2.3	23	W	9 58	10 11	p2.2	3 41	3 51	2.3
24	T	10 45	11 07	p2.0	4 31	4 42	p2.2	24	T	10 54	11 00	p2.2	4 28	4 46	a2.3
25	W	11 44	11 58	p2.1	5 24	5 41	2.2	25	F	11 52	11 53	p2.2	5 19	5 46	a2.4
26	T	...	12 41	1.9	6 15	6 39	a2.4	26	S	...	12 51	1.9	6 13	6 46	a2.6
27	F	12 48	1 34	a2.2	7 04	7 33	a2.6	27	S	12 47	1 47	a2.4	7 07	7 45	a2.8
28	S	1 36	2 23	a2.4	7 51	8 23	a2.9	28	M	1 42	2 40	a2.6	8 00	8 40	a3.1
29	S	2 21	3 09	a2.7	8 36	9 10	a3.2	29	T	2 35	3 31	a2.8	8 53	9 33	a3.4
30	M	3 06	3 54	a2.9	9 20	9 56	a3.5	30	W	3 28	4 20	a3.1	9 44	10 23	a3.7
								31	T	4 20	5 08	3.3	10 34	11 13	a3.9

***Standard Time starts**

The Kts. (knots) columns show the **maximum** predicted velocities of the stronger one of the Flood Currents and the stronger one of the Ebb Currents for each day. The letter "a" means the velocity shown should occur **after** the **a.m.** Current Change. The letter "p" means the velocity shown should occur **after** the **p.m.** Current Change (even if next morning). No "a" or "p" means a.m. and p.m. velocities are the same for that day.

Av. Max. Vel.: Fl. 2.9 Kts., Ebb 3.5 Kts.

Max. Flood 2 hrs. 45 min. after Flood Starts, ±15 min.

Max. Ebb 3 hrs. 25 min. after Ebb Starts, ±15 min.

Use THE RACE tables with current charts pp. 92-97

See pp. 22-29 for Current Change at other points.

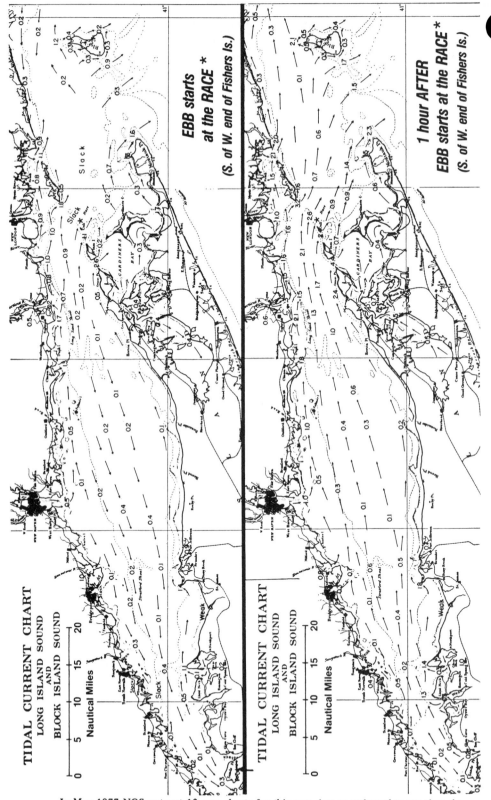

TIDAL CURRENT CHART
LONG ISLAND SOUND
AND
BLOCK ISLAND SOUND

Nautical Miles

**EBB starts
at the RACE** *
(S. of W. end of Fishers Is.)

TIDAL CURRENT CHART
LONG ISLAND SOUND
AND
BLOCK ISLAND SOUND

Nautical Miles

**1 hour AFTER
EBB starts at the RACE** *
(S. of W. end of Fishers Is.)

In May 1977 NOS put out 13 new charts for this area, but so awkward to use that, despite some differences (rarely major) in current directions and velocities, we believe the old presentation comprehensibly more useful.

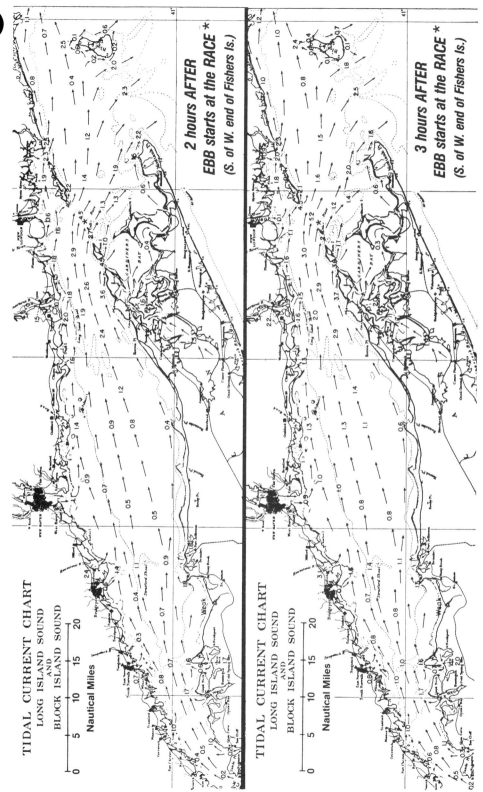

TIDAL CURRENT CHART
LONG ISLAND SOUND
AND
BLOCK ISLAND SOUND

Nautical Miles

2 hours AFTER
EBB starts at the RACE ✱
(S. of W. end of Fishers Is.)

TIDAL CURRENT CHART
LONG ISLAND SOUND
AND
BLOCK ISLAND SOUND

Nautical Miles

3 hours AFTER
EBB starts at the RACE ✱
(S. of W. end of Fishers Is.)

In May 1977 NOS put out 13 new charts for this area, but so awkward to use that, despite some differences (rarely major) in current directions and velocities, we believe the old presentation comprehensibly more useful.

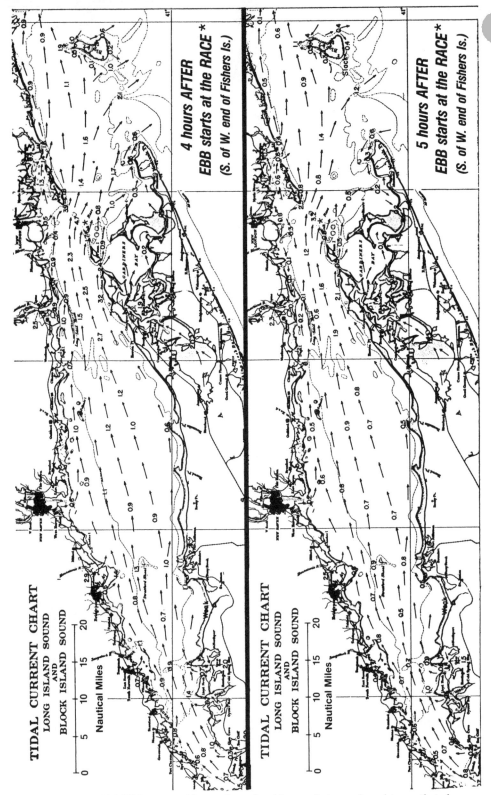

TIDAL CURRENT CHART
LONG ISLAND SOUND
AND
BLOCK ISLAND SOUND

Nautical Miles

4 hours AFTER
EBB starts at the RACE *
(S. of W. end of Fishers Is.)

TIDAL CURRENT CHART
LONG ISLAND SOUND
AND
BLOCK ISLAND SOUND

Nautical Miles

5 hours AFTER
EBB starts at the RACE *
(S. of W. end of Fishers Is.)

In May 1977 NOS put out 13 new charts for this area, but so awkward to use that, despite some differences (rarely major) in current directions and velocities, we believe the old presentation comprehensibly more useful.

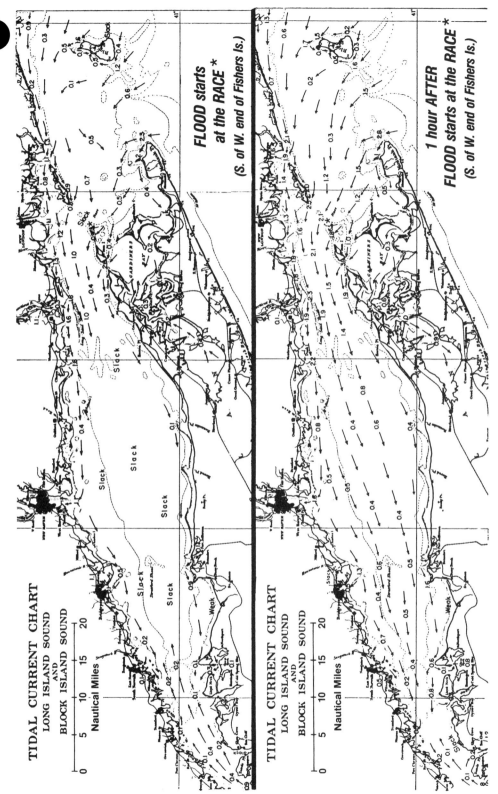

TIDAL CURRENT CHART
LONG ISLAND SOUND
AND
BLOCK ISLAND SOUND

Nautical Miles

FLOOD starts at the RACE *
(S. of W. end of Fishers Is.)

1 hour AFTER FLOOD starts at the RACE *
(S. of W. end of Fishers Is.)

In May 1977 NOS put out 13 new charts for this area, but so awkward to use that, despite some differences (rarely major) in current directions and velocities, we believe the old presentation comprehensibly more useful.

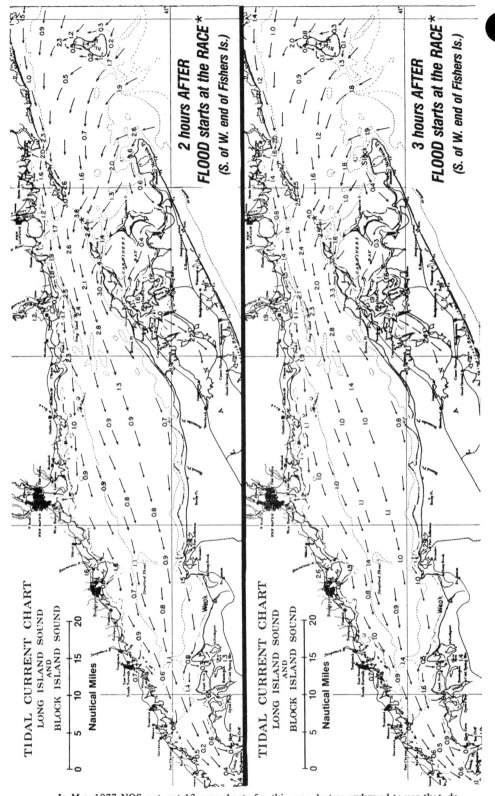

TIDAL CURRENT CHART
LONG ISLAND SOUND
AND
BLOCK ISLAND SOUND

Nautical Miles

2 hours AFTER
FLOOD starts at the RACE*
(S. of W. end of Fishers Is.)

TIDAL CURRENT CHART
LONG ISLAND SOUND
AND
BLOCK ISLAND SOUND

Nautical Miles

3 hours AFTER
FLOOD starts at the RACE*
(S. of W. end of Fishers Is.)

In May 1977 NOS put out 13 new charts for this area, but so awkward to use that, despite some differences (rarely major) in current directions and velocities, we believe the old presentation comprehensibly more useful.

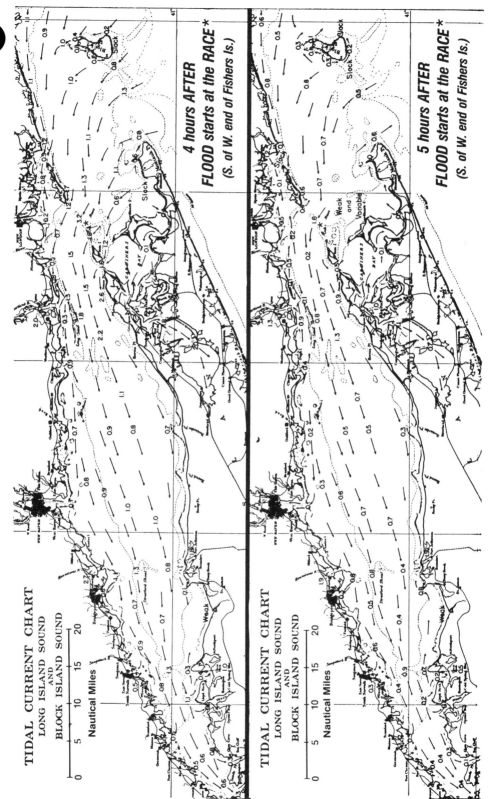

TIDAL CURRENT CHART
LONG ISLAND SOUND
AND
BLOCK ISLAND SOUND

4 hours AFTER
FLOOD starts at the RACE*
(S. of W. end of Fishers Is.)

5 hours AFTER
FLOOD starts at the RACE*
(S. of W. end of Fishers Is.)

In May 1977 NOS put out 13 new charts for this area, but so awkward to use that, despite some differences (rarely major) in current directions and velocities, we believe the old presentation comprehensibly more useful.

2009 HIGH & LOW WATER
BRIDGEPORT, CT
41°10.4'N, 73°10.9'W

						Standard Time								Standard Time			

DAY OF MONTH	DAY OF WEEK	JANUARY						DAY OF MONTH	DAY OF WEEK	FEBRUARY					
		HIGH				LOW				HIGH				LOW	
		a.m.	Ht.	p.m.	Ht.	a.m.	p.m.			a.m.	Ht.	p.m.	Ht.	a.m.	p.m.
1	T	1 55	6.4	2 08	6.4	8 04	8 26	1	S	2 46	6.9	3 17	6.1	9 15	9 25
2	F	2 36	6.5	2 54	6.2	8 52	9 10	2	M	3 36	6.8	4 15	5.9	10 12	10 21
3	S	3 21	6.5	3 45	6.0	9 44	9 58	3	T	4 34	6.8	5 19	5.7	11 17	11 24
4	S	4 11	6.6	4 43	5.9	10 41	10 52	4	W	5 39	6.8	6 26	5.8	...	12 24
5	M	5 06	6.8	5 45	5.8	11 43	11 51	5	T	6 47	7.0	7 32	6.0	12 32	1 30
6	T	6 05	6.9	6 48	5.9	...	12 46	6	F	7 53	7.2	8 33	6.4	1 38	2 31
7	W	7 06	7.2	7 50	6.1	12 52	1 48	7	S	8 55	7.5	9 29	6.8	2 41	3 27
8	T	8 07	7.5	8 49	6.4	1 53	2 47	8	S	9 51	7.7	10 22	7.2	3 39	4 19
9	F	9 07	7.8	9 45	6.7	2 52	3 43	9	M	10 44	7.8	11 12	7.5	4 33	5 07
10	S	10 02	8.0	10 38	7.0	3 49	4 36	10	T	11 33	7.8	11 59	7.7	5 25	5 53
11	S	10 57	8.1	11 30	7.2	4 45	5 27	11	W	...	...	12 21	7.6	6 15	6 38
12	M	11 49	8.0	...	...	5 39	6 16	12	T	12 45	7.7	1 08	7.3	7 04	7 22
13	T	12 20	7.4	12 40	7.8	6 32	7 05	13	F	1 31	7.6	1 56	6.8	7 52	8 07
14	W	1 11	7.4	1 32	7.4	7 26	7 53	14	S	2 17	7.3	2 45	6.4	8 42	8 54
15	T	2 01	7.4	2 24	7.0	8 20	8 42	15	S	3 05	7.0	3 37	6.0	9 34	9 44
16	F	2 53	7.2	3 18	6.5	9 15	9 33	16	M	3 57	6.6	4 33	5.6	10 29	10 39
17	S	3 45	7.0	4 14	6.1	10 13	10 26	17	T	4 53	6.3	5 32	5.5	11 28	11 38
18	S	4 40	6.7	5 13	5.7	11 11	11 22	18	W	5 54	6.1	6 33	5.4	...	12 27
19	M	5 36	6.5	6 12	5.6	...	12 11	19	T	6 54	6.0	7 30	5.6	12 39	1 24
20	T	6 34	6.4	7 11	5.6	12 19	1 08	20	F	7 51	6.1	8 22	5.8	1 37	2 16
21	W	7 29	6.4	8 05	5.7	1 15	2 02	21	S	8 41	6.3	9 09	6.1	2 29	3 02
22	T	8 21	6.4	8 54	5.8	2 08	2 50	22	S	9 26	6.5	9 51	6.4	3 15	3 43
23	F	9 09	6.5	9 39	6.0	2 57	3 34	23	M	10 07	6.7	10 29	6.7	3 58	4 21
24	S	9 53	6.6	10 20	6.2	3 42	4 14	24	T	10 45	6.9	11 06	6.9	4 38	4 58
25	S	10 33	6.7	10 59	6.4	4 23	4 52	25	W	11 22	6.9	11 41	7.1	5 17	5 34
26	M	11 10	6.8	11 35	6.5	5 02	5 28	26	T	11 59	6.9	...	...	5 55	6 09
27	T	11 47	6.8	...	...	5 40	6 04	27	F	12 16	7.2	12 39	6.9	6 35	6 47
28	W	12 11	6.7	12 24	6.8	6 19	6 39	28	S	12 53	7.3	1 20	6.7	7 16	7 27
29	T	12 46	6.8	1 01	6.7	6 58	7 15								
30	F	1 23	6.8	1 42	6.6	7 39	7 54								
31	S	2 02	6.9	2 27	6.3	8 24	8 37								

Dates when Ht. of **Low** Water is below Mean Lower Low with Ht. of lowest given for each period and Date of lowest in ( ):

7th - 15th: -1.1' (11th - 12th)   6th - 14th: -1.1' (10th)
26th - 29th: -0.3' (27th - 28th)   24th - 28th: -0.3' (25th - 28th)

Average Rise and Fall 6.8 ft.

When a high tide exceeds av. ht., the *following* low tide will be lower than av.

2009 HIGH & LOW WATER
BRIDGEPORT, CT
41°10.4'N, 73°10.9'W

Daylight Time starts March 8 at 2 a.m. **Daylight Saving Time**

Day of Month	Day of Week	MARCH HIGH a.m.	Ht.	HIGH p.m.	Ht.	LOW a.m.	LOW p.m.	Day of Month	Day of Week	APRIL HIGH a.m.	Ht.	HIGH p.m.	Ht.	LOW a.m.	LOW p.m.
1	S	1 34	7.3	2 06	6.5	8 02	8 11	1	W	3 59	7.3	4 44	6.3	10 39	10 52
2	M	2 19	7.2	2 57	6.2	8 54	9 02	2	T	5 03	7.0	5 50	6.2	11 44	...
3	T	3 12	7.1	3 56	6.0	9 52	10 02	3	F	6 13	6.8	6 57	6.4	12 01	12 51
4	W	4 14	6.9	5 02	5.9	10 58	11 09	4	S	7 23	6.8	8 01	6.7	1 12	1 55
5	T	5 23	6.8	6 11	5.9	...	12 07	5	S	8 28	6.9	9 00	7.1	2 19	2 53
6	F	6 35	6.9	7 17	6.2	12 20	1 14	6	M	9 27	7.1	9 53	7.5	3 19	3 46
7	S	7 42	7.1	8 18	6.6	1 29	2 15	7	T	10 20	7.2	10 41	7.8	4 14	4 34
8	S	*9 42	7.3	*10 13	7.1	*3 31	*4 09	8	W	11 08	7.3	11 25	8.0	5 04	5 18
9	M	10 38	7.5	11 04	7.5	4 28	4 58	9	T	11 55	7.2	...	...	5 50	6 01
10	T	11 27	7.6	11 50	7.8	5 20	5 44	10	F	12 08	8.0	12 37	7.1	6 33	6 41
11	W	...	...	12 14	7.5	6 09	6 28	11	S	12 48	7.8	1 19	6.9	7 15	7 22
12	T	12 34	7.9	12 59	7.4	6 55	7 09	12	S	1 29	7.6	2 02	6.7	7 56	8 02
13	F	1 16	7.9	1 43	7.1	7 39	7 51	13	M	2 10	7.3	2 45	6.4	8 38	8 45
14	S	1 59	7.7	2 28	6.8	8 24	8 33	14	T	2 54	6.9	3 30	6.2	9 22	9 31
15	S	2 42	7.3	3 13	6.4	9 09	9 17	15	W	3 41	6.6	4 19	6.0	10 09	10 22
16	M	3 27	6.9	4 02	6.1	9 56	10 05	16	T	4 33	6.3	5 12	5.9	11 00	11 19
17	T	4 16	6.5	4 54	5.8	10 47	10 58	17	F	5 29	6.1	6 08	5.9	11 54	...
18	W	5 11	6.2	5 51	5.6	11 42	11 58	18	S	6 28	6.0	7 04	6.1	12 18	12 49
19	T	6 11	6.0	6 51	5.6	...	12 41	19	S	7 26	6.0	7 57	6.3	1 18	1 43
20	F	7 13	5.9	7 49	5.7	1 00	1 39	20	M	8 20	6.1	8 46	6.6	2 14	2 32
21	S	8 12	6.0	8 43	6.0	2 00	2 33	21	T	9 10	6.3	9 31	7.0	3 05	3 19
22	S	9 04	6.2	9 31	6.3	2 54	3 21	22	W	9 58	6.6	10 14	7.4	3 53	4 03
23	M	9 51	6.4	10 14	6.7	3 43	4 04	23	T	10 43	6.8	10 55	7.7	4 38	4 46
24	T	10 35	6.7	10 54	7.0	4 28	4 44	24	F	11 28	6.9	11 37	7.9	5 23	5 29
25	W	11 15	6.9	11 31	7.3	5 10	5 23	25	S	...	...	12 12	7.0	6 07	6 13
26	T	11 55	7.0	...	...	5 50	6 01	26	S	12 21	8.1	12 58	7.1	6 53	6 59
27	F	12 08	7.6	12 36	7.0	6 31	6 40	27	M	1 07	8.1	1 47	7.0	7 42	7 48
28	S	12 47	7.7	1 18	7.0	7 13	7 21	28	T	1 56	8.0	2 38	6.9	8 33	8 42
29	S	1 27	7.8	2 02	6.9	7 57	8 05	29	W	2 51	7.7	3 34	6.8	9 28	9 41
30	M	2 12	7.7	2 51	6.7	8 46	8 54	30	T	3 50	7.5	4 34	6.7	10 27	10 45
31	T	3 02	7.5	3 44	6.5	9 39	9 49								

*Daylight Saving Time starts

Dates when Ht. of **Low** Water is below Mean Lower Low with Ht. of lowest given for each period and Date of lowest in ( ):

1st: -0.2'

8th - 14th: -0.9' (11th - 12th)

26th - 30th: -0.5' (28th - 29th)

7th - 12th: -0.6' (9th - 10th)

24th - 29th: -0.6' (26th - 27th)

Average Rise and Fall 6.8 ft.

When a high tide exceeds av. ht., the *following* low tide will be lower than av.

2009 HIGH & LOW WATER
BRIDGEPORT, CT
41°10.4'N, 73°10.9'W

Daylight Saving Time   Daylight Saving Time

DAY OF MONTH	DAY OF WEEK	MAY HIGH a.m.	Ht.	MAY HIGH p.m.	Ht.	MAY LOW a.m.	MAY LOW p.m.	DAY OF MONTH	DAY OF WEEK	JUNE HIGH a.m.	Ht.	JUNE HIGH p.m.	Ht.	JUNE LOW a.m.	JUNE LOW p.m.
1	F	4 53	7.2	5 37	6.8	11 29	11 52	1	M	6 44	6.7	7 15	7.4	12 42	1 02
2	S	6 00	6.9	6 40	6.9	...	12 31	2	T	7 45	6.5	8 11	7.5	1 44	1 57
3	S	7 06	6.8	7 41	7.2	1 00	1 31	3	W	8 43	6.5	9 02	7.6	2 42	2 50
4	M	8 09	6.8	8 37	7.4	2 04	2 27	4	T	9 36	6.5	9 51	7.6	3 35	3 40
5	T	9 07	6.8	9 29	7.7	3 03	3 19	5	F	10 26	6.5	10 36	7.5	4 24	4 27
6	W	9 59	6.8	10 16	7.8	3 57	4 07	6	S	11 12	6.5	11 20	7.5	5 09	5 11
7	T	10 48	6.9	11 00	7.8	4 45	4 52	7	S	11 54	6.6	...	...	5 50	5 53
8	F	11 33	6.9	11 42	7.8	5 30	5 35	8	M	12 01	7.4	12 35	6.6	6 30	6 34
9	S	...	...	12 17	6.8	6 12	6 16	9	T	12 43	7.3	1 16	6.6	7 08	7 14
10	S	12 23	7.6	12 57	6.7	6 52	6 56	10	W	1 22	7.1	1 55	6.6	7 46	7 55
11	M	1 03	7.4	1 38	6.6	7 31	7 37	11	T	2 02	7.0	2 35	6.6	8 25	8 37
12	T	1 44	7.2	2 19	6.5	8 11	8 18	12	F	2 43	6.8	3 16	6.6	9 04	9 21
13	W	2 26	7.0	3 02	6.4	8 52	9 03	13	S	3 25	6.6	3 59	6.6	9 45	10 08
14	T	3 11	6.7	3 47	6.3	9 35	9 51	14	S	4 11	6.5	4 43	6.7	10 28	10 59
15	F	3 58	6.5	4 35	6.3	10 21	10 43	15	M	5 00	6.3	5 30	6.8	11 14	11 53
16	S	4 48	6.3	5 25	6.3	11 10	11 38	16	T	5 53	6.2	6 20	6.9	...	12 04
17	S	5 42	6.1	6 17	6.4	11 59	...	17	W	6 49	6.1	7 11	7.1	12 49	12 56
18	M	6 37	6.1	7 08	6.6	12 34	12 51	18	T	7 47	6.1	8 05	7.4	1 46	1 50
19	T	7 33	6.1	7 57	6.9	1 30	1 42	19	F	8 45	6.3	8 59	7.7	2 43	2 45
20	W	8 27	6.2	8 46	7.2	2 24	2 32	20	S	9 41	6.5	9 54	7.9	3 39	3 41
21	T	9 20	6.4	9 34	7.6	3 16	3 22	21	S	10 36	6.7	10 49	8.2	4 33	4 36
22	F	10 11	6.6	10 21	7.9	4 07	4 11	22	M	11 30	7.0	11 43	8.3	5 27	5 31
23	S	11 00	6.8	11 10	8.1	4 56	5 00	23	T	...	...	12 23	7.2	6 19	6 27
24	S	11 50	7.0	...	...	5 46	5 50	24	W	12 37	8.4	1 15	7.4	7 11	7 22
25	M	12 01	8.3	12 40	7.1	6 36	6 42	25	T	1 32	8.2	2 09	7.6	8 03	8 19
26	T	12 51	8.3	1 32	7.2	7 27	7 35	26	F	2 26	8.0	3 02	7.6	8 54	9 16
27	W	1 45	8.2	2 25	7.2	8 20	8 32	27	S	3 22	7.7	3 57	7.7	9 47	10 16
28	T	2 40	7.9	3 21	7.2	9 14	9 31	28	S	4 18	7.2	4 53	7.6	10 40	11 16
29	F	3 38	7.6	4 19	7.2	10 10	10 34	29	M	5 17	6.8	5 49	7.5	11 34	...
30	S	4 39	7.3	5 18	7.3	11 07	11 38	30	T	6 17	6.5	6 45	7.4	12 17	12 30
31	S	5 41	6.9	6 17	7.4	...	12 05								

Dates when Ht. of **Low** Water is below Mean Lower Low with Ht. of lowest given for each period and Date of lowest in ( ):

7th - 9th: -0.3' (8th - 9th)   21st - 27th: -0.7' (24th - 25th)
23rd - 29th: -0.6' (25th - 26th)

Average Rise and Fall 6.8 ft.

When a high tide exceeds av. ht., the *following* low tide will be lower than av.

2009 HIGH & LOW WATER
BRIDGEPORT, CT

41°10.4'N, 73°10.9'W

Daylight Saving Time   **Daylight Saving Time**

DAY OF MONTH	DAY OF WEEK	JULY HIGH a.m.	Ht.	JULY HIGH p.m.	Ht.	JULY LOW a.m.	JULY LOW p.m.	DAY OF MONTH	DAY OF WEEK	AUGUST HIGH a.m.	Ht.	AUGUST HIGH p.m.	Ht.	AUGUST LOW a.m.	AUGUST LOW p.m.
1	W	7 17	6.3	7 41	7.4	1 18	1 26	1	S	8 43	6.1	9 01	6.9	2 41	2 47
2	T	8 16	6.2	8 35	7.3	2 16	2 21	2	S	9 35	6.2	9 51	7.0	3 33	3 39
3	F	9 11	6.2	9 26	7.2	3 10	3 13	3	M	10 22	6.4	10 36	7.0	4 18	4 25
4	S	10 02	6.3	10 14	7.2	4 00	4 03	4	T	11 05	6.6	11 18	7.1	5 00	5 08
5	S	10 49	6.4	10 59	7.2	4 46	4 49	5	W	11 45	6.7	11 57	7.1	5 38	5 49
6	M	11 31	6.5	11 41	7.2	5 27	5 32	6	T	...	...	12 22	6.9	6 14	6 27
7	T	...	...	12 12	6.6	6 06	6 12	7	F	12 34	7.1	12 58	7.0	6 49	7 06
8	W	12 21	7.2	12 51	6.7	6 44	6 52	8	S	1 10	7.1	1 33	7.1	7 24	7 44
9	T	1 00	7.1	1 29	6.7	7 20	7 31	9	S	1 48	7.0	2 10	7.2	7 59	8 24
10	F	1 37	7.0	2 06	6.8	7 56	8 11	10	M	2 26	6.9	2 46	7.2	8 36	9 06
11	S	2 15	6.9	2 43	6.9	8 32	8 52	11	T	3 08	6.7	3 26	7.3	9 16	9 53
12	S	2 55	6.8	3 22	6.9	9 10	9 36	12	W	3 55	6.5	4 12	7.3	10 01	10 46
13	M	3 37	6.6	4 03	7.0	9 50	10 24	13	T	4 48	6.3	5 05	7.2	10 52	11 46
14	T	4 24	6.4	4 48	7.1	10 34	11 16	14	F	5 47	6.1	6 06	7.2	11 51	...
15	W	5 16	6.2	5 38	7.1	11 23	...	15	S	6 51	6.1	7 11	7.3	12 50	12 56
16	T	6 13	6.1	6 33	7.2	12 13	12 18	16	S	7 56	6.3	8 17	7.5	1 55	2 02
17	F	7 14	6.1	7 32	7.4	1 14	1 17	17	M	8 59	6.6	9 20	7.8	2 57	3 06
18	S	8 16	6.2	8 34	7.6	2 15	2 18	18	T	9 57	7.0	10 18	8.1	3 55	4 06
19	S	9 17	6.4	9 34	7.9	3 16	3 19	19	W	10 51	7.5	11 13	8.2	4 48	5 03
20	M	10 15	6.8	10 32	8.1	4 13	4 19	20	T	11 43	7.9	...	...	5 38	5 57
21	T	11 10	7.1	11 28	8.3	5 08	5 16	21	F	12 05	8.2	12 32	8.1	6 26	6 49
22	W	...	...	12 04	7.5	6 00	6 12	22	S	12 55	8.1	1 21	8.3	7 13	7 40
23	T	12 22	8.3	12 55	7.8	6 51	7 07	23	S	1 44	7.8	2 08	8.2	7 59	8 31
24	F	1 15	8.2	1 47	8.0	7 40	8 01	24	M	2 34	7.4	2 57	8.0	8 46	9 23
25	S	2 07	8.0	2 38	8.0	8 29	8 56	25	T	3 25	7.0	3 47	7.6	9 35	10 17
26	S	2 59	7.6	3 29	7.9	9 18	9 52	26	W	4 18	6.6	4 40	7.3	10 27	11 13
27	M	3 53	7.1	4 22	7.7	10 08	10 49	27	T	5 15	6.2	5 37	6.9	11 22	...
28	T	4 49	6.7	5 16	7.5	11 01	11 47	28	F	6 14	6.0	6 36	6.7	12 11	12 21
29	W	5 47	6.3	6 12	7.2	11 57	...	29	S	7 14	6.0	7 36	6.6	1 11	1 21
30	T	6 47	6.1	7 10	7.0	12 47	12 54	30	S	8 11	6.1	8 32	6.7	2 07	2 19
31	F	7 47	6.0	8 07	6.9	1 46	1 52	31	M	9 04	6.3	9 23	6.8	2 59	3 11

Dates when Ht. of **Low** Water is below Mean Lower Low with Ht. of lowest given for each period and Date of lowest in ( ):

21st - 26th: -0.7' (23rd - 24th)   19th - 23rd: -0.7' (21st)

Average Rise and Fall 6.8 ft.

When a high tide exceeds av. ht., the *following* low tide will be lower than av.

101

2009 HIGH & LOW WATER
BRIDGEPORT, CT
41°10.4'N, 73°10.9'W

Daylight Saving Time · · · · · · · · · · · · Daylight Saving Time

DAY OF MONTH	DAY OF WEEK	SEPTEMBER HIGH a.m.	Ht.	HIGH p.m.	Ht.	LOW a.m.	LOW p.m.	DAY OF MONTH	DAY OF WEEK	OCTOBER HIGH a.m.	Ht.	HIGH p.m.	Ht.	LOW a.m.	LOW p.m.
1	T	9 51	6.5	10 08	6.9	3 45	3 58	1	T	9 56	7.0	10 17	6.9	3 46	4 10
2	W	10 33	6.8	10 50	7.0	4 26	4 41	2	F	10 35	7.2	10 57	7.0	4 25	4 51
3	T	11 12	7.0	11 28	7.1	5 04	5 22	3	S	11 13	7.5	11 36	7.1	5 03	5 31
4	F	11 49	7.2	...	...	5 40	6 00	4	S	11 49	7.6	...	...	5 40	6 10
5	S	12 06	7.2	12 24	7.4	6 16	6 38	5	M	12 15	7.1	12 25	7.7	6 18	6 51
6	S	12 43	7.1	12 59	7.5	6 51	7 17	6	T	12 55	7.0	1 04	7.8	6 58	7 34
7	M	1 20	7.1	1 34	7.5	7 27	7 57	7	W	1 38	6.9	1 47	7.7	7 40	8 20
8	T	2 00	6.9	2 13	7.5	8 05	8 40	8	T	2 24	6.7	2 34	7.6	8 26	9 12
9	W	2 45	6.7	2 57	7.5	8 48	9 29	9	F	3 17	6.5	3 30	7.4	9 19	10 10
10	T	3 32	6.5	3 46	7.4	9 36	10 25	10	S	4 14	6.4	4 31	7.2	10 20	11 13
11	F	4 28	6.3	4 44	7.2	10 32	11 27	11	S	5 18	6.4	5 39	7.1	11 27	...
12	S	5 30	6.2	5 50	7.2	11 36	...	12	M	6 23	6.5	6 47	7.1	12 18	12 37
13	S	6 36	6.2	6 59	7.2	12 33	12 44	13	T	7 27	6.8	7 53	7.2	1 21	1 43
14	M	7 42	6.5	8 05	7.4	1 39	1 52	14	W	8 26	7.2	8 53	7.3	2 20	2 45
15	T	8 43	6.9	9 07	7.6	2 40	2 56	15	T	9 21	7.7	9 47	7.4	3 13	3 42
16	W	9 40	7.4	10 04	7.8	3 35	3 54	16	F	10 11	8.0	10 38	7.5	4 03	4 34
17	T	10 32	7.8	10 56	7.9	4 26	4 49	17	S	10 58	8.2	11 26	7.4	4 50	5 23
18	F	11 20	8.2	11 45	7.9	5 14	5 40	18	S	11 42	8.3	...	...	5 34	6 09
19	S	...	...	12 07	8.4	6 00	6 29	19	M	12 12	7.3	12 26	8.1	6 18	6 54
20	S	12 33	7.7	12 53	8.3	6 45	7 17	20	T	12 57	7.1	1 09	7.9	7 01	7 38
21	M	1 20	7.5	1 38	8.1	7 29	8 05	21	W	1 42	6.9	1 54	7.5	7 45	8 23
22	T	2 07	7.1	2 24	7.8	8 14	8 53	22	T	2 28	6.6	2 40	7.1	8 30	9 09
23	W	2 56	6.8	3 12	7.4	9 01	9 43	23	F	3 15	6.3	3 29	6.8	9 19	9 58
24	T	3 47	6.4	4 04	7.0	9 52	10 36	24	S	4 06	6.1	4 23	6.5	10 12	10 51
25	F	4 41	6.1	5 00	6.6	10 47	11 32	25	S	5 00	6.0	5 19	6.2	11 08	11 45
26	S	5 38	6.0	6 00	6.4	11 47	...	26	M	5 56	6.0	6 17	6.1	...	12 07
27	S	6 37	6.0	7 00	6.4	12 30	12 47	27	T	6 51	6.2	7 13	6.1	12 38	1 05
28	M	7 34	6.1	7 57	6.4	1 26	1 45	28	W	7 43	6.4	8 06	6.2	1 29	1 59
29	T	8 27	6.4	8 48	6.6	2 18	2 38	29	T	8 31	6.7	8 55	6.4	2 17	2 49
30	W	9 14	6.7	9 34	6.7	3 04	3 26	30	F	9 15	7.0	9 40	6.6	3 02	3 35
								31	S	9 56	7.3	10 24	6.7	3 44	4 19

Dates when Ht. of **Low** Water is below Mean Lower Low with Ht. of lowest given for each period and Date of lowest in ( ):

17th - 21st: -0.6' (18th - 19th) · · · · · · · 15th - 19th: -0.6' (17th)

Average Rise and Fall 6.8 ft.

When a high tide exceeds av. ht., the *following* low tide will be lower than av.

2009 HIGH & LOW WATER
BRIDGEPORT, CT
41°10.4'N, 73°10.9'W

Standard Time starts Nov. 1 at 2 a.m.   Standard Time

DAY OF MONTH	DAY OF WEEK	NOVEMBER HIGH a.m.	Ht.	HIGH p.m.	Ht.	LOW a.m.	LOW p.m.	DAY OF MONTH	DAY OF WEEK	DECEMBER HIGH a.m.	Ht.	HIGH p.m.	Ht.	LOW a.m.	LOW p.m.
1	S	*9 36	7.6	*10 06	6.8	*3 26	*4 02	1	T	9 48	7.7	10 26	6.7	3 38	4 22
2	M	10 16	7.7	10 49	6.9	4 07	4 44	2	W	10 35	7.9	11 13	6.8	4 26	5 10
3	T	10 57	7.9	11 33	6.9	4 49	5 29	3	T	11 24	7.9	...	...	5 15	5 59
4	W	11 41	7.9	...	...	5 33	6 15	4	F	12 03	6.9	12 15	7.9	6 06	6 50
5	T	12 19	6.8	12 28	7.8	6 20	7 04	5	S	12 54	6.9	1 09	7.7	7 00	7 43
6	F	1 09	6.7	1 21	7.6	7 12	7 58	6	S	1 48	6.9	2 06	7.4	7 58	8 38
7	S	2 03	6.6	2 18	7.4	8 09	8 56	7	M	2 45	6.9	3 05	7.1	8 59	9 34
8	S	3 01	6.6	3 20	7.2	9 11	9 57	8	T	3 44	7.0	4 08	6.8	10 03	10 33
9	M	4 05	6.6	4 27	7.0	10 18	10 58	9	W	4 45	7.1	5 13	6.5	11 09	11 31
10	T	5 07	6.8	5 32	6.8	11 26	11 59	10	T	5 44	7.2	6 15	6.4	...	12 13
11	W	6 09	7.1	6 36	6.8	...	12 31	11	F	6 42	7.3	7 15	6.3	12 29	1 14
12	T	7 06	7.4	7 35	6.8	12 56	1 32	12	S	7 37	7.4	8 11	6.3	1 24	2 10
13	F	8 00	7.7	8 30	6.9	1 49	2 27	13	S	8 29	7.4	9 03	6.4	2 17	3 01
14	S	8 49	7.9	9 21	6.9	2 39	3 18	14	M	9 17	7.4	9 51	6.4	3 07	3 48
15	S	9 36	7.9	10 08	6.9	3 27	4 06	15	T	10 02	7.3	10 35	6.4	3 53	4 32
16	M	10 20	7.9	10 53	6.8	4 12	4 50	16	W	10 45	7.3	11 16	6.4	4 37	5 13
17	T	11 03	7.7	11 36	6.7	4 55	5 33	17	T	11 26	7.1	11 57	6.4	5 18	5 52
18	W	11 46	7.5	...	...	5 38	6 14	18	F	...	...	12 07	7.0	5 59	6 31
19	T	12 19	6.6	12 28	7.2	6 20	6 56	19	S	12 36	6.4	12 47	6.8	6 39	7 09
20	F	1 01	6.4	1 12	6.9	7 03	7 38	20	S	1 16	6.3	1 28	6.6	7 20	7 47
21	S	1 45	6.3	1 57	6.6	7 48	8 22	21	M	1 57	6.3	2 10	6.4	8 04	8 27
22	S	2 31	6.2	2 46	6.4	8 37	9 08	22	T	2 40	6.3	2 55	6.1	8 50	9 10
23	M	3 20	6.1	3 36	6.1	9 29	9 57	23	W	3 24	6.3	3 43	5.9	9 40	9 55
24	T	4 11	6.1	4 30	6.0	10 24	10 46	24	T	4 11	6.3	4 35	5.7	10 33	10 44
25	W	5 02	6.2	5 25	5.9	11 20	11 37	25	F	5 00	6.4	5 31	5.7	11 29	11 37
26	T	5 53	6.4	6 19	5.9	...	12 15	26	S	5 52	6.5	6 28	5.7	...	12 25
27	F	6 42	6.6	7 12	6.0	12 27	1 08	27	S	6 45	6.7	7 25	5.8	12 31	1 22
28	S	7 30	6.9	8 02	6.2	1 16	1 58	28	M	7 39	7.0	8 20	6.0	1 26	2 17
29	S	8 16	7.2	8 51	6.4	2 03	2 47	29	T	8 33	7.3	9 13	6.3	2 20	3 10
30	M	9 02	7.5	9 38	6.5	2 51	3 34	30	W	9 26	7.6	10 05	6.6	3 14	4 02
								31	T	10 19	7.8	10 55	6.8	4 07	4 52

*Standard Time starts

Dates when Ht. of **Low** Water is below Mean Lower Low with Ht. of lowest given for each period and Date of lowest in ( ):

2nd - 5th: -0.4' (3rd)   1st - 7th: -0.7' (3rd)
13th - 17th: -0.5' (15th)   12th - 16th: -0.3' (13th - 15th)
30th: -0.3'   29th - 31st: -0.8' (31st)

Average Rise and Fall 6.8 ft.

When a high tide exceeds av. ht., the *following* low tide will be lower than av.

2009 HIGH & LOW WATER
KINGS POINT, NY
40°48.6'N, 73°45.9'W

DAY OF MONTH	DAY OF WEEK	JANUARY Standard Time						DAY OF MONTH	DAY OF WEEK	FEBRUARY Standard Time					
		HIGH		LOW		LOW				HIGH		HIGH		LOW	LOW
		a.m.	Ht.	p.m.	Ht.	a.m.	p.m.			a.m.	Ht.	p.m.	Ht.	a.m.	p.m.
1	T	1 33	7.0	1 47	7.2	7 49	8 11	1	S	2 28	7.7	2 57	6.9	8 59	9 12
2	F	2 14	7.1	2 33	7.0	8 35	8 55	2	M	3 17	7.7	3 51	6.7	9 54	10 06
3	S	2 59	7.3	3 23	6.8	9 26	9 43	3	T	4 12	7.5	4 53	6.4	11 00	11 07
4	S	3 48	7.4	4 18	6.6	10 22	10 36	4	W	5 15	7.4	6 06	6.3	...	12 31
5	M	4 42	7.5	5 18	6.5	11 26	11 33	5	T	6 29	7.4	7 35	6.5	12 18	2 08
6	T	5 42	7.6	6 27	6.5	...	12 41	6	F	7 55	7.7	8 49	6.9	1 49	3 13
7	W	6 47	7.7	7 41	6.6	12 36	2 10	7	S	9 08	8.0	9 48	7.5	3 13	4 08
8	T	7 56	8.0	8 50	7.0	1 45	3 20	8	S	10 06	8.4	10 40	7.9	4 15	4 59
9	F	9 03	8.3	9 52	7.3	3 00	4 19	9	M	11 00	8.5	11 30	8.2	5 10	5 46
10	S	10 03	8.6	10 47	7.6	4 09	5 12	10	T	11 48	8.5	...	...	6 00	6 30
11	S	11 00	8.7	11 41	7.9	5 10	6 02	11	W	12 16	8.4	12 37	8.3	6 49	7 13
12	M	11 55	8.6	...	...	6 06	6 51	12	T	1 02	8.4	1 25	7.9	7 38	7 56
13	T	12 34	8.0	12 50	8.4	7 02	7 40	13	F	1 47	8.2	2 14	7.4	8 28	8 38
14	W	1 27	8.1	1 45	8.0	7 58	8 29	14	S	2 33	7.9	3 05	6.9	9 22	9 23
15	T	2 21	8.0	2 42	7.5	8 57	9 21	15	S	3 21	7.5	4 02	6.5	10 20	10 18
16	F	3 15	7.8	3 41	7.0	9 58	10 15	16	M	4 16	7.0	5 06	6.1	11 21	11 26
17	S	4 11	7.6	4 44	6.6	11 00	11 14	17	T	5 24	6.7	6 13	6.0	...	12 22
18	S	5 11	7.3	5 49	6.3	...	12 01	18	W	6 34	6.5	7 16	6.0	12 33	1 20
19	M	6 13	7.1	6 52	6.2	12 13	1 00	19	T	7 37	6.6	8 12	6.2	1 33	2 15
20	T	7 13	7.0	7 51	6.2	1 11	1 55	20	F	8 32	6.8	9 01	6.5	2 28	3 04
21	W	8 09	7.0	8 43	6.4	2 07	2 48	21	S	9 19	7.0	9 44	6.7	3 16	3 48
22	T	9 00	7.1	9 30	6.5	2 58	3 36	22	S	9 59	7.2	10 22	7.0	3 59	4 27
23	F	9 45	7.2	10 13	6.7	3 45	4 20	23	M	10 32	7.3	10 51	7.2	4 37	4 59
24	S	10 25	7.3	10 52	6.8	4 27	4 59	24	T	10 57	7.4	11 10	7.4	5 07	5 21
25	S	10 58	7.4	11 24	6.9	5 03	5 33	25	W	11 17	7.5	11 28	7.6	5 30	5 37
26	M	11 23	7.4	11 47	7.0	5 31	5 58	26	T	11 45	7.6	11 57	7.9	5 55	6 03
27	T	11 41	7.4	...	...	5 49	6 10	27	F	...	...	12 20	7.6	6 28	6 37
28	W	12 02	7.1	12 07	7.4	6 13	6 32	28	S	12 34	8.1	1 01	7.5	7 06	7 16
29	T	12 28	7.3	12 42	7.4	6 47	7 04								
30	F	1 03	7.5	1 23	7.4	7 26	7 42								
31	S	1 43	7.6	2 08	7.2	8 10	8 25								

Dates when Ht. of **Low** Water is below Mean Lower Low with Ht. of lowest given for each period and Date of lowest in ( ):

8th - 16th: -1.4' (11th - 12th)   6th - 14th: -1.4' (9th - 10th)
25th - 30th: -0.3' (28th - 30th)   24th - 28th: -0.5' (27th - 28th)

Average Rise and Fall 7.1 ft.

When a high tide exceeds av. ht., the *following* low tide will be lower than av.

104

2009 HIGH & LOW WATER
KINGS POINT, NY
40°48.6'N, 73°45.9'W

Daylight Time starts March 8 at 2 a.m.   **Daylight Saving Time**

D A Y O F M O N T H	D A Y O F W E E K	MARCH HIGH a.m.	Ht.	HIGH p.m.	Ht.	LOW a.m.	LOW p.m.	D A Y O F M O N T H	D A Y O F W E E K	APRIL HIGH a.m.	Ht.	HIGH p.m.	Ht.	LOW a.m.	LOW p.m.
1	S	1 16	8.1	1 47	7.3	7 50	8 00	1	W	3 38	7.9	4 24	6.9	10 33	10 38
2	M	2 02	8.1	2 37	7.1	8 39	8 49	2	T	4 41	7.5	5 39	6.7	...	12 13
3	T	2 54	7.8	3 32	6.7	9 36	9 46	3	F	6 01	7.2	7 17	6.9	12 09	1 36
4	W	3 52	7.5	4 38	6.5	10 50	10 54	4	S	7 47	7.2	8 32	7.3	1 55	2 41
5	T	5 00	7.3	6 04	6.4	...	12 44	5	S	9 01	7.5	9 31	7.8	3 04	3 38
6	F	6 31	7.2	7 39	6.7	12 31	1 59	6	M	9 58	7.8	10 22	8.3	4 03	4 29
7	S	8 05	7.5	8 45	7.3	2 09	2 58	7	T	10 49	8.0	11 07	8.6	4 56	5 17
8	S	*10 09	7.9	*10 38	7.9	*4 14	*4 51	8	W	11 35	8.0	11 49	8.8	5 45	6 01
9	M	11 03	8.2	11 27	8.3	5 09	5 40	9	T	...	...	12 20	8.0	6 31	6 42
10	T	11 50	8.3	...	...	6 00	6 24	10	F	12 29	8.7	1 01	7.8	7 14	7 20
11	W	12 11	8.6	12 35	8.3	6 48	7 06	11	S	1 05	8.5	1 41	7.5	7 54	7 51
12	T	12 53	8.7	1 19	8.1	7 33	7 46	12	S	1 39	8.2	2 20	7.3	8 32	8 12
13	F	1 33	8.5	2 02	7.7	8 16	8 22	13	M	2 12	7.8	2 58	7.0	9 03	8 36
14	S	2 12	8.3	2 45	7.3	8 59	8 52	14	T	2 46	7.4	3 37	6.7	9 25	9 13
15	S	2 50	7.9	3 29	6.9	9 42	9 18	15	W	3 27	7.0	4 20	6.4	9 57	10 00
16	M	3 29	7.4	4 17	6.5	10 29	9 54	16	T	4 13	6.7	5 15	6.3	10 44	10 55
17	T	4 13	7.0	5 14	6.2	11 27	10 42	17	F	5 08	6.5	6 28	6.2	11 47	...
18	W	5 10	6.6	6 24	6.0	...	12 34	18	S	6 22	6.3	7 36	6.4	12 04	1 19
19	T	6 38	6.3	7 33	6.0	12 31	1 36	19	S	7 51	6.4	8 29	6.7	1 48	2 15
20	F	7 55	6.3	8 33	6.2	1 51	2 33	20	M	8 48	6.6	9 10	7.0	2 49	2 58
21	S	8 54	6.5	9 24	6.5	2 49	3 23	21	T	9 32	6.9	9 41	7.4	3 38	3 34
22	S	9 43	6.8	10 07	6.9	3 40	4 07	22	W	10 09	7.1	10 09	7.9	4 21	4 08
23	M	10 23	7.0	10 41	7.2	4 24	4 44	23	T	10 42	7.4	10 41	8.3	4 59	4 45
24	T	10 56	7.3	11 06	7.5	5 03	5 14	24	F	11 18	7.6	11 20	8.6	5 36	5 25
25	W	11 23	7.4	11 25	7.9	5 36	5 35	25	S	11 58	7.7	...	...	6 15	6 07
26	T	11 49	7.6	11 52	8.2	6 04	6 00	26	S	12 02	8.8	12 42	7.8	6 56	6 52
27	F	...	...	12 21	7.7	6 34	6 34	27	M	12 48	8.9	1 30	7.7	7 42	7 40
28	S	12 28	8.5	1 00	7.7	7 10	7 12	28	T	1 38	8.7	2 22	7.6	8 33	8 33
29	S	1 08	8.6	1 43	7.7	7 50	7 55	29	W	2 32	8.4	3 20	7.4	9 33	9 33
30	M	1 54	8.5	2 31	7.5	8 36	8 42	30	T	3 32	8.0	4 29	7.3	10 50	10 54
31	T	2 43	8.3	3 24	7.2	9 28	9 35								

*Daylight Saving Time starts

Dates when Ht. of **Low** Water is below Mean Lower Low with Ht. of lowest given for each period and Date of lowest in ( ):

1st: -0.4'   5th - 12th: -1.0' (8th - 9th)
7th - 14th: -1.2' (11th)   24th - 28th: -0.7' (26th - 27th)
26th - 30th: -0.7 (28th)

Average Rise and Fall 7.1 ft.

When a high tide exceeds av. ht., the *following* low tide will be lower than av.

2009 HIGH & LOW WATER
KINGS POINT, NY
40°48.6'N, 73°45.9'W

		Daylight Saving Time								Daylight Saving Time				
DAY OF MONTH	DAY OF WEEK	**MAY**					DAY OF MONTH	DAY OF WEEK	**JUNE**					
		HIGH		LOW					HIGH		LOW			
		a.m.	Ht.	**p.m.**	Ht.	a.m.	**p.m.**			a.m.	Ht.	**p.m.**	Ht.	a.m.	**p.m.**
1	F	4 43	7.6	**5 50**	7.3	...	**12 09**	1	M	7 18	7.2	**7 48**	8.1	1 27	**1 48**
2	S	6 15	7.3	**7 08**	7.5	12 33	**1 16**	2	T	8 22	7.2	**8 45**	8.3	2 28	**2 44**
3	S	7 38	7.3	**8 14**	7.8	1 46	**2 17**	3	W	9 19	7.2	**9 36**	8.4	3 25	**3 38**
4	M	8 45	7.4	**9 10**	8.2	2 49	**3 12**	4	T	10 11	7.3	**10 24**	8.4	4 17	**4 28**
5	T	9 41	7.6	**10 00**	8.5	3 46	**4 04**	5	F	10 59	7.3	**11 08**	8.3	5 07	**5 16**
6	W	10 32	7.7	**10 46**	8.6	4 39	**4 53**	6	S	11 43	7.4	**11 48**	8.2	5 52	**6 00**
7	T	11 18	7.7	**11 28**	8.6	5 27	**5 38**	7	S	...	...	**12 24**	7.3	6 35	**6 39**
8	F	...	...	**12 01**	7.6	6 13	**6 20**	8	M	12 25	8.0	**1 03**	7.2	7 14	**7 12**
9	S	12 07	8.5	**12 44**	7.5	6 55	**6 58**	9	T	12 58	7.8	**1 40**	7.1	7 48	**7 32**
10	S	12 42	8.2	**1 22**	7.4	7 34	**7 29**	10	W	1 24	7.6	**2 09**	7.1	8 11	**7 49**
11	M	1 14	8.0	**1 58**	7.2	8 09	**7 46**	11	T	1 51	7.5	**2 33**	7.0	8 22	**8 22**
12	T	1 43	7.7	**2 32**	7.0	8 34	**8 09**	12	F	2 25	7.4	**3 01**	7.0	8 49	**9 03**
13	W	2 15	7.4	**3 04**	6.8	8 48	**8 45**	13	S	3 04	7.2	**3 37**	7.1	9 26	**9 48**
14	T	2 53	7.2	**3 38**	6.7	9 19	**9 29**	14	S	3 48	7.1	**4 18**	7.2	10 09	**10 38**
15	F	3 36	7.0	**4 18**	6.7	10 00	**10 19**	15	M	4 35	6.9	**5 03**	7.3	10 55	**11 32**
16	S	4 23	6.8	**5 04**	6.7	10 48	**11 15**	16	T	5 27	6.8	**5 53**	7.5	11 45	...
17	S	5 15	6.6	**5 54**	6.8	11 39	...	17	W	6 24	6.7	**6 46**	7.7	12 30	**12 38**
18	M	6 13	6.5	**6 47**	7.0	12 16	**12 32**	18	T	7 25	6.7	**7 42**	7.9	1 33	**1 33**
19	T	7 15	6.6	**7 40**	7.3	1 22	**1 26**	19	F	8 29	6.9	**8 40**	8.2	2 41	**2 31**
20	W	8 18	6.7	**8 30**	7.7	2 28	**2 19**	20	S	9 31	7.1	**9 38**	8.6	3 51	**3 32**
21	T	9 14	7.0	**9 18**	8.1	3 28	**3 11**	21	S	10 28	7.4	**10 35**	8.8	4 55	**4 33**
22	F	10 03	7.3	**10 05**	8.5	4 21	**4 03**	22	M	11 24	7.7	**11 31**	9.0	5 51	**5 35**
23	S	10 50	7.5	**10 53**	8.8	5 11	**4 55**	23	T	...	...	**12 19**	7.9	6 45	**6 35**
24	S	11 38	7.7	**11 42**	9.0	6 00	**5 46**	24	W	12 27	9.0	**1 16**	8.1	7 37	**7 36**
25	M	...	...	**12 28**	7.8	6 50	**6 39**	25	T	1 26	8.8	**2 14**	8.2	8 30	**8 39**
26	T	12 34	9.0	**1 21**	7.8	7 42	**7 34**	26	F	2 27	8.6	**3 14**	8.3	9 24	**9 45**
27	W	1 29	8.8	**2 19**	7.8	8 38	**8 34**	27	S	3 30	8.2	**4 14**	8.3	10 21	**10 53**
28	T	2 29	8.5	**3 23**	7.8	9 39	**9 45**	28	S	4 36	7.8	**5 15**	8.2	11 19	...
29	F	3 34	8.1	**4 31**	7.8	10 44	**11 07**	29	M	5 44	7.4	**6 16**	8.2	12 01	**12 19**
30	S	4 48	7.7	**5 40**	7.8	11 48	...	30	T	6 52	7.1	**7 18**	8.1	1 03	**1 18**
31	S	6 06	7.4	**6 46**	8.0	12 21	**12 50**								

Dates when Ht. of **Low** Water is below Mean Lower Low with Ht. of lowest given for each period and Date of lowest in ( ):

5th - 10th: -0.7' (7th - 8th)   3rd - 7th: -0.4' (5th - 6th)
23rd - 29th: -0.7' (25th - 26th)   21st - 27th: -0.9' (24th)

Average Rise and Fall 7.1 ft.

When a high tide exceeds av. ht., the *following* low tide will be lower than av.

2009 HIGH & LOW WATER
KINGS POINT, NY
40°48.6'N, 73°45.9'W

		Daylight Saving Time								Daylight Saving Time					

DAY OF MONTH	DAY OF WEEK	JULY						DAY OF MONTH	DAY OF WEEK	AUGUST					
		HIGH				LOW				HIGH				LOW	
		a.m.	Ht.	p.m.	Ht.	a.m.	p.m.			a.m.	Ht.	p.m.	Ht.	a.m.	p.m.
1	W	7 56	6.9	8 17	8.0	2 03	2 16	1	S	9 23	6.8	9 40	7.6	3 27	3 39
2	T	8 55	6.9	9 12	8.0	3 00	3 11	2	S	10 13	7.0	10 28	7.7	4 17	4 29
3	F	9 48	7.0	10 02	8.0	3 53	4 04	3	M	10 58	7.2	11 11	7.7	5 03	5 15
4	S	10 37	7.1	10 49	7.9	4 43	4 54	4	T	11 39	7.3	11 48	7.8	5 45	5 55
5	S	11 22	7.2	11 31	7.9	5 30	5 39	5	W	...	...	12 15	7.4	6 22	6 30
6	M	...	...	12 04	7.2	6 12	6 19	6	T	12 20	7.7	12 45	7.5	6 51	6 55
7	T	12 10	7.8	12 42	7.3	6 51	6 54	7	F	12 42	7.7	1 03	7.6	7 07	7 11
8	W	12 42	7.7	1 16	7.2	7 23	7 17	8	S	1 01	7.7	1 19	7.7	7 19	7 37
9	T	1 07	7.6	1 42	7.2	7 44	7 32	9	S	1 31	7.6	1 49	7.9	7 47	8 11
10	F	1 27	7.6	1 58	7.3	7 53	8 00	10	M	2 06	7.6	2 24	8.0	8 22	8 52
11	S	1 57	7.5	2 24	7.4	8 19	8 37	11	T	2 48	7.5	3 06	8.1	9 03	9 37
12	S	2 34	7.4	2 59	7.6	8 54	9 19	12	W	3 34	7.3	3 52	8.1	9 48	10 27
13	M	3 16	7.3	3 40	7.7	9 34	10 06	13	T	4 25	7.1	4 43	8.0	10 38	11 25
14	T	4 02	7.2	4 25	7.8	10 19	10 57	14	F	5 21	6.9	5 41	7.9	11 35	...
15	W	4 53	7.0	5 14	7.9	11 08	11 53	15	S	6 26	6.8	6 47	7.9	12 34	12 38
16	T	5 48	6.8	6 09	7.9	...	12 02	16	S	7 42	6.8	8 02	8.0	2 10	1 52
17	F	6 50	6.7	7 09	8.0	12 57	1 00	17	M	9 04	7.2	9 20	8.3	3 33	3 20
18	S	7 59	6.8	8 15	8.2	2 12	2 04	18	T	10 10	7.7	10 27	8.6	4 33	4 37
19	S	9 10	7.1	9 22	8.4	3 38	3 14	19	W	11 06	8.2	11 23	8.8	5 26	5 37
20	M	10 16	7.4	10 26	8.7	4 45	4 28	20	T	11 57	8.6	...	...	6 15	6 31
21	T	11 15	7.9	11 26	8.9	5 42	5 36	21	F	12 16	8.9	12 45	8.9	7 01	7 23
22	W	...	...	12 10	8.2	6 33	6 37	22	S	1 07	8.7	1 34	9.0	7 46	8 14
23	T	12 24	9.0	1 04	8.5	7 22	7 34	23	S	1 58	8.4	2 22	8.9	8 30	9 07
24	F	1 20	8.8	1 58	8.6	8 11	8 31	24	M	2 51	8.0	3 11	8.6	9 16	10 02
25	S	2 16	8.5	2 51	8.7	9 00	9 29	25	T	3 46	7.5	4 02	8.2	10 06	11 01
26	S	3 14	8.1	3 46	8.6	9 51	10 30	26	W	4 45	7.1	5 00	7.8	11 05	...
27	M	4 13	7.7	4 42	8.3	10 45	11 33	27	T	5 49	6.8	6 06	7.4	12 02	12 11
28	T	5 16	7.2	5 41	8.1	11 44	...	28	F	6 54	6.6	7 14	7.2	1 03	1 16
29	W	6 22	6.9	6 44	7.8	12 35	12 46	29	S	7 57	6.6	8 17	7.2	2 01	2 16
30	T	7 27	6.7	7 47	7.6	1 35	1 47	30	S	8 53	6.8	9 13	7.3	2 55	3 11
31	F	8 28	6.7	8 46	7.6	2 32	2 45	31	M	9 43	7.0	10 01	7.5	3 45	4 01

Dates when Ht. of **Low** Water is below Mean Lower Low with Ht. of lowest given for each period and Date of lowest in ( ):

20th - 26th: -1.0' (23rd)   19th - 23rd: -0.9' (21st)

Average Rise and Fall 7.1 ft.

When a high tide exceeds av. ht., the *following* low tide will be lower than av.

2009 HIGH & LOW WATER
KINGS POINT, NY
40°48.6'N, 73°45.9'W

Daylight Saving Time   Daylight Saving Time

D A Y O F M O N T H	D A Y O F W E E K	SEPTEMBER HIGH a.m.	Ht.	HIGH p.m.	Ht.	LOW a.m.	LOW p.m.	D A Y O F M O N T H	D A Y O F W E E K	OCTOBER HIGH a.m.	Ht.	HIGH p.m.	Ht.	LOW a.m.	LOW p.m.
1	T	10 28	7.3	10 43	7.6	4 31	4 46	1	T	10 29	7.6	10 44	7.5	4 28	4 51
2	W	11 08	7.5	11 20	7.7	5 11	5 26	2	F	10 58	7.8	11 13	7.6	4 59	5 26
3	T	11 41	7.7	11 50	7.7	5 46	6 00	3	S	11 15	8.1	11 35	7.6	5 19	5 53
4	F	...	...	12 05	7.8	6 11	6 26	4	S	11 35	8.3	...	...	5 40	6 18
5	S	12 10	7.7	12 18	8.0	6 23	6 45	5	M	12 02	7.7	12 06	8.5	6 10	6 49
6	S	12 32	7.7	12 40	8.1	6 44	7 12	6	T	12 37	7.7	12 44	8.6	6 47	7 26
7	M	1 02	7.7	1 13	8.3	7 16	7 47	7	W	1 18	7.7	1 27	8.6	7 29	8 09
8	T	1 40	7.7	1 53	8.4	7 53	8 27	8	T	2 03	7.5	2 15	8.5	8 14	8 58
9	W	2 25	7.5	2 38	8.4	8 35	9 13	9	F	2 55	7.3	3 09	8.2	9 05	9 56
10	T	3 11	7.3	3 26	8.2	9 23	10 06	10	S	3 51	7.1	4 06	7.9	10 03	11 12
11	F	4 04	7.1	4 21	8.0	10 16	11 09	11	S	4 56	7.0	5 14	7.6	11 14	...
12	S	5 03	6.9	5 23	7.8	11 18	...	12	M	6 22	7.0	6 43	7.5	12 52	1 01
13	S	6 15	6.8	6 36	7.7	12 36	12 32	13	T	7 50	7.4	8 16	7.6	2 03	2 25
14	M	7 46	7.0	8 08	7.8	2 15	2 13	14	W	8 55	7.9	9 21	7.9	3 02	3 29
15	T	9 04	7.5	9 25	8.1	3 21	3 34	15	T	9 48	8.5	10 15	8.1	3 55	4 25
16	W	10 02	8.1	10 24	8.4	4 16	4 35	16	F	10 36	8.9	11 04	8.2	4 45	5 17
17	T	10 53	8.6	11 15	8.6	5 06	5 29	17	S	11 20	9.1	11 50	8.2	5 31	6 05
18	F	11 39	9.0	...		5 53	6 20	18	S	...	...	12 02	9.1	6 15	6 51
19	S	12 04	8.6	12 23	9.2	6 37	7 08	19	M	12 35	8.0	12 42	8.9	6 56	7 35
20	S	12 51	8.4	1 06	9.1	7 19	7 55	20	T	1 18	7.7	1 21	8.5	7 34	8 18
21	M	1 38	8.1	1 49	8.8	8 00	8 42	21	W	2 02	7.4	2 00	8.1	8 07	9 01
22	T	2 26	7.7	2 33	8.4	8 39	9 32	22	T	2 48	7.1	2 41	7.7	8 36	9 46
23	W	3 16	7.3	3 20	8.0	9 20	10 26	23	F	3 36	6.8	3 26	7.3	9 11	10 35
24	T	4 11	7.0	4 13	7.5	10 10	11 25	24	S	4 30	6.6	4 19	6.9	9 58	11 35
25	F	5 12	6.7	5 19	7.1	11 26	...	25	S	5 30	6.5	5 27	6.6	11 16	...
26	S	6 17	6.5	6 34	6.9	12 25	12 39	26	M	6 33	6.5	6 44	6.5	12 32	12 51
27	S	7 20	6.5	7 40	6.8	1 23	1 41	27	T	7 31	6.6	7 47	6.6	1 25	1 51
28	M	8 18	6.7	8 37	7.0	2 17	2 36	28	W	8 22	6.9	8 40	6.7	2 12	2 43
29	T	9 08	7.0	9 26	7.2	3 06	3 26	29	T	9 05	7.2	9 24	6.9	2 53	3 29
30	W	9 52	7.3	10 08	7.3	3 50	4 11	30	F	9 38	7.5	10 01	7.1	3 26	4 11
								31	S	10 02	7.8	10 31	7.3	3 53	4 47

Dates when Ht. of **Low** Water is below Mean Lower Low with Ht. of lowest given for each period and Date of lowest in ( ):

16th - 21st: -0.9' (19th)   5th - 6th: -0.2'
                                             15th - 20th: -0.9' (17th - 18th)

Average Rise and Fall 7.1 ft.

When a high tide exceeds av. ht., the *following* low tide will be lower than av.

108

2009 HIGH & LOW WATER
KINGS POINT, NY
40°48.6'N, 73°45.9'W

Standard Time starts Nov. 1 at 2 a.m.   **Standard Time**

DAY OF MONTH	DAY OF WEEK	NOVEMBER HIGH a.m.	Ht.	HIGH p.m.	Ht.	LOW a.m.	LOW p.m.	DAY OF MONTH	DAY OF WEEK	DECEMBER HIGH a.m.	Ht.	HIGH p.m.	Ht.	LOW a.m.	LOW p.m.
1	S	*9 27	8.2	*10 01	7.4	*3 24	*4 20	1	T	9 31	8.4	10 13	7.3	3 30	4 35
2	M	9 59	8.5	10 36	7.6	4 01	4 53	2	W	10 17	8.6	10 59	7.5	4 19	5 21
3	T	10 38	8.7	11 15	7.6	4 41	5 31	3	T	11 06	8.7	11 48	7.5	5 08	6 09
4	W	11 21	8.7	...	...	5 24	6 13	4	F	11 58	8.6	...	...	6 00	7 00
5	T	12 01	7.6	12 09	8.7	6 10	7 00	5	S	12 41	7.5	12 53	8.4	6 55	7 56
6	F	12 49	7.5	1 00	8.4	7 00	7 54	6	S	1 40	7.5	1 52	8.1	7 56	8 59
7	S	1 44	7.3	1 57	8.1	7 56	9 00	7	M	2 44	7.5	2 58	7.7	9 13	10 07
8	S	2 46	7.2	3 00	7.7	9 02	10 23	8	T	3 54	7.5	4 15	7.3	10 40	11 13
9	M	4 01	7.2	4 17	7.4	10 38	11 38	9	W	5 07	7.6	5 38	7.0	11 53	...
10	T	5 24	7.3	5 50	7.3	...	12 08	10	T	6 14	7.8	6 48	6.9	12 15	12 58
11	W	6 37	7.7	7 06	7.3	12 41	1 16	11	F	7 15	8.0	7 50	7.0	1 14	1 57
12	T	7 37	8.1	8 08	7.5	1 39	2 15	12	S	8 11	8.2	8 46	7.1	2 10	2 52
13	F	8 30	8.5	9 01	7.6	2 33	3 10	13	S	9 01	8.2	9 36	7.2	3 04	3 44
14	S	9 18	8.7	9 50	7.7	3 23	4 01	14	M	9 48	8.2	10 22	7.2	3 54	4 32
15	S	10 03	8.8	10 36	7.7	4 11	4 49	15	T	10 32	8.1	11 05	7.2	4 40	5 17
16	M	10 44	8.7	11 20	7.5	4 56	5 34	16	W	11 12	8.0	11 46	7.1	5 23	5 58
17	T	11 24	8.4	...	...	5 38	6 17	17	T	11 48	7.7	...	...	6 01	6 36
18	W	12 02	7.4	12 01	8.1	6 16	6 57	18	F	12 24	7.0	12 20	7.5	6 32	7 09
19	T	12 42	7.2	12 36	7.8	6 46	7 34	19	S	12 58	6.9	12 48	7.3	6 49	7 27
20	F	1 22	6.9	1 11	7.5	7 06	8 05	20	S	1 27	6.8	1 17	7.1	7 11	7 40
21	S	2 01	6.7	1 48	7.2	7 36	8 24	21	M	1 53	6.7	1 52	7.0	7 47	8 11
22	S	2 41	6.6	2 29	6.9	8 18	8 56	22	T	2 25	6.7	2 33	6.8	8 30	8 51
23	M	3 22	6.5	3 16	6.6	9 07	9 39	23	W	3 03	6.8	3 18	6.6	9 18	9 35
24	T	4 08	6.5	4 08	6.5	10 04	10 28	24	T	3 45	6.8	4 08	6.4	10 11	10 24
25	W	4 58	6.6	5 06	6.3	11 10	11 19	25	F	4 33	6.9	5 02	6.2	11 08	11 16
26	T	5 48	6.7	6 12	6.3	...	12 25	26	S	5 24	7.1	6 03	6.2	...	12 09
27	F	6 36	7.0	7 13	6.4	12 11	1 28	27	S	6 20	7.3	7 07	6.3	12 11	1 17
28	S	7 21	7.4	8 03	6.6	1 02	2 19	28	M	7 18	7.5	8 08	6.6	1 08	2 28
29	S	8 03	7.7	8 47	6.9	1 51	3 06	29	T	8 16	7.9	9 04	6.9	2 07	3 31
30	M	8 47	8.1	9 29	7.1	2 41	3 51	30	W	9 11	8.2	9 56	7.2	3 07	4 25
								31	T	10 05	8.5	10 48	7.5	4 05	5 16

*Standard Time starts

Dates when Ht. of **Low** Water is below Mean Lower Low with Ht. of lowest given for each period and Date of lowest in ( ):

     2nd - 5th: -0.4' (3rd - 4th)   1st - 7th: -0.7' (2nd - 4th)
     12th - 18th: -0.9' (15th)   11th - 17th: -0.8' (14th)
     30th: -0.2'   29th - 31st: -0.9' (31st)

Average Rise and Fall 7.1 ft.

When a high tide exceeds av. ht., the *following* low tide will be lower than av.

2009 CURRENT TABLE
HELL GATE, NY (EAST RIVER)
40°46.7'N, 73°56.3'W Off Mill Rock

Standard Time   Standard Time

DAY OF MONTH	DAY OF WEEK	JANUARY NORTHEAST Flood Starts a.m.	p.m.	Kts.	SOUTHWEST Ebb Starts a.m.	p.m.	Kts.	DAY OF MONTH	DAY OF WEEK	FEBRUARY NORTHEAST Flood Starts a.m.	p.m.	Kts.	SOUTHWEST Ebb Starts a.m.	p.m.	Kts.
1	T	6 50	7 17	a3.3	12 27	12 50	4.7	1	S	7 50	8 08	3.2	1 23	1 53	a4.8
2	F	7 34	7 59	a3.2	1 08	1 34	4.6	2	M	8 45	9 03	3.1	2 14	2 48	a4.7
3	S	8 23	8 47	a3.1	1 53	2 22	a4.6	3	T	9 50	10 09	p3.1	3 13	3 50	a4.6
4	S	9 19	9 41	3.0	2 44	3 16	a4.6	4	W	11 01	11 20	p3.1	4 20	4 58	a4.5
5	M	10 21	10 40	p3.1	3 41	4 16	a4.6	5	T	...	12 12	3.1	5 30	6 07	a4.5
6	T	11 26	11 44	p3.2	4 43	5 19	a4.6	6	F	12 30	1 17	3.3	6 39	7 12	4.6
7	W	...	12 31	3.1	5 47	6 23	a4.7	7	S	1 35	2 17	3.5	7 42	8 12	4.8
8	T	12 47	1 33	a3.4	6 50	7 25	a4.8	8	S	2 35	3 11	a3.8	8 41	9 07	5.0
9	F	1 49	2 32	3.5	7 51	8 23	a5.0	9	M	3 31	4 03	3.9	9 35	9 58	5.1
10	S	2 46	3 27	3.7	8 50	9 19	a5.1	10	T	4 21	4 50	a4.0	10 26	10 47	5.1
11	S	3 42	4 20	a3.9	9 45	10 13	a5.2	11	W	5 11	5 38	a4.0	11 15	11 35	5.1
12	M	4 37	5 12	a3.9	10 39	11 05	a5.2	12	T	6 00	6 24	a3.9	...	12 03	5.0
13	T	5 31	6 03	a3.9	11 32	11 56	a5.1	13	F	6 49	7 11	a3.7	12 22	12 50	a4.9
14	W	6 24	6 54	a3.8	...	12 24	5.0	14	S	7 38	7 59	a3.4	1 09	1 37	a4.7
15	T	7 18	7 46	a3.6	12 47	1 16	a4.9	15	S	8 28	8 49	a3.1	1 58	2 26	a4.5
16	F	8 13	8 39	a3.4	1 39	2 08	a4.7	16	M	9 22	9 42	a2.9	2 48	3 18	a4.2
17	S	9 09	9 33	a3.1	2 31	3 01	a4.5	17	T	10 19	10 39	2.7	3 42	4 12	a4.0
18	S	10 07	10 29	2.9	3 25	3 55	a4.3	18	W	11 16	11 35	p2.7	4 38	5 08	a3.9
19	M	11 04	11 24	2.8	4 20	4 50	a4.2	19	T	...	12 12	2.7	5 34	6 03	a3.9
20	T	11 59	...	2.8	5 14	5 43	a4.1	20	F	12 28	1 02	2.8	6 28	6 54	4.0
21	W	12 17	12 52	a2.9	6 07	6 35	a4.2	21	S	1 17	1 48	3.0	7 17	7 40	4.2
22	T	1 06	1 39	2.9	6 57	7 23	a4.3	22	S	2 00	2 29	3.2	8 03	8 22	4.4
23	F	1 51	2 22	a3.1	7 44	8 07	a4.4	23	M	2 41	3 08	3.4	8 45	9 02	4.6
24	S	2 33	3 03	3.2	8 28	8 49	4.5	24	T	3 20	3 44	a3.6	9 25	9 40	p4.8
25	S	3 12	3 41	3.3	9 09	9 29	a4.7	25	W	3 57	4 20	a3.7	10 04	10 18	p4.9
26	M	3 50	4 18	a3.5	9 49	10 07	a4.8	26	T	4 34	4 54	a3.7	10 42	10 55	p5.0
27	T	4 27	4 54	a3.5	10 28	10 44	4.8	27	F	5 12	5 30	a3.7	11 21	11 34	p5.0
28	W	5 04	5 29	a3.6	11 06	11 21	4.9	28	S	5 52	6 08	a3.7	...	12 02	4.8
29	T	5 41	6 04	a3.5	11 44	11 59	4.9								
30	F	6 20	6 41	a3.5	...	12 24	4.8								
31	S	7 02	7 21	a3.4	12 39	1 06	a4.8								

The Kts. (knots) columns show the **maximum** predicted velocities of the stronger one of the Flood Currents and the stronger one of the Ebb Currents for each day. The letter "a" means the velocity shown should occur **after** the **a.m.** Current Change. The letter "p" means the velocity shown should occur **after** the **p.m.** Current Change (even if next morning). No "a" or "p" means a.m. and p.m. velocities are the same for that day.

Av. Max. Vel.: Fl. 3.4 Kts., Ebb 4.6 Kts.

Max. Flood & Max. Ebb 3 hrs. after Flood Starts & Ebb Starts, ±10 min.

At **City Island** the Current turns 2 hours before Hell Gate. At **Throg's Neck** the Current turns 1 hour before Hell Gate. At **Whitestone Pt.** the Current turns 35 min. before Hell Gate. At **College Pt.** the Current turns 20 min. before Hell Gate.

2009 CURRENT TABLE
HELL GATE, NY (EAST RIVER)

40°46.7'N, 73°56.3'W Off Mill Rock

Daylight Time starts March 8 at 2 a.m.   Daylight Saving Time

MARCH							APRIL						
DAY OF MONTH	DAY OF WEEK	CURRENT TURNS TO						DAY OF MONTH	DAY OF WEEK	CURRENT TURNS TO			
		NORTHEAST Flood Starts			SOUTHWEST Ebb Starts					NORTHEAST Flood Starts			SOUTHWEST Ebb Starts		
		a.m.	p.m.	Kts.	a.m.	p.m.	Kts.			a.m.	p.m.	Kts.	a.m.	p.m.	Kts.
1	S	6 35	6 50	3.5	12 16	12 45	a4.9	1	W	9 09	9 28	3.2	2 44	3 21	a4.6
2	M	7 24	7 39	a3.4	1 02	1 34	a4.8	2	T	10 18	10 41	3.1	3 50	4 28	a4.4
3	T	8 21	8 38	3.2	1 56	2 31	a4.6	3	F	11 31	11 57	p3.2	5 01	5 38	a4.3
4	W	9 28	9 48	p3.1	2 58	3 36	a4.5	4	S	...	12 40	3.2	6 12	6 45	4.3
5	T	10 42	11 05	p3.1	4 08	4 47	a4.3	5	S	1 08	1 43	p3.4	7 19	7 46	p4.5
6	F	11 55	...	3.1	5 22	5 58	4.3	6	M	2 09	2 38	3.6	8 18	8 41	p4.7
7	S	12 18	1 01	3.3	6 31	7 02	4.5	7	T	3 04	3 28	p3.8	9 12	9 31	p4.9
8	S	1 23	*2 59	3.6	*8 33	*8 59	4.7	8	W	3 54	4 14	3.9	10 00	10 17	p5.0
9	M	3 21	3 52	3.8	9 30	9 52	p5.0	9	T	4 41	4 58	3.9	10 45	11 01	p5.0
10	T	4 13	4 39	3.9	10 21	10 40	p5.1	10	F	5 24	5 39	a3.9	11 28	11 44	p4.9
11	W	5 01	5 24	a4.0	11 09	11 26	p5.1	11	S	6 06	6 20	a3.8	...	12 10	4.7
12	T	5 48	6 08	a4.0	11 54	...	5.0	12	S	6 48	7 01	a3.6	12 26	12 52	a4.8
13	F	6 33	6 51	a3.9	12 11	12 38	a5.0	13	M	7 30	7 43	a3.4	1 08	1 34	a4.6
14	S	7 17	7 34	a3.7	12 55	1 22	a4.9	14	T	8 14	8 28	a3.2	1 51	2 18	a4.5
15	S	8 02	8 18	a3.4	1 39	2 06	a4.7	15	W	9 00	9 16	2.9	2 37	3 05	a4.2
16	M	8 48	9 05	a3.2	2 24	2 52	a4.4	16	T	9 51	10 09	2.8	3 26	3 56	a4.1
17	T	9 38	9 56	a2.9	3 12	3 42	a4.2	17	F	10 46	11 06	2.7	4 19	4 50	a3.9
18	W	10 32	10 52	2.7	4 04	4 35	a4.0	18	S	11 41	...	2.7	5 15	5 44	a3.9
19	T	11 30	11 50	p2.7	5 00	5 31	a3.8	19	S	12 02	12 34	2.8	6 11	6 36	4.0
20	F	...	12 27	2.7	5 58	6 27	a3.8	20	M	12 55	1 23	3.0	7 03	7 24	p4.2
21	S	12 46	1 20	2.8	6 53	7 19	3.9	21	T	1 44	2 08	3.2	7 52	8 10	p4.5
22	S	1 37	2 07	3.0	7 45	8 06	p4.2	22	W	2 30	2 50	3.4	8 38	8 54	p4.7
23	M	2 23	2 50	3.2	8 32	8 50	p4.4	23	T	3 14	3 31	3.6	9 23	9 37	p4.9
24	T	3 06	3 30	3.4	9 15	9 31	p4.7	24	F	3 58	4 13	3.7	10 06	10 20	p5.1
25	W	3 47	4 07	3.6	9 56	10 10	p4.9	25	S	4 43	4 55	3.8	10 50	11 05	p5.1
26	T	4 26	4 44	3.7	10 37	10 50	p5.0	26	S	5 28	5 40	3.8	11 36	11 52	p5.1
27	F	5 06	5 22	3.8	11 17	11 30	p5.1	27	M	6 16	6 28	3.7	...	12 23	4.8
28	S	5 47	6 01	3.8	11 58	...	4.9	28	T	7 08	7 21	3.6	12 42	1 15	a5.0
29	S	6 30	6 43	3.7	12 12	12 42	a5.1	29	W	8 04	8 20	3.4	1 36	2 10	a4.8
30	M	7 17	7 29	3.6	12 57	1 29	a5.0	30	T	9 05	9 27	p3.3	2 36	3 11	a4.6
31	T	8 09	8 23	3.4	1 47	2 21	a4.8								

*Daylight Saving Time starts

The Kts. (knots) columns show the **maximum** predicted velocities of the stronger one of the Flood Currents and the stronger one of the Ebb Currents for each day. The letter "a" means the velocity shown should occur **after** the **a.m.** Current Change. The letter "p" means the velocity shown should occur **after** the **p.m.** Current Change (even if next morning). No "a" or "p" means a.m. and p.m. velocities are the same for that day.

Av. Max. Vel.: Fl. 3.4 Kts., Ebb 4.6 Kts.

Max. Flood & Max. Ebb 3 hrs. after Flood Starts & Ebb Starts, ±10 min.

See pp. 22-29 for Current Change at other points.

2009 CURRENT TABLE
HELL GATE, NY (EAST RIVER)
40°46.7'N, 73°56.3'W Off Mill Rock

Daylight Saving Time | Daylight Saving Time

		MAY							JUNE					
		CURRENT TURNS TO								CURRENT TURNS TO				
D A Y O F M O N T H	D A Y O F W E E K	NORTHEAST Flood Starts			SOUTHWEST Ebb Starts			D A Y O F M O N T H	D A Y O F W E E K	NORTHEAST Flood Starts			SOUTHWEST Ebb Starts	
		a.m.	p.m.	Kts.	a.m.	p.m.	Kts.			a.m.	p.m.	Kts.	a.m.	p.m.	Kts.
1	F	10 12	10 38	p3.2	3 41	4 16	a4.4	1	M	...	12 02	3.2	5 33	6 00	4.4
2	S	11 19	11 49	p3.2	4 49	5 23	4.3	2	T	12 35	12 59	p3.3	6 33	6 57	p4.5
3	S	...	12 24	3.2	5 56	6 26	p4.4	3	W	1 32	1 51	p3.4	7 28	7 49	p4.6
4	M	12 54	1 23	p3.4	6 59	7 24	p4.5	4	T	2 24	2 40	p3.4	8 18	8 37	p4.7
5	T	1 53	2 16	3.5	7 56	8 17	p4.7	5	F	3 12	3 25	p3.5	9 05	9 23	p4.7
6	W	2 46	3 04	3.6	8 47	9 05	p4.8	6	S	3 56	4 07	p3.5	9 49	10 06	p4.7
7	T	3 33	3 49	3.7	9 34	9 50	p4.9	7	S	4 38	4 48	p3.5	10 31	10 47	p4.7
8	F	4 18	4 32	3.7	10 18	10 33	p4.9	8	M	5 19	5 28	3.4	11 11	11 28	p4.7
9	S	5 02	5 14	3.6	11 00	11 15	p4.8	9	T	5 59	6 09	p3.4	11 51	...	4.5
10	S	5 42	5 53	a3.6	11 40	11 56	p4.8	10	W	6 38	6 48	3.3	12 08	12 31	a4.7
11	M	6 22	6 33	3.4	...	12 21	4.5	11	T	7 18	7 28	3.2	12 49	1 12	a4.6
12	T	7 03	7 14	3.3	12 37	1 02	a4.7	12	F	7 58	8 10	3.1	1 30	1 53	a4.5
13	W	7 45	7 57	3.1	1 18	1 44	a4.5	13	S	8 40	8 54	3.0	2 13	2 35	a4.5
14	T	8 29	8 43	3.0	2 02	2 28	a4.4	14	S	9 23	9 41	p3.0	2 58	3 20	a4.4
15	F	9 16	9 31	2.9	2 48	3 15	a4.3	15	M	10 09	10 33	p3.0	3 45	4 07	4.3
16	S	10 05	10 24	2.8	3 37	4 04	a4.2	16	T	10 57	11 28	p3.0	4 36	4 58	p4.4
17	S	10 56	11 18	p2.9	4 29	4 55	4.1	17	W	11 49	...	3.0	5 30	5 52	p4.5
18	M	11 47	...	2.9	5 22	5 46	p4.2	18	T	12 24	12 42	p3.2	6 25	6 48	p4.6
19	T	12 12	12 36	3.0	6 15	6 37	p4.4	19	F	1 22	1 36	p3.3	7 21	7 43	p4.8
20	W	1 04	1 24	p3.2	7 07	7 27	p4.6	20	S	2 18	2 31	p3.5	8 17	8 39	p4.9
21	T	1 55	2 12	p3.4	7 58	8 16	p4.8	21	S	3 13	3 26	p3.7	9 12	9 35	p5.0
22	F	2 45	2 59	p3.6	8 47	9 05	p5.0	22	M	4 08	4 21	p3.8	10 06	10 30	p5.1
23	S	3 34	3 46	p3.7	9 36	9 54	p5.1	23	T	5 02	5 16	p3.8	11 00	11 25	p5.1
24	S	4 24	4 35	p3.8	10 25	10 44	p5.1	24	W	5 56	6 12	p3.8	11 54	...	5.0
25	M	5 15	5 26	p3.8	11 16	11 36	p5.1	25	T	6 50	7 09	p3.8	12 20	12 48	a5.1
26	T	6 07	6 20	3.7	...	12 08	4.9	26	F	7 45	8 07	3.6	1 16	1 43	a5.0
27	W	7 01	7 17	3.6	12 30	1 02	a5.0	27	S	8 41	9 07	3.5	2 12	2 39	a4.8
28	T	7 58	8 18	3.5	1 27	1 58	a4.9	28	S	9 38	10 08	a3.4	3 10	3 37	4.6
29	F	8 58	9 22	p3.4	2 26	2 58	a4.7	29	M	10 36	11 09	a3.3	4 08	4 35	4.5
30	S	10 00	10 28	3.3	3 28	3 59	a4.6	30	T	11 33	...	3.2	5 07	5 32	p4.4
31	S	11 02	11 34	3.2	4 31	5 00	4.4								

The Kts. (knots) columns show the **maximum** predicted velocities of the stronger one of the Flood Currents and the stronger one of the Ebb Currents for each day. The letter "a" means the velocity shown should occur **after** the **a.m.** Current Change. The letter "p" means the velocity shown should occur **after** the **p.m.** Current Change (even if next morning). No "a" or "p" means a.m. and p.m. velocities are the same for that day.

Av. Max. Vel.: Fl. 3.4 Kts., Ebb 4.6 Kts.

Max. Flood & Max. Ebb 3 hrs. after Flood Starts & Ebb Starts, ±10 min.

At **City Island** the Current turns 2 hours before Hell Gate. At **Throg's Neck** the Current turns 1 hour before Hell Gate. At **Whitestone Pt.** the Current turns 35 min. before Hell Gate. At **College Pt.** the Current turns 20 min. before Hell Gate.

2009 CURRENT TABLE
HELL GATE, NY (EAST RIVER)
40°46.7'N, 73°56.3'W Off Mill Rock

Daylight Saving Time · Daylight Saving Time

JULY							AUGUST						
DAY OF MONTH	DAY OF WEEK	CURRENT TURNS TO						DAY OF MONTH	DAY OF WEEK	CURRENT TURNS TO					
		NORTHEAST Flood Starts			SOUTHWEST Ebb Starts					NORTHEAST Flood Starts			SOUTHWEST Ebb Starts		
		a.m.	p.m.	Kts.	a.m.	p.m.	Kts.			a.m.	p.m.	Kts.	a.m.	p.m.	Kts.
1	W	12 09	12 30	p3.2	6 04	6 28	p4.4	1	S	1 26	1 42	p3.1	7 24	7 47	p4.1
2	T	1 06	1 23	p3.2	6 59	7 22	p4.4	2	S	2 16	2 29	p3.2	8 13	8 35	p4.2
3	F	1 59	2 13	p3.2	7 50	8 11	p4.4	3	M	3 01	3 13	p3.3	8 58	9 20	p4.4
4	S	2 47	2 59	p3.3	8 38	8 58	p4.5	4	T	3 43	3 53	p3.4	9 41	10 01	p4.5
5	S	3 31	3 42	p3.3	9 22	9 42	p4.5	5	W	4 22	4 32	p3.5	10 20	10 41	p4.6
6	M	4 13	4 23	p3.4	10 05	10 23	p4.6	6	T	4 59	5 09	p3.6	10 59	11 20	p4.7
7	T	4 53	5 02	p3.4	10 45	11 04	p4.7	7	F	5 34	5 45	p3.6	11 36	11 58	4.7
8	W	5 31	5 40	p3.5	11 25	11 43	p4.7	8	S	6 09	6 21	p3.6	...	12 12	4.7
9	T	6 10	6 19	p3.5	...	12 03	4.6	9	S	6 44	6 59	p3.6	12 35	12 49	4.7
10	F	6 46	6 56	p3.4	12 23	12 42	a4.7	10	M	7 18	7 37	3.5	1 14	1 28	4.7
11	S	7 23	7 34	3.3	1 02	1 20	a4.7	11	T	7 55	8 20	3.4	1 54	2 09	4.6
12	S	8 00	8 15	p3.3	1 42	1 59	4.6	12	W	8 36	9 10	a3.3	2 38	2 57	4.5
13	M	8 38	8 59	3.2	2 23	2 41	4.5	13	T	9 25	10 08	a3.2	3 29	3 51	p4.4
14	T	9 21	9 49	3.1	3 08	3 27	p4.5	14	F	10 25	11 15	a3.2	4 26	4 54	p4.4
15	W	10 08	10 45	a3.1	3 57	4 19	p4.5	15	S	11 33	...	3.2	5 31	6 02	p4.4
16	T	11 03	11 46	a3.1	4 52	5 17	p4.5	16	S	12 26	12 45	p3.3	6 38	7 10	p4.5
17	F	...	12 03	3.2	5 52	6 19	p4.5	17	M	1 34	1 53	p3.5	7 43	8 14	p4.6
18	S	12 51	1 06	p3.3	6 54	7 22	p4.6	18	T	2 36	2 55	p3.8	8 44	9 14	p4.8
19	S	1 54	2 09	p3.5	7 56	8 23	p4.8	19	W	3 33	3 53	p4.0	9 40	10 10	p5.0
20	M	2 54	3 09	p3.7	8 55	9 23	p4.9	20	T	4 26	4 47	p4.1	10 33	11 02	5.0
21	T	3 51	4 07	p3.9	9 52	10 20	p5.0	21	F	5 16	5 39	p4.1	11 24	11 53	a5.1
22	W	4 46	5 03	p4.0	10 47	11 15	p5.1	22	S	6 05	6 29	4.0	...	12 13	5.1
23	T	5 39	5 58	p4.0	11 40	...	5.0	23	S	6 53	7 19	3.9	12 43	1 03	4.9
24	F	6 30	6 52	3.9	12 08	12 33	a5.1	24	M	7 41	8 10	a3.7	1 33	1 52	4.7
25	S	7 22	7 46	3.8	1 01	1 25	a5.0	25	T	8 31	9 02	a3.5	2 23	2 44	4.5
26	S	8 14	8 41	a3.7	1 54	2 18	4.8	26	W	9 22	9 57	a3.3	3 15	3 37	4.2
27	M	9 07	9 37	a3.5	2 48	3 11	4.6	27	T	10 17	10 54	a3.1	4 09	4 33	p4.0
28	T	10 01	10 36	a3.3	3 43	4 07	4.4	28	F	11 15	11 53	2.9	5 05	5 31	p3.8
29	W	10 57	11 34	a3.1	4 39	5 04	p4.2	29	S	...	12 12	2.9	6 02	6 28	p3.8
30	T	11 54	...	3.0	5 35	6 00	p4.1	30	S	12 49	1 07	p3.0	6 57	7 22	p3.9
31	F	12 32	12 50	p3.0	6 31	6 56	p4.1	31	M	1 40	1 56	p3.2	7 47	8 10	p4.1

The Kts. (knots) columns show the **maximum** predicted velocities of the stronger one of the Flood Currents and the stronger one of the Ebb Currents for each day. The letter "a" means the velocity shown should occur **after** the **a.m.** Current Change. The letter "p" means the velocity shown should occur **after** the **p.m.** Current Change (even if next morning). No "a" or "p" means a.m. and p.m. velocities are the same for that day.
Av. Max. Vel.: Fl. 3.4 Kts., Ebb 4.6 Kts.
Max. Flood & Max. Ebb 3 hrs. after Flood Starts & Ebb Starts, ±10 min.

See pp. 22-29 for Current Change at other points.

2009 CURRENT TABLE
HELL GATE, NY (EAST RIVER)
40°46.7'N, 73°56.3'W Off Mill Rock

		SEPTEMBER									OCTOBER					
		Daylight Saving Time									Daylight Saving Time					
DAY OF MONTH	DAY OF WEEK	CURRENT TURNS TO						DAY OF MONTH	DAY OF WEEK	CURRENT TURNS TO						
		NORTHEAST Flood Starts			SOUTHWEST Ebb Starts					NORTHEAST Flood Starts			SOUTHWEST Ebb Starts			
		a.m.	p.m.	Kts.	a.m.	p.m.	Kts.			a.m.	p.m.	Kts.	a.m.	p.m.	Kts.	
1	T	2 26	2 40	p3.3	8 32	8 55	p4.2	1	T	2 30	2 47	p3.5	8 39	9 03	p4.4	
2	W	3 08	3 21	p3.5	9 14	9 36	p4.4	2	F	3 09	3 27	p3.6	9 18	9 43	p4.6	
3	T	3 46	3 59	p3.6	9 53	10 15	p4.6	3	S	3 46	4 05	p3.7	9 56	10 22	4.7	
4	F	4 22	4 36	p3.7	10 30	10 53	p4.7	4	S	4 22	4 43	p3.8	10 34	11 01	a4.9	
5	S	4 57	5 12	p3.8	11 06	11 31	a4.8	5	M	4 58	5 22	3.8	11 12	11 40	a4.9	
6	S	5 31	5 48	p3.8	11 43	...	4.8	6	T	5 34	6 03	3.8	11 52	...	5.0	
7	M	6 05	6 26	3.7	12 08	12 20	p4.8	7	W	6 13	6 46	a3.7	12 21	12 35	p4.9	
8	T	6 40	7 06	3.6	12 47	1 00	p4.8	8	T	6 56	7 35	a3.6	1 05	1 22	p4.8	
9	W	7 20	7 52	a3.6	1 28	1 43	p4.7	9	F	7 47	8 31	a3.5	1 55	2 16	p4.6	
10	T	8 03	8 42	a3.4	2 14	2 33	p4.5	10	S	8 45	9 35	a3.3	2 51	3 18	p4.4	
11	F	8 56	9 44	a3.3	3 07	3 32	p4.4	11	S	9 56	10 46	a3.2	3 55	4 27	p4.2	
12	S	10 02	10 55	a3.2	4 09	4 39	p4.2	12	M	11 13	11 58	3.2	5 04	5 39	p4.2	
13	S	11 17	...	3.2	5 18	5 51	p4.2	13	T	...	12 27	3.3	6 12	6 47	p4.3	
14	M	12 09	12 33	p3.3	6 27	7 01	p4.3	14	W	1 04	1 32	p3.6	7 15	7 48	p4.5	
15	T	1 17	1 42	p3.6	7 32	8 04	p4.5	15	T	2 02	2 30	p3.8	8 12	8 44	4.7	
16	W	2 19	2 43	p3.8	8 31	9 02	p4.7	16	F	2 55	3 23	p3.9	9 04	9 35	a4.9	
17	T	3 13	3 38	p4.0	9 25	9 55	4.9	17	S	3 43	4 12	p4.0	9 53	10 22	a5.0	
18	F	4 04	4 29	p4.1	10 15	10 45	5.0	18	S	4 30	4 59	4.0	10 39	11 08	a5.0	
19	S	4 52	5 18	4.1	11 03	11 33	a5.1	19	M	5 14	5 44	3.9	11 24	11 52	a5.0	
20	S	5 38	6 05	4.0	11 50	...	5.0	20	T	5 58	6 28	a3.8	...	12 08	4.9	
21	M	6 23	6 52	a3.9	12 20	12 37	p4.9	21	W	6 41	7 12	a3.6	12 36	12 53	p4.7	
22	T	7 09	7 39	a3.7	1 06	1 24	p4.7	22	T	7 26	7 58	a3.4	1 21	1 38	p4.5	
23	W	7 56	8 28	a3.5	1 54	2 12	4.4	23	F	8 13	8 47	a3.2	2 07	2 26	4.2	
24	T	8 45	9 19	a3.2	2 43	3 03	4.1	24	S	9 03	9 38	a3.0	2 55	3 17	4.0	
25	F	9 37	10 14	a3.0	3 35	3 58	3.9	25	S	9 57	10 33	a2.8	3 47	4 10	p3.9	
26	S	10 35	11 12	a2.9	4 30	4 55	3.7	26	M	10 54	11 28	a2.8	4 41	5 06	p3.8	
27	S	11 33	...	2.8	5 27	5 52	p3.7	27	T	11 49	...	2.8	5 34	6 00	p3.9	
28	M	12 09	12 29	p2.9	6 21	6 47	p3.8	28	W	12 20	12 41	p3.0	6 25	6 51	p4.0	
29	T	1 01	1 20	p3.1	7 12	7 36	p4.0	29	T	1 08	1 29	p3.1	7 12	7 38	p4.2	
30	W	1 48	2 05	p3.3	7 57	8 21	p4.2	30	F	1 52	2 14	p3.3	7 56	8 23	4.4	
								31	S	2 33	2 57	p3.5	8 38	9 06	4.6	

The Kts. (knots) columns show the **maximum** predicted velocities of the stronger one of the Flood Currents and the stronger one of the Ebb Currents for each day. The letter "a" means the velocity shown should occur **after** the **a.m.** Current Change. The letter "p" means the velocity shown should occur **after** the **p.m.** Current Change (even if next morning). No "a" or "p" means a.m. and p.m. velocities are the same for that day.

Av. Max. Vel.: Fl. 3.4 Kts., Ebb 4.6 Kts.

Max. Flood & Max. Ebb 3 hrs. after Flood Starts & Ebb Starts, ±10 min.

At **City Island** the Current turns 2 hours before Hell Gate. At **Throg's Neck** the Current turns 1 hour before Hell Gate. At **Whitestone Pt.** the Current turns 35 min. before Hell Gate. At **College Pt.** the Current turns 20 min. before Hell Gate.

2009 CURRENT TABLE
HELL GATE, NY (EAST RIVER)
40°46.7'N, 73°56.3'W Off Mill Rock

Standard Time starts Nov. 1 at 2 a.m.   Standard Time

NOVEMBER							DECEMBER					
D A Y O F M O N T H	D A Y O F W E E K	CURRENT TURNS TO					D A Y O F M O N T H	D A Y O F W E E K	CURRENT TURNS TO			
		NORTHEAST Flood Starts		SOUTHWEST Ebb Starts					NORTHEAST Flood Starts		SOUTHWEST Ebb Starts	
		a.m.	**p.m.**	Kts.	a.m.	**p.m.** Kts.			a.m.	**p.m.** Kts.	a.m.	**p.m.** Kts.
1	S	*2 12	*2 39	p3.6	*8 19	*8 47 a4.8	1	T	2 26	3 03 3.6	8 31	9 00 a5.0
2	M	2 52	3 21	3.7	9 01	9 29 a5.0	2	W	3 13	3 51 a3.7	9 19	9 48 a5.1
3	T	3 31	4 04	a3.8	9 43	10 12 a5.1	3	T	4 01	4 40 a3.7	10 08	10 37 a5.1
4	W	4 13	4 49	a3.8	10 27	10 57 a5.1	4	F	4 51	5 31 a3.7	10 59	11 28 a5.1
5	T	4 58	5 37	a3.7	11 14	11 45 a5.0	5	S	5 45	6 25 a3.7	11 53	... 5.0
6	F	5 48	6 29	a3.6	...	12 05 4.9	6	S	6 43	7 22 a3.5	12 22	12 49 p4.9
7	S	6 43	7 28	a3.5	12 38	1 02 p4.7	7	M	7 45	8 23 a3.4	1 19	1 49 4.7
8	S	7 47	8 32	a3.3	1 36	2 04 p4.5	8	T	8 51	9 26 a3.3	2 19	2 52 a4.6
9	M	8 58	9 40	a3.2	2 39	3 11 4.3	9	W	9 59	10 30 3.2	3 22	3 56 a4.5
10	T	10 10	10 46	3.2	3 45	4 20 4.3	10	T	11 04	11 29 3.2	4 24	4 58 a4.5
11	W	11 19	11 49	3.3	4 51	5 25 4.4	11	F	...	12 06 3.2	5 24	5 58 a4.5
12	T	...	12 21	3.4	5 52	6 25 4.5	12	S	12 26	1 02 3.3	6 21	6 53 a4.6
13	F	12 45	1 18	p3.6	6 48	7 20 a4.7	13	S	1 19	1 54 3.4	7 14	7 43 a4.7
14	S	1 37	2 09	p3.7	7 39	8 10 a4.8	14	M	2 08	2 41 a3.5	8 03	8 30 a4.8
15	S	2 25	2 57	3.7	8 28	8 56 a4.9	15	T	2 53	3 25 a3.5	8 48	9 14 a4.8
16	M	3 10	3 42	a3.8	9 13	9 40 a5.0	16	W	3 36	4 07 a3.5	9 32	9 55 a4.8
17	T	3 54	4 25	a3.7	9 57	10 23 a4.9	17	T	4 18	4 48 a3.5	10 13	10 36 a4.8
18	W	4 37	5 08	a3.6	10 40	11 05 a4.8	18	F	4 58	5 28 a3.4	10 54	11 16 a4.8
19	T	5 19	5 50	a3.5	11 22	11 47 a4.7	19	S	5 38	6 07 a3.4	11 34	11 55 a4.7
20	F	6 02	6 33	a3.3	...	12 04 4.6	20	S	6 18	6 47 a3.3	...	12 14 4.6
21	S	6 46	7 18	a3.2	12 30	12 48 4.4	21	M	7 00	7 28 a3.1	12 36	12 56 4.5
22	S	7 32	8 04	a3.0	1 14	1 34 p4.3	22	T	7 43	8 11 a3.0	1 17	1 38 4.4
23	M	8 21	8 53	a2.9	2 00	2 22 p4.2	23	W	8 28	8 55 a2.9	1 59	2 24 4.4
24	T	9 13	9 44	a2.8	2 49	3 13 p4.1	24	T	9 18	9 42 2.8	2 45	3 12 4.3
25	W	10 06	10 34	2.8	3 39	4 06 4.1	25	F	10 11	10 33 2.8	3 34	4 04 4.3
26	T	10 59	11 23	p2.9	4 30	4 58 4.1	26	S	11 07	11 25 p2.9	4 27	4 59 a4.4
27	F	11 50	...	3.0	5 20	5 49 4.3	27	S	...	12 04 2.9	5 22	5 55 a4.5
28	S	12 10	12 40	p3.1	6 08	6 38 a4.5	28	M	12 18	1 00 3.1	6 18	6 50 a4.6
29	S	12 56	1 28	p3.3	6 56	7 26 a4.7	29	T	1 12	1 54 3.3	7 13	7 45 a4.8
30	M	1 41	2 15	3.4	7 43	8 13 a4.9	30	W	2 05	2 46 a3.5	8 08	8 38 a5.0
							31	T	2 58	3 38 a3.7	9 02	9 30 a5.1

*Standard Time starts

The Kts. (knots) columns show the **maximum** predicted velocities of the stronger one of the Flood Currents and the stronger one of the Ebb Currents for each day. The letter "a" means the velocity shown should occur **after** the **a.m.** Current Change. The letter "p" means the velocity shown should occur **after** the **p.m.** Current Change (even if next morning). No "a" or "p" means a.m. and p.m. velocities are the same for that day.

Av. Max. Vel.: Fl. 3.4 Kts., Ebb 4.6 Kts.

Max. Flood & Max. Ebb 3 hrs. after Flood Starts & Ebb Starts, ±10 min.

See pp. 22-29 for Current Change at other points.

2009 CURRENT TABLE
THE NARROWS, NY HARBOR
40°36.56'N, 74°02.77'W Mid-Channel

JANUARY

Standard Time

DAY OF MONTH	DAY OF WEEK	NORTH Flood Starts a.m.	NORTH Flood Starts p.m.	NORTH Flood Starts Kts.	SOUTH Ebb Starts a.m.	SOUTH Ebb Starts p.m.	SOUTH Ebb Starts Kts.
1	T	6 54	7 27	a1.6	12 33	12 50	p2.0
2	F	7 49	8 14	1.6	1 21	1 38	p2.0
3	S	8 47	9 04	p1.8	2 10	2 27	p2.0
4	S	9 48	9 55	p1.9	3 01	3 19	p2.0
5	M	10 49	10 47	p2.0	3 56	4 15	p2.0
6	T	11 49	11 40	p2.0	4 54	5 15	2.0
7	W	...	12 48	1.5	5 53	6 16	a2.1
8	T	12 34	1 46	a2.1	6 51	7 15	a2.2
9	F	1 30	2 42	a2.2	7 46	8 11	a2.3
10	S	2 26	3 32	a2.2	8 39	9 05	a2.4
11	S	3 22	4 22	a2.2	9 31	9 59	a2.4
12	M	4 18	5 12	a2.1	10 24	10 54	a2.3
13	T	5 16	6 06	a2.0	11 19	11 52	a2.2
14	W	6 17	7 04	a1.8	...	12 16	2.1
15	T	7 23	8 04	a1.7	12 52	1 14	p2.0
16	F	8 31	9 04	p1.6	1 52	2 11	1.8
17	S	9 39	10 03	p1.6	2 52	3 08	1.7
18	S	10 45	10 59	p1.5	3 54	4 09	1.6
19	M	11 50	11 51	p1.5	4 58	5 14	a1.7
20	T	...	12 51	1.1	6 00	6 18	a1.7
21	W	12 41	1 50	a1.4	6 53	7 14	a1.7
22	T	1 28	2 40	a1.4	7 38	8 01	a1.8
23	F	2 10	3 18	a1.4	8 15	8 38	a1.7
24	S	2 48	3 47	a1.4	8 47	9 09	a1.7
25	S	3 21	4 11	a1.4	9 16	9 37	a1.7
26	M	3 52	4 33	a1.5	9 45	10 06	a1.8
27	T	4 24	4 57	a1.5	10 17	10 38	a1.9
28	W	4 59	5 27	a1.6	10 53	11 16	a2.0
29	T	5 39	6 03	a1.6	11 34	11 59	a2.0
30	F	6 25	6 44	p1.7	...	12 17	2.1
31	S	7 17	7 29	p1.7	12 45	1 04	p2.1

FEBRUARY

Standard Time

DAY OF MONTH	DAY OF WEEK	NORTH Flood Starts a.m.	NORTH Flood Starts p.m.	NORTH Flood Starts Kts.	SOUTH Ebb Starts a.m.	SOUTH Ebb Starts p.m.	SOUTH Ebb Starts Kts.
1	S	8 14	8 18	p1.8	1 33	1 51	p2.0
2	M	9 16	9 11	p1.8	2 23	2 42	p1.9
3	T	10 20	10 08	p1.8	3 17	3 38	a1.8
4	W	11 25	11 09	p1.8	4 16	4 42	a1.8
5	T	...	12 27	1.3	5 21	5 50	a1.8
6	F	12 11	1 26	a1.8	6 26	6 55	a2.0
7	S	1 14	2 21	a1.9	7 26	7 54	a2.1
8	S	2 15	3 12	a2.0	8 22	8 49	2.2
9	M	3 14	4 01	a2.1	9 15	9 41	2.3
10	T	4 07	4 48	a2.1	10 07	10 34	2.3
11	W	5 01	5 37	a2.0	10 58	11 28	a2.3
12	T	5 56	6 30	a1.9	11 51	...	2.2
13	F	6 56	7 26	1.7	12 24	12 46	2.1
14	S	8 01	8 24	p1.6	1 21	1 41	1.9
15	S	9 08	9 22	p1.6	2 18	2 36	1.7
16	M	10 16	10 19	p1.5	3 15	3 36	a1.6
17	T	11 21	11 15	p1.4	4 15	4 41	a1.6
18	W	...	12 21	1.1	5 17	5 48	a1.7
19	T	12 07	1 15	a1.3	6 14	6 46	a1.7
20	F	12 57	2 02	a1.3	7 03	7 33	a1.7
21	S	1 43	2 39	a1.4	7 44	8 11	a1.8
22	S	2 24	3 08	a1.4	8 19	8 42	a1.8
23	M	3 00	3 33	a1.5	8 50	9 11	a1.9
24	T	3 34	3 58	a1.6	9 22	9 41	a1.9
25	W	4 07	4 24	a1.7	9 54	10 13	a2.0
26	T	4 42	4 54	1.7	10 29	10 49	a2.1
27	F	5 20	5 28	p1.8	11 07	11 29	a2.1
28	S	6 03	6 07	p1.8	11 49	...	2.1

The Kts. (knots) columns show the **maximum** predicted velocities of the stronger one of the Flood Currents and the stronger one of the Ebb Currents for each day. The letter "a" means the velocity shown should occur **after** the **a.m.** Current Change. The letter "p" means the velocity shown should occur **after** the **p.m.** Current Change (even if next morning). No "a" or "p" means a.m. and p.m. velocities are the same for that day.

Av. Max. Vel.: Fl. 1.7 Kts., Ebb 2.0 Kts.

Max. Flood 2 hrs. 25 min. after Flood Starts, ±30 min.

Max. Ebb 3 hrs. 15 min. after Ebb Starts, ±10 min.

At **The Battery, Desbrosses St., & Chelsea Dock** Current turns 1 1/2 hrs. after the Narrows. At **42nd St.** and the **George Washington Br.**, the Current turns 1 3/4 hrs. after the Narrows.

2009 CURRENT TABLE
THE NARROWS, NY HARBOR
40°36.56'N, 74°02.77'W Mid-Channel

Daylight Time starts March 8 at 2 a.m. Daylight Saving Time

		MARCH						APRIL			
		CURRENT TURNS TO						CURRENT TURNS TO			
		NORTH Flood Starts		SOUTH Ebb Starts				NORTH Flood Starts		SOUTH Ebb Starts	
DAY OF MONTH	DAY OF WEEK	a.m.	p.m.	Kts.	a.m.	p.m.	Kts.	DAY OF MONTH	DAY OF WEEK	a.m.	p.m.	Kts.	a.m.	p.m.	Kts.

D.O.M.	D.O.W.	a.m.	p.m.	Kts.	a.m.	p.m.	Kts.	D.O.M.	D.O.W.	a.m.	p.m.	Kts.	a.m.	p.m.	Kts.
1	S	6 53	6 51	p1.8	12 14	12 34	2.0	1	W	9 43	9 20	p1.4	2 30	3 01	a1.7
2	M	7 50	7 40	p1.7	1 01	1 23	1.9	2	T	10 52	10 36	p1.3	3 26	4 02	a1.5
3	T	8 54	8 37	p1.7	1 52	2 15	a1.8	3	F	11 57	11 53	p1.2	4 30	5 13	a1.4
4	W	10 01	9 42	p1.6	2 45	3 13	a1.6	4	S	...	12 57	1.1	5 42	6 26	1.4
5	T	11 08	10 53	p1.5	3 47	4 21	a1.5	5	S	1 04	1 52	1.2	6 55	7 32	p1.7
6	F	...	12 10	1.1	4 57	5 34	a1.6	6	M	2 06	2 43	p1.5	7 58	8 29	p1.9
7	S	12 03	1 08	a1.5	6 07	6 41	1.7	7	T	3 03	3 30	p1.7	8 51	9 20	p2.1
8	S	1 08	*3 01	a1.6	*8 11	*8 40	p2.0	8	W	3 54	4 14	p1.8	9 39	10 06	2.1
9	M	3 10	3 52	a1.8	9 07	9 34	p2.2	9	T	4 42	4 56	p1.9	10 24	10 51	p2.2
10	T	4 03	4 37	1.9	9 58	10 23	p2.3	10	F	5 27	5 36	p1.8	11 08	11 35	2.1
11	W	4 54	5 21	a2.0	10 46	11 12	2.3	11	S	6 14	6 18	p1.7	11 54	...	2.0
12	T	5 43	6 07	1.9	11 34	...	2.3	12	S	7 05	7 03	p1.6	12 20	12 43	a2.0
13	F	6 34	6 54	1.8	12 02	12 23	2.2	13	M	8 02	7 53	p1.5	1 08	1 35	a1.8
14	S	7 30	7 45	p1.7	12 53	1 14	a2.1	14	T	9 02	8 49	p1.4	1 58	2 30	a1.8
15	S	8 31	8 39	p1.6	1 46	2 08	a1.9	15	W	10 02	9 50	p1.3	2 48	3 24	a1.6
16	M	9 36	9 37	p1.5	2 39	3 04	a1.8	16	T	10 58	10 50	p1.3	3 40	4 20	a1.6
17	T	10 42	10 36	p1.4	3 32	4 01	a1.6	17	F	11 48	11 47	p1.3	4 34	5 17	a1.6
18	W	11 44	11 34	p1.3	4 27	5 03	a1.6	18	S	...	12 32	1.2	5 29	6 12	a1.6
19	T	...	12 39	1.1	5 25	6 07	a1.6	19	S	12 40	1 13	a1.4	6 24	7 02	a1.8
20	F	12 30	1 28	a1.3	6 24	7 04	a1.6	20	M	1 29	1 52	1.5	7 15	7 47	1.9
21	S	1 21	2 10	a1.3	7 17	7 52	a1.7	21	T	2 15	2 30	p1.7	8 00	8 29	2.1
22	S	2 09	2 47	a1.4	8 02	8 32	1.8	22	W	2 59	3 06	p1.9	8 42	9 07	2.2
23	M	2 52	3 20	a1.5	8 42	9 07	1.9	23	T	3 42	3 42	p2.1	9 22	9 45	2.3
24	T	3 32	3 50	a1.7	9 19	9 41	2.0	24	F	4 24	4 18	p2.2	10 01	10 23	2.3
25	W	4 10	4 20	1.8	9 53	10 14	2.1	25	S	5 06	4 54	p2.2	10 41	11 02	p2.3
26	T	4 46	4 51	p1.9	10 28	10 48	a2.2	26	S	5 50	5 33	p2.1	11 23	11 44	p2.2
27	F	5 24	5 23	p2.0	11 04	11 25	a2.2	27	M	6 37	6 16	p1.9	...	12 09	1.9
28	S	6 04	5 58	p2.0	11 43	...	2.1	28	T	7 30	7 06	p1.7	12 31	1 00	a2.0
29	S	6 48	6 37	p1.9	12 05	12 26	a2.1	29	W	8 30	8 08	p1.4	1 22	1 57	a1.8
30	M	7 39	7 22	p1.7	12 49	1 13	a2.0	30	T	9 36	9 22	p1.3	2 18	2 57	a1.6
31	T	8 38	8 15	p1.6	1 37	2 05	a1.9								

*Daylight Saving Time starts

The Kts. (knots) columns show the **maximum** predicted velocities of the stronger one of the Flood Currents and the stronger one of the Ebb Currents for each day. The letter "a" means the velocity shown should occur **after** the **a.m.** Current Change. The letter "p" means the velocity shown should occur **after** the **p.m.** Current Change (even if next morning). No "a" or "p" means a.m. and p.m. velocities are the same for that day.

Av. Max. Vel.: Fl. 1.7 Kts., Ebb 2.0 Kts.
Max. Flood 2 hrs. 25 min. after Flood Starts, ±30 min.
Max. Ebb 3 hrs. 15 min. after Ebb Starts, ±10 min.

See pp. 22-29 for Current Change at other points.

2009 CURRENT TABLE
THE NARROWS, NY HARBOR

40°36.56'N, 74°02.77'W Mid-Channel

Daylight Saving Time   Daylight Saving Time

		MAY								JUNE					
		CURRENT TURNS TO								CURRENT TURNS TO					
		NORTH Flood Starts			SOUTH Ebb Starts					NORTH Flood Starts			SOUTH Ebb Starts		
DAY OF MONTH	DAY OF WEEK	a.m.	p.m.	Kts.	a.m.	p.m.	Kts.	DAY OF MONTH	DAY OF WEEK	a.m.	p.m.	Kts.	a.m.	p.m.	Kts.
1	F	10 43	10 43	p1.2	3 18	4 01	a1.4	1	M	...	12 23	1.3	5 21	6 11	1.5
2	S	11 45	11 58	1.1	4 23	5 10	1.3	2	T	12 56	1 16	p1.4	6 26	7 15	1.6
3	S	...	12 43	1.2	5 34	6 21	1.4	3	W	1 55	2 05	p1.5	7 27	8 10	p1.7
4	M	1 04	1 35	p1.3	6 43	7 25	p1.6	4	T	2 52	2 51	p1.6	8 20	8 57	p1.8
5	T	2 03	2 25	p1.5	7 43	8 20	p1.8	5	F	3 44	3 34	p1.6	9 08	9 37	p1.8
6	W	2 57	3 10	p1.6	8 35	9 08	p1.9	6	S	4 30	4 12	p1.6	9 51	10 14	p1.8
7	T	3 47	3 52	p1.7	9 21	9 51	p2.0	7	S	5 11	4 47	p1.5	10 31	10 48	p1.8
8	F	4 33	4 31	p1.7	10 04	10 31	p2.0	8	M	5 48	5 21	p1.5	11 10	11 23	p1.8
9	S	5 17	5 09	p1.7	10 46	11 10	p1.9	9	T	6 26	5 58	p1.4	11 49	...	1.4
10	S	6 00	5 45	p1.6	11 29	11 50	p1.9	10	W	7 01	6 37	p1.4	12 01	12 30	a1.7
11	M	6 45	6 25	p1.5	...	12 14	1.6	11	T	7 39	7 25	p1.4	12 42	1 14	a1.7
12	T	7 33	7 11	p1.4	12 32	1 03	a1.8	12	F	8 20	8 18	p1.4	1 27	2 01	a1.8
13	W	8 23	8 04	p1.3	1 18	1 53	a1.7	13	S	9 04	9 15	p1.5	2 14	2 48	a1.8
14	T	9 14	9 03	p1.3	2 06	2 44	a1.7	14	S	9 49	10 12	p1.5	3 01	3 35	a1.8
15	F	10 02	10 02	p1.4	2 55	3 34	a1.6	15	M	10 35	11 08	1.6	3 50	4 25	a1.9
16	S	10 49	11 00	p1.4	3 45	4 24	a1.6	16	T	11 22	...	1.7	4 41	5 19	a1.9
17	S	11 34	11 54	p1.5	4 37	5 17	a1.7	17	W	12 04	12 09	p1.9	5 35	6 13	2.0
18	M	...	12 17	1.5	5 31	6 10	1.8	18	T	12 59	12 56	p2.0	6 30	7 07	p2.2
19	T	12 46	1 00	p1.7	6 24	7 01	2.0	19	F	1 53	1 43	p2.1	7 25	7 58	p2.3
20	W	1 36	1 42	p1.9	7 15	7 48	p2.2	20	S	2 47	2 32	p2.3	8 18	8 47	p2.4
21	T	2 25	2 23	p2.1	8 03	8 33	p2.3	21	S	3 39	3 21	p2.3	9 09	9 35	p2.5
22	F	3 14	3 05	p2.2	8 49	9 16	p2.4	22	M	4 29	4 11	p2.3	10 00	10 24	p2.4
23	S	4 02	3 47	p2.3	9 34	9 59	p2.5	23	T	5 18	5 03	p2.2	10 50	11 13	p2.3
24	S	4 48	4 30	p2.3	10 19	10 42	p2.4	24	W	6 07	5 58	p2.0	11 43	...	2.0
25	M	5 35	5 15	p2.2	11 06	11 28	p2.3	25	T	6 59	6 58	p1.8	12 06	12 39	a2.2
26	T	6 24	6 04	p1.9	11 56	...	1.9	26	F	7 57	8 05	p1.6	1 03	1 39	a2.0
27	W	7 18	7 02	p1.7	12 18	12 51	a2.1	27	S	8 58	9 17	1.4	2 02	2 40	a1.8
28	T	8 18	8 10	p1.4	1 13	1 51	a1.9	28	S	10 00	10 28	a1.5	3 00	3 41	a1.7
29	F	9 22	9 27	p1.3	2 13	2 53	a1.7	29	M	11 00	11 36	a1.5	3 59	4 45	a1.6
30	S	10 26	10 42	1.2	3 13	3 56	a1.5	30	T	11 58	...	1.5	5 00	5 53	1.6
31	S	11 27	11 52	a1.3	4 15	5 02	a1.5								

The Kts. (knots) columns show the **maximum** predicted velocities of the stronger one of the Flood Currents and the stronger one of the Ebb Currents for each day. The letter "a" means the velocity shown should occur **after** the **a.m.** Current Change. The letter "p" means the velocity shown should occur **after** the **p.m.** Current Change (even if next morning). No "a" or "p" means a.m. and p.m. velocities are the same for that day.

Av. Max. Vel.: Fl. 1.7 Kts., Ebb 2.0 Kts.

Max. Flood 2 hrs. 25 min. after Flood Starts, ±30 min.

Max. Ebb 3 hrs. 15 min. after Ebb Starts, ±10 min.

At **The Battery, Desbrosses St., & Chelsea Dock** Current turns 1 1/2 hrs. after the Narrows. At **42nd St.** and the **George Washington Br.**, the Current turns 1 3/4 hrs. after the Narrows.

2009 CURRENT TABLE
THE NARROWS, NY HARBOR
40°36.56'N, 74°02.77'W Mid-Channel

Daylight Saving Time	Daylight Saving Time

		JULY								AUGUST					
		CURRENT TURNS TO								CURRENT TURNS TO					
		NORTH Flood Starts			SOUTH Ebb Starts					NORTH Flood Starts			SOUTH Ebb Starts		
DAY OF MONTH	DAY OF WEEK	a.m.	p.m.	Kts.	a.m.	p.m.	Kts.	DAY OF MONTH	DAY OF WEEK	a.m.	p.m.	Kts.	a.m.	p.m.	Kts.
1	W	12 42	12 52	p1.4	6 05	6 58	1.6	1	S	2 30	2 13	p1.4	7 57	8 23	p1.7
2	T	1 44	1 44	p1.4	7 10	7 55	p1.7	2	S	3 25	3 01	p1.4	8 50	9 06	p1.8
3	F	2 45	2 33	p1.5	8 08	8 44	p1.7	3	M	4 09	3 43	p1.4	9 33	9 41	p1.7
4	S	3 40	3 18	p1.5	9 00	9 24	p1.8	4	T	4 42	4 18	p1.4	10 06	10 11	p1.7
5	S	4 26	3 57	p1.4	9 44	9 59	p1.7	5	W	5 07	4 49	p1.4	10 33	10 39	p1.7
6	M	5 03	4 32	p1.4	10 21	10 30	p1.7	6	T	5 28	5 19	p1.5	11 00	11 09	p1.8
7	T	5 34	5 04	p1.4	10 54	11 01	p1.7	7	F	5 49	5 50	p1.5	11 28	11 41	p1.9
8	W	6 00	5 37	p1.4	11 25	11 34	p1.7	8	S	6 14	6 26	p1.6	...	12 01	1.7
9	T	6 27	6 13	p1.4	11 58	...	1.4	9	S	6 46	7 08	p1.6	12 17	12 40	a2.0
10	F	6 55	6 53	p1.5	12 10	12 36	a1.8	10	M	7 22	7 55	1.6	12 58	1 24	a2.1
11	S	7 29	7 39	p1.5	12 50	1 18	a1.9	11	T	8 05	8 49	a1.7	1 42	2 10	a2.1
12	S	8 09	8 31	p1.5	1 34	2 03	a1.9	12	W	8 52	9 48	a1.8	2 29	2 59	a2.0
13	M	8 54	9 26	1.6	2 20	2 50	a2.0	13	T	9 44	10 51	a1.9	3 18	3 50	a1.9
14	T	9 41	10 24	a1.7	3 07	3 39	a2.0	14	F	10 40	11 55	a1.9	4 12	4 47	1.8
15	W	10 30	11 24	a1.9	3 56	4 31	a2.0	15	S	11 41	...	1.8	5 14	5 50	p1.8
16	T	11 22	...	2.0	4 50	5 28	1.9	16	S	12 57	12 44	p1.8	6 22	6 55	p2.0
17	F	12 24	12 15	p2.0	5 49	6 27	p2.0	17	M	1 56	1 47	p1.9	7 27	7 58	p2.1
18	S	1 23	1 09	p2.0	6 51	7 25	p2.2	18	T	2 52	2 48	p2.0	8 27	8 56	p2.2
19	S	2 21	2 05	p2.1	7 51	8 21	p2.3	19	W	3 44	3 47	p2.1	9 22	9 49	p2.3
20	M	3 16	3 01	p2.2	8 47	9 14	p2.4	20	T	4 32	4 41	p2.1	10 15	10 40	p2.4
21	T	4 07	3 58	p2.2	9 41	10 07	p2.4	21	F	5 19	5 34	p2.1	11 06	11 31	a2.4
22	W	4 56	4 53	p2.2	10 33	10 59	p2.3	22	S	6 07	6 28	2.0	11 58	...	2.3
23	T	5 44	5 48	p2.1	11 26	11 51	p2.3	23	S	6 57	7 26	a1.9	12 23	12 53	a2.3
24	F	6 35	6 46	p1.9	...	12 21	2.2	24	M	7 52	8 30	a1.8	1 16	1 51	a2.1
25	S	7 29	7 49	1.7	12 46	1 19	a2.2	25	T	8 51	9 39	a1.7	2 12	2 49	a1.9
26	S	8 27	8 56	a1.7	1 42	2 18	a2.0	26	W	9 52	10 49	a1.6	3 09	3 47	1.7
27	M	9 27	10 05	a1.6	2 39	3 18	a1.9	27	T	10 54	11 58	a1.5	4 10	4 49	p1.6
28	T	10 28	11 14	a1.6	3 35	4 19	a1.7	28	F	11 55	...	1.4	5 17	5 54	p1.7
29	W	11 28	...	1.5	4 36	5 25	1.6	29	S	1 02	12 53	p1.3	6 30	6 57	p1.7
30	T	12 22	12 25	p1.4	5 42	6 31	p1.7	30	S	2 00	1 47	p1.3	7 35	7 52	p1.7
31	F	1 28	1 20	p1.4	6 52	7 32	p1.7	31	M	2 51	2 37	p1.4	8 27	8 37	p1.8

The Kts. (knots) columns show the **maximum** predicted velocities of the stronger one of the Flood Currents and the stronger one of the Ebb Currents for each day. The letter "a" means the velocity shown should occur **after** the a.m. Current Change. The letter "p" means the velocity shown should occur **after** the p.m. Current Change (even if next morning). No "a" or "p" means a.m. and p.m. velocities are the same for that day.

Av. Max. Vel.: Fl. 1.7 Kts., Ebb 2.0 Kts.
Max. Flood 2 hrs. 25 min. after Flood Starts, ±30 min.
Max. Ebb 3 hrs. 15 min. after Ebb Starts, ±10 min.

See pp. 22-29 for Current Change at other points.

2009 CURRENT TABLE
THE NARROWS, NY HARBOR
40°36.56'N, 74°02.77'W Mid-Channel

Daylight Saving Time   Daylight Saving Time

		SEPTEMBER CURRENT TURNS TO								OCTOBER CURRENT TURNS TO					
		NORTH Flood Starts			SOUTH Ebb Starts					NORTH Flood Starts			SOUTH Ebb Starts		
DAY OF MONTH	DAY OF WEEK	a.m.	**p.m.**	Kts.	a.m.	**p.m.**	Kts.	DAY OF MONTH	DAY OF WEEK	a.m.	**p.m.**	Kts.	a.m.	**p.m.**	Kts.
1	T	3 32	3 20	p1.4	9 07	9 13	p1.8	1	T	3 12	3 25	p1.6	9 00	9 09	1.9
2	W	4 03	3 56	p1.5	9 38	9 44	p1.8	2	F	3 41	4 01	p1.6	9 31	9 41	p2.0
3	T	4 28	4 28	p1.5	10 05	10 13	p1.8	3	S	4 07	4 35	1.7	10 01	10 13	p2.1
4	F	4 49	4 58	p1.6	10 31	10 42	p1.9	4	S	4 34	5 09	a1.8	10 32	10 46	2.1
5	S	5 11	5 29	p1.6	11 00	11 13	p2.0	5	M	5 04	5 46	a1.9	11 06	11 22	2.1
6	S	5 37	6 04	1.6	11 32	11 48	p2.1	6	T	5 36	6 26	a2.0	11 42	...	2.1
7	M	6 07	6 43	a1.8	...	12 08	2.0	7	W	6 12	7 12	a1.9	12 02	12 23	2.0
8	T	6 43	7 29	a1.8	12 27	12 50	a2.1	8	T	6 55	8 06	a1.8	12 47	1 09	p2.0
9	W	7 26	8 23	a1.8	1 10	1 36	a2.0	9	F	7 46	9 07	a1.7	1 38	2 00	p1.8
10	T	8 12	9 22	a1.8	1 58	2 24	1.9	10	S	8 45	10 12	a1.5	2 33	2 54	p1.6
11	F	9 07	10 27	a1.7	2 49	3 17	1.7	11	S	9 56	11 17	a1.4	3 31	3 54	p1.5
12	S	10 09	11 33	a1.7	3 46	4 14	1.6	12	M	11 14	...	1.3	4 36	5 01	p1.4
13	S	11 19	...	1.6	4 49	5 20	p1.6	13	T	12 18	12 26	p1.3	5 46	6 14	p1.6
14	M	12 36	12 30	p1.6	6 00	6 32	p1.7	14	W	1 14	1 31	p1.4	6 54	7 20	p1.7
15	T	1 34	1 37	p1.6	7 09	7 38	p1.9	15	T	2 06	2 30	1.5	7 55	8 18	1.9
16	W	2 28	2 39	p1.8	8 10	8 37	p2.1	16	F	2 56	3 25	1.7	8 49	9 09	2.1
17	T	3 19	3 35	p1.9	9 05	9 30	2.2	17	S	3 42	4 15	a1.9	9 38	9 57	2.2
18	F	4 07	4 28	p2.0	9 56	10 19	2.3	18	S	4 27	5 03	a2.0	10 25	10 43	a2.2
19	S	4 52	5 18	2.0	10 45	11 07	a2.4	19	M	5 10	5 52	a2.0	11 11	11 30	a2.2
20	S	5 38	6 09	a2.0	11 34	11 56	a2.3	20	T	5 54	6 44	a1.9	11 58	...	2.1
21	M	6 25	7 04	a2.0	...	12 26	2.2	21	W	6 41	7 41	a1.7	12 21	12 48	p2.0
22	T	7 16	8 05	a1.8	12 48	1 20	2.1	22	T	7 34	8 43	a1.6	1 16	1 40	p1.9
23	W	8 12	9 11	a1.7	1 44	2 15	1.9	23	F	8 33	9 45	a1.4	2 14	2 33	p1.6
24	T	9 13	10 20	a1.5	2 42	3 11	p1.7	24	S	9 37	10 45	a1.3	3 12	3 27	p1.6
25	F	10 17	11 26	a1.4	3 43	4 08	p1.6	25	S	10 40	11 37	a1.3	4 10	4 22	p1.6
26	S	11 20	...	1.4	4 48	5 09	p1.6	26	M	11 38	...	1.3	5 08	5 18	p1.6
27	S	12 25	12 20	p1.3	5 55	6 11	p1.7	27	T	12 23	12 31	p1.3	6 03	6 13	p1.7
28	M	1 17	1 13	p1.3	6 57	7 08	p1.7	28	W	1 04	1 20	p1.4	6 54	7 03	p1.8
29	T	2 02	2 02	p1.4	7 46	7 55	p1.8	29	T	1 42	2 05	p1.5	7 38	7 47	p1.9
30	W	2 40	2 46	p1.5	8 26	8 34	p1.8	30	F	2 18	2 48	p1.6	8 17	8 27	2.0
								31	S	2 52	3 30	1.7	8 54	9 06	2.1

The Kts. (knots) columns show the **maximum** predicted velocities of the stronger one of the Flood Currents and the stronger one of the Ebb Currents for each day. The letter "a" means the velocity shown should occur **after** the a.m. Current Change. The letter "p" means the velocity shown should occur **after** the p.m. Current Change (even if next morning). No "a" or "p" means a.m. and p.m. velocities are the same for that day.

Av. Max. Vel.: Fl. 1.7 Kts., Ebb 2.0 Kts.

Max. Flood 2 hrs. 25 min. after Flood Starts, ±30 min.

Max. Ebb 3 hrs. 15 min. after Ebb Starts, ±10 min.

At **The Battery, Desbrosses St., & Chelsea Dock** Current turns 1 1/2 hrs. after the Narrows. At **42nd St.** and the **George Washington Br.**, the Current turns 1 3/4 hrs. after the Narrows.

THE NARROWS, NY HARBOR
40°36.56'N, 74°02.77'W Mid-Channel

Standard Time starts Nov. 1 at 2 a.m.　　　　　Standard Time

		NOVEMBER						DECEMBER			
DAY OF MONTH	DAY OF WEEK	CURRENT TURNS TO				DAY OF MONTH	DAY OF WEEK	CURRENT TURNS TO			
		NORTH Flood Starts		SOUTH Ebb Starts				NORTH Flood Starts		SOUTH Ebb Starts	
		a.m.	p.m.	Kts.	a.m.	p.m.	Kts.	a.m.	p.m.	Kts.	a.m.	p.m.	Kts.
1	S	*2 25	*3 10	a1.9	*8 30	*8 43	a2.2	1	T	2 27	3 30	a2.2	8 39	8 59	a2.4
2	M	2 59	3 50	a2.1	9 06	9 21	a2.3	2	W	3 09	4 14	a2.2	9 21	9 44	a2.4
3	T	3 34	4 30	a2.1	9 42	10 01	a2.2	3	T	3 53	5 00	a2.2	10 04	10 32	a2.3
4	W	4 11	5 13	a2.1	10 21	10 45	a2.2	4	F	4 40	5 49	a2.0	10 51	11 23	a2.1
5	T	4 51	6 01	a2.0	11 04	11 33	a2.1	5	S	5 32	6 43	a1.8	11 43	...	2.0
6	F	5 38	6 56	a1.8	11 53	...	1.9	6	S	6 34	7 42	a1.5	12 19	12 40	p1.8
7	S	6 34	7 56	a1.6	12 27	12 46	p1.7	7	M	7 44	8 43	a1.4	1 18	1 38	p1.6
8	S	7 42	9 00	a1.4	1 25	1 44	p1.5	8	T	8 59	9 44	p1.3	2 17	2 36	p1.5
9	M	8 59	10 04	a1.2	2 25	2 44	p1.4	9	W	10 11	10 43	p1.3	3 19	3 36	1.4
10	T	10 15	11 01	1.2	3 28	3 49	p1.4	10	T	11 17	11 36	p1.3	4 25	4 40	p1.5
11	W	11 24	11 56	p1.3	4 36	4 58	p1.5	11	F	...	12 20	1.1	5 32	5 44	1.5
12	T	...	12 26	1.2	5 43	6 02	1.6	12	S	12 28	1 19	a1.4	6 33	6 44	1.6
13	F	12 47	1 23	a1.4	6 43	6 59	p1.8	13	S	1 18	2 15	a1.5	7 26	7 38	a1.7
14	S	1 35	2 17	a1.6	7 36	7 50	1.9	14	M	2 05	3 06	a1.6	8 12	8 27	a1.8
15	S	2 21	3 07	a1.7	8 23	8 38	a2.0	15	T	2 50	3 52	a1.6	8 54	9 13	a1.9
16	M	3 05	3 54	a1.8	9 07	9 24	a2.0	16	W	3 31	4 33	a1.6	9 33	9 56	a1.9
17	T	3 46	4 40	a1.8	9 50	10 10	a2.0	17	T	4 10	5 12	a1.5	10 11	10 39	a1.8
18	W	4 28	5 27	a1.7	10 32	10 58	a2.0	18	F	4 49	5 51	a1.5	10 50	11 22	a1.8
19	T	5 11	6 17	a1.6	11 17	11 49	a1.9	19	S	5 31	6 31	a1.4	11 33	...	1.8
20	F	6 00	7 10	a1.5	...	12 05	1.8	20	S	6 18	7 12	a1.4	12 06	12 17	p1.8
21	S	6 54	8 03	a1.4	12 43	12 55	p1.7	21	M	7 10	7 54	a1.4	12 51	1 03	p1.8
22	S	7 54	8 53	a1.3	1 35	1 45	p1.7	22	T	8 04	8 37	a1.4	1 37	1 49	p1.8
23	M	8 53	9 40	a1.3	2 25	2 35	p1.6	23	W	8 58	9 20	1.5	2 22	2 35	p1.8
24	T	9 50	10 23	a1.4	3 14	3 25	p1.7	24	T	9 53	10 05	p1.7	3 10	3 23	p1.9
25	W	10 43	11 05	1.4	4 05	4 16	p1.8	25	F	10 47	10 50	p1.8	4 00	4 14	p1.9
26	T	11 34	11 45	p1.6	4 56	5 08	p1.9	26	S	11 41	11 36	p1.9	4 53	5 09	p2.0
27	F	...	12 22	1.5	5 46	5 57	p2.0	27	S	...	12 34	1.5	5 46	6 04	2.0
28	S	12 25	1 10	a1.7	6 32	6 45	2.1	28	M	12 23	1 27	a2.0	6 38	6 57	a2.2
29	S	1 05	1 58	a1.9	7 16	7 30	2.2	29	T	1 11	2 19	a2.1	7 27	7 49	a2.3
30	M	1 46	2 44	a2.1	7 58	8 15	a2.3	30	W	2 00	3 08	a2.2	8 14	8 39	a2.4
								31	T	2 50	3 55	a2.3	9 02	9 28	a2.4

*Standard Time starts

The Kts. (knots) columns show the **maximum** predicted velocities of the stronger one of the Flood Currents and the stronger one of the Ebb Currents for each day. The letter "a" means the velocity shown should occur **after** the **a.m.** Current Change. The letter "p" means the velocity shown should occur **after** the **p.m.** Current Change (even if next morning). No "a" or "p" means a.m. and p.m. velocities are the same for that day.

Av. Max. Vel.: Fl. 1.7 Kts., Ebb 2.0 Kts.
Max. Flood 2 hrs. 25 min. after Flood Starts, ±30 min.
Max. Ebb 3 hrs. 15 min. after Ebb Starts, ±10 min.

See pp. 22-29 for Current Change at other points.

2009 HIGH & LOW WATER
THE BATTERY, NY HARBOR
40°42'N, 74°00.9'W

Standard Time   Standard Time

D A Y O F M O N T H	D A Y O F W E E K	JANUARY						D A Y O F M O N T H	D A Y O F W E E K	FEBRUARY					
		HIGH				LOW				HIGH				LOW	
		a.m.	Ht.	p.m.	Ht.	a.m.	p.m.			a.m.	Ht.	p.m.	Ht.	a.m.	p.m.
1	T	10 50	4.3	11 30	3.9	4 46	5 21	1	S	11 54	4.0	...	...	5 58	6 00
2	F	11 32	4.2	...	...	5 25	5 56	2	M	12 20	4.5	12 50	3.8	7 11	7 00
3	S	12 10	4.0	12 20	4.0	6 19	6 41	3	T	1 15	4.5	1 53	3.7	8 33	8 24
4	S	12 56	4.2	1 13	3.9	7 41	7 44	4	W	2 19	4.6	3 05	3.7	9 43	9 39
5	M	1 47	4.4	2 13	3.8	8 58	8 54	5	T	3 33	4.7	4 24	3.8	10 44	10 44
6	T	2 46	4.6	3 22	3.8	10 03	9 58	6	F	4 50	4.9	5 33	4.1	11 41	11 43
7	W	3 54	4.8	4 36	3.9	11 02	10 57	7	S	5 56	5.2	6 32	4.5	...	12 36
8	T	5 02	5.1	5 43	4.1	11 58	11 55	8	S	6 52	5.5	7 24	4.8	12 40	1 27
9	F	6 06	5.4	6 43	4.4	...	12 53	9	M	7 43	5.6	8 14	5.1	1 35	2 16
10	S	7 00	5.6	7 36	4.6	12 52	1 46	10	T	8 30	5.5	9 02	5.2	2 26	3 02
11	S	7 53	5.7	8 29	4.8	1 48	2 37	11	W	9 18	5.3	9 51	5.2	3 15	3 46
12	M	8 45	5.7	9 23	4.9	2 41	3 25	12	T	10 07	5.0	10 40	5.0	4 03	4 28
13	T	9 38	5.5	10 18	4.9	3 32	4 12	13	F	10 56	4.7	11 29	4.9	4 50	5 11
14	W	10 32	5.2	11 13	4.8	4 23	4 59	14	S	11 46	4.3	...	...	5 39	5 55
15	T	11 25	4.8	...	...	5 14	5 47	15	S	12 17	4.6	12 36	3.9	6 33	6 45
16	F	12 05	4.7	12 17	4.4	6 09	6 37	16	M	1 05	4.4	1 28	3.6	7 34	7 43
17	S	12 56	4.6	1 09	4.1	7 10	7 32	17	T	1 55	4.2	2 24	3.4	8 38	8 46
18	S	1 46	4.4	2 01	3.7	8 14	8 29	18	W	2 51	4.0	3 26	3.3	9 38	9 44
19	M	2 38	4.3	2 59	3.5	9 16	9 25	19	T	3 53	4.0	4 30	3.4	10 32	10 37
20	T	3 34	4.2	4 01	3.4	10 12	10 17	20	F	4 54	4.1	5 27	3.6	11 20	11 26
21	W	4 32	4.2	5 02	3.4	11 03	11 05	21	S	5 46	4.3	6 15	3.8	...	12 05
22	T	5 26	4.3	5 55	3.6	11 51	11 52	22	S	6 30	4.5	6 55	4.1	12 12	12 48
23	F	6 14	4.5	6 41	3.7	...	12 36	23	M	7 08	4.7	7 31	4.3	12 56	1 28
24	S	6 56	4.6	7 22	3.9	12 38	1 19	24	T	7 43	4.8	8 03	4.5	1 39	2 06
25	S	7 34	4.7	8 00	4.0	1 21	2 00	25	W	8 15	4.8	8 33	4.6	2 20	2 42
26	M	8 09	4.8	8 35	4.1	2 03	2 39	26	T	8 47	4.8	9 02	4.7	2 59	3 15
27	T	8 42	4.7	9 07	4.1	2 42	3 14	27	F	9 21	4.6	9 35	4.8	3 38	3 48
28	W	9 13	4.7	9 38	4.2	3 20	3 47	28	S	10 00	4.5	10 14	4.9	4 17	4 20
29	T	9 45	4.5	10 10	4.2	3 55	4 18								
30	F	10 21	4.4	10 47	4.3	4 31	4 47								
31	S	11 04	4.2	11 30	4.4	5 09	5 19								

Dates when Ht. of **Low** Water is below Mean Lower Low with Ht. of lowest given for each period and Date of lowest in ( ):

7th - 16th: -1.2' (12th)   6th - 13th: -1.1' (9th - 10th)
25th - 30th: -0.3' (26th - 29th)   24th - 28th: -0.4' (26th)

Average Rise and Fall 4.6 ft.

When a high tide exceeds av. ht., the *following* low tide will be lower than av.

2009 HIGH & LOW WATER
THE BATTERY, NY HARBOR
40°42'N, 74°00.9'W

Daylight Time starts March 8 at 2 a.m.   **Daylight Saving Time**

DAY OF MONTH	DAY OF WEEK	MARCH HIGH a.m.	Ht.	MARCH HIGH p.m.	Ht.	MARCH LOW a.m.	MARCH LOW p.m.	DAY OF MONTH	DAY OF WEEK	APRIL HIGH a.m.	Ht.	APRIL HIGH p.m.	Ht.	APRIL LOW a.m.	APRIL LOW p.m.
1	S	10 47	4.3	11 01	4.9	4 58	4 55	1	W	12 46	5.1	1 42	4.2	7 47	7 42
2	M	11 41	4.1	11 56	4.8	5 48	5 39	2	T	1 51	4.9	2 47	4.2	8 59	9 05
3	T	...	...	12 41	3.9	6 57	6 43	3	F	3 00	4.8	3 55	4.3	10 07	10 18
4	W	12 57	4.7	1 47	3.8	8 15	8 12	4	S	4 13	4.7	5 03	4.5	11 06	11 20
5	T	2 05	4.6	3 00	3.9	9 26	9 28	5	S	5 24	4.8	6 06	4.8	11 59	...
6	F	3 23	4.7	4 15	4.1	10 27	10 33	6	M	6 26	4.9	6 59	5.2	12 18	12 50
7	S	4 40	4.8	5 22	4.4	11 23	11 31	7	T	7 18	5.0	7 46	5.5	1 11	1 37
8	S	*6 44	5.1	*7 18	4.8	...	*1 15	8	W	8 05	5.1	8 29	5.6	2 02	2 22
9	M	7 39	5.3	8 08	5.2	1 27	2 04	9	T	8 50	5.0	9 11	5.7	2 51	3 06
10	T	8 25	5.4	8 53	5.4	2 19	2 51	10	F	9 32	4.9	9 50	5.6	3 37	3 47
11	W	9 10	5.3	9 37	5.5	3 09	3 35	11	S	10 16	4.7	10 31	5.4	4 20	4 27
12	T	9 55	5.1	10 21	5.4	3 57	4 17	12	S	11 02	4.4	11 13	5.1	5 02	5 05
13	F	10 40	4.9	11 05	5.3	4 42	4 57	13	M	11 51	4.2	11 58	4.8	5 44	5 42
14	S	11 28	4.5	11 50	5.0	5 26	5 36	14	T	...	...	12 41	4.0	6 27	6 20
15	S	...	...	12 17	4.2	6 10	6 15	15	W	12 46	4.6	1 32	3.8	7 15	7 06
16	M	12 36	4.7	1 07	3.9	6 58	6 58	16	T	1 35	4.3	2 22	3.7	8 12	8 13
17	T	1 23	4.5	1 58	3.7	7 53	7 53	17	F	2 25	4.2	3 13	3.7	9 14	9 26
18	W	2 13	4.2	2 52	3.5	8 56	9 03	18	S	3 19	4.1	4 07	3.8	10 10	10 28
19	T	3 07	4.1	3 50	3.5	9 59	10 08	19	S	4 16	4.1	5 02	4.0	10 59	11 21
20	F	4 08	4.0	4 52	3.6	10 54	11 05	20	M	5 13	4.2	5 52	4.3	11 43	...
21	S	5 11	4.1	5 50	3.8	11 43	11 55	21	T	6 06	4.3	6 34	4.7	12 09	12 25
22	S	6 07	4.2	6 39	4.1	...	12 27	22	W	6 52	4.5	7 12	5.1	12 57	1 06
23	M	6 54	4.5	7 20	4.4	12 42	1 09	23	T	7 34	4.6	7 48	5.4	1 44	1 48
24	T	7 34	4.6	7 55	4.7	1 28	1 49	24	F	8 15	4.7	8 25	5.7	2 30	2 31
25	W	8 10	4.8	8 27	5.0	2 12	2 28	25	S	8 58	4.8	9 04	5.8	3 17	3 14
26	T	8 45	4.8	8 57	5.2	2 55	3 06	26	S	9 43	4.8	9 48	5.8	4 04	3 59
27	F	9 21	4.8	9 30	5.4	3 38	3 44	27	M	10 35	4.7	10 39	5.7	4 51	4 45
28	S	10 00	4.7	10 07	5.4	4 20	4 21	28	T	11 35	4.6	11 38	5.5	5 40	5 35
29	S	10 45	4.6	10 52	5.4	5 03	4 59	29	W	...	...	12 38	4.5	6 34	6 31
30	M	11 39	4.4	11 45	5.2	5 49	5 42	30	T	12 44	5.3	1 41	4.5	7 35	7 39
31	T	...	...	12 39	4.3	6 42	6 33								

*Daylight Saving Time starts

Dates when Ht. of **Low** Water is below Mean Lower Low with Ht. of lowest given for each period and Date of lowest in ( ):

7th - 14th: -0.8' (10th - 11th)   6th - 11th: -0.4' (7th - 10th)
26th - 29th: -0.3' (28th - 29th)   25th - 28th: -0.4' (26th - 27th)

Average Rise and Fall 4.6 ft.

When a high tide exceeds av. ht., the *following* low tide will be lower than av.

Daylight Saving Time   Daylight Saving Time

DAY OF MONTH	DAY OF WEEK	MAY HIGH a.m.	Ht.	HIGH p.m.	Ht.	LOW a.m.	LOW p.m.	DAY OF MONTH	DAY OF WEEK	JUNE HIGH a.m.	Ht.	HIGH p.m.	Ht.	LOW a.m.	LOW p.m.
1	F	1 48	5.1	2 41	4.6	8 40	8 53	1	M	3 33	4.7	4 19	5.1	10 11	10 45
2	S	2 52	4.9	3 43	4.7	9 44	10 03	2	T	4 33	4.5	5 15	5.2	11 03	11 41
3	S	3 56	4.8	4 44	4.9	10 41	11 05	3	W	5 33	4.3	6 08	5.3	11 52	...
4	M	5 01	4.7	5 43	5.1	11 33	...	4	T	6 30	4.3	6 56	5.4	12 32	12 38
5	T	6 01	4.7	6 36	5.4	12 01	12 21	5	F	7 21	4.3	7 40	5.5	1 22	1 24
6	W	6 55	4.7	7 22	5.6	12 53	1 08	6	S	8 07	4.3	8 21	5.4	2 09	2 09
7	T	7 43	4.7	8 04	5.7	1 43	1 53	7	S	8 51	4.3	9 00	5.4	2 54	2 53
8	F	8 28	4.6	8 44	5.6	2 31	2 36	8	M	9 34	4.3	9 39	5.2	3 37	3 35
9	S	9 12	4.6	9 24	5.5	3 16	3 19	9	T	10 19	4.2	10 19	5.1	4 18	4 15
10	S	9 54	4.4	10 02	5.3	3 59	3 59	10	W	11 03	4.2	10 59	4.9	4 57	4 53
11	M	10 39	4.3	10 42	5.1	4 40	4 38	11	T	11 49	4.1	11 40	4.8	5 35	5 30
12	T	11 27	4.1	11 26	4.9	5 20	5 15	12	F	...	...	12 33	4.1	6 11	6 06
13	W	...	...	12 17	4.0	6 00	5 52	13	S	12 21	4.6	1 13	4.1	6 48	6 47
14	T	12 12	4.7	1 05	3.9	6 43	6 32	14	S	1 01	4.4	1 50	4.2	7 27	7 44
15	F	12 58	4.5	1 50	3.9	7 29	7 25	15	M	1 42	4.3	2 27	4.3	8 12	8 57
16	S	1 43	4.3	2 34	4.0	8 22	8 37	16	T	2 26	4.2	3 07	4.6	9 05	10 04
17	S	2 28	4.2	3 18	4.1	9 17	9 45	17	W	3 17	4.1	3 55	4.8	9 59	11 02
18	M	3 16	4.2	4 04	4.3	10 07	10 43	18	T	4 17	4.1	4 49	5.1	10 53	11 57
19	T	4 10	4.1	4 52	4.6	10 53	11 35	19	F	5 24	4.1	5 48	5.4	11 46	...
20	W	5 08	4.2	5 40	4.9	11 38	...	20	S	6 29	4.3	6 45	5.7	12 51	12 40
21	T	6 06	4.3	6 27	5.3	12 25	12 23	21	S	7 27	4.5	7 40	5.9	1 45	1 36
22	F	6 59	4.5	7 13	5.6	1 16	1 10	22	M	8 22	4.7	8 33	6.1	2 38	2 33
23	S	7 49	4.6	7 58	5.9	2 06	2 00	23	T	9 17	4.9	9 27	6.1	3 30	3 28
24	S	8 38	4.7	8 45	6.0	2 57	2 51	24	W	10 14	5.0	10 23	6.0	4 21	4 22
25	M	9 30	4.8	9 36	6.0	3 47	3 42	25	T	11 13	5.1	11 23	5.8	5 10	5 15
26	T	10 26	4.8	10 32	5.9	4 37	4 34	26	F	...	...	12 13	5.1	6 00	6 10
27	W	11 28	4.8	11 34	5.7	5 27	5 27	27	S	12 22	5.5	1 09	5.2	6 51	7 09
28	T	...	...	12 30	4.8	6 20	6 24	28	S	1 18	5.2	2 03	5.2	7 45	8 13
29	F	12 37	5.4	1 30	4.9	7 16	7 27	29	M	2 12	4.8	2 56	5.2	8 42	9 19
30	S	1 37	5.2	2 26	4.9	8 16	8 36	30	T	3 06	4.5	3 49	5.1	9 38	10 21
31	S	2 35	4.9	3 22	5.0	9 15	9 44								

Dates when Ht. of **Low** Water is below Mean Lower Low with Ht. of lowest given for each period and Date of lowest in ( ):

24th - 29th: -0.5' (25th - 27th)   22nd - 27th: -0.7' (24th - 25th)

Average Rise and Fall 4.6 ft.

When a high tide exceeds av. ht., the *following* low tide will be lower than av.

2009 HIGH & LOW WATER
THE BATTERY, NY HARBOR
40°42'N, 74°00.9'W

			Daylight Saving Time							Daylight Saving Time			
DAY OF MONTH	DAY OF WEEK	**JULY**				DAY OF MONTH	DAY OF WEEK	**AUGUST**					
		HIGH				LOW					
		a.m.	Ht.	**p.m.**	Ht.	a.m.	**p.m.**			a.m.	Ht.	**p.m.**	Ht.	a.m.	**p.m.**

DAY OF MONTH	DAY OF WEEK	a.m.	Ht.	p.m.	Ht.	a.m.	p.m.	DAY OF MONTH	DAY OF WEEK	a.m.	Ht.	p.m.	Ht.	a.m.	p.m.
1	W	4 04	4.2	4 44	5.1	10 32	11 18	1	S	5 39	3.9	6 04	4.9	11 45	...
2	T	5 05	4.0	5 38	5.1	11 22	...	2	S	6 36	4.0	6 54	5.0	12 33	12 33
3	F	6 05	4.0	6 30	5.1	12 09	12 11	3	M	7 25	4.2	7 38	5.1	1 19	1 20
4	S	6 59	4.0	7 17	5.2	12 58	12 58	4	T	8 08	4.3	8 18	5.2	2 03	2 05
5	S	7 47	4.1	8 00	5.2	1 45	1 44	5	W	8 47	4.4	8 54	5.2	2 44	2 48
6	M	8 31	4.2	8 40	5.2	2 30	2 29	6	T	9 24	4.5	9 28	5.2	3 23	3 29
7	T	9 12	4.3	9 18	5.2	3 13	3 12	7	F	9 59	4.6	10 00	5.1	3 59	4 07
8	W	9 53	4.3	9 55	5.1	3 53	3 52	8	S	10 31	4.6	10 32	4.9	4 33	4 44
9	T	10 35	4.3	10 32	5.0	4 30	4 31	9	S	11 03	4.7	11 06	4.7	5 03	5 20
10	F	11 14	4.3	11 06	4.8	5 06	5 07	10	M	11 33	4.7	11 43	4.5	5 31	5 56
11	S	11 51	4.3	11 42	4.7	5 38	5 42	11	T	...	...	12 11	4.8	5 59	6 38
12	S	...	...	12 26	4.4	6 08	6 19	12	W	12 29	4.3	12 55	4.9	6 33	7 39
13	M	12 19	4.5	1 00	4.5	6 38	7 04	13	T	1 22	4.2	1 47	5.0	7 20	9 00
14	T	1 01	4.3	1 37	4.6	7 12	8 11	14	F	2 21	4.1	2 44	5.1	8 32	10 12
15	W	1 48	4.2	2 21	4.8	8 00	9 27	15	S	3 28	4.0	3 51	5.2	9 58	11 15
16	T	2 41	4.1	3 12	5.0	9 07	10 34	16	S	4 43	4.1	5 06	5.3	11 08	...
17	F	3 44	4.0	4 13	5.2	10 17	11 33	17	M	5 56	4.4	6 17	5.6	12 12	12 10
18	S	4 56	4.1	5 21	5.4	11 21	...	18	T	6 59	4.8	7 17	5.9	1 06	1 08
19	S	6 08	4.3	6 28	5.7	12 30	12 21	19	W	7 54	5.2	8 09	6.0	1 58	2 04
20	M	7 12	4.6	7 27	5.9	1 25	1 21	20	T	8 45	5.5	8 59	6.0	2 48	2 59
21	T	8 08	4.9	8 22	6.1	2 19	2 18	21	F	9 35	5.7	9 48	5.9	3 35	3 50
22	W	9 02	5.1	9 15	6.1	3 11	3 14	22	S	10 25	5.8	10 39	5.6	4 21	4 41
23	T	9 56	5.3	10 08	6.0	4 00	4 07	23	S	11 16	5.7	11 31	5.2	5 05	5 30
24	F	10 52	5.4	11 03	5.7	4 48	4 59	24	M	...	...	12 08	5.6	5 49	6 21
25	S	11 47	5.5	11 58	5.4	5 34	5 52	25	T	12 25	4.8	12 59	5.3	6 35	7 16
26	S	...	...	12 42	5.4	6 21	6 46	26	W	1 19	4.5	1 50	5.1	7 25	8 17
27	M	12 53	5.0	1 34	5.3	7 11	7 46	27	T	2 13	4.2	2 41	4.9	8 23	9 21
28	T	1 46	4.7	2 24	5.2	8 05	8 50	28	F	3 09	4.0	3 36	4.7	9 27	10 22
29	W	2 40	4.3	3 16	5.0	9 02	9 54	29	S	4 09	3.8	4 34	4.6	10 26	11 16
30	T	3 36	4.0	4 10	4.9	10 00	10 52	30	S	5 11	3.9	5 33	4.7	11 19	...
31	F	4 37	3.9	5 07	4.8	10 54	11 44	31	M	6 09	4.0	6 26	4.8	12 04	12 08

Dates when Ht. of **Low** Water is below Mean Lower Low with Ht. of lowest given for each period and Date of lowest in ( ):

21st - 26th: -0.8' (23rd - 24th)   19th - 24th: -0.7' (21st - 22nd)

Average Rise and Fall 4.6 ft.

When a high tide exceeds av. ht., the *following* low tide will be lower than av.

2009 HIGH & LOW WATER
THE BATTERY, NY HARBOR
40°42'N, 74°00.9'W

Daylight SavingTime · Daylight Saving Time

DAY OF MONTH	DAY OF WEEK	SEPTEMBER HIGH a.m.	Ht.	HIGH p.m.	Ht.	LOW a.m.	LOW p.m.	DAY OF MONTH	DAY OF WEEK	OCTOBER HIGH a.m.	Ht.	HIGH p.m.	Ht.	LOW a.m.	LOW p.m.
1	T	6 58	4.3	7 11	5.0	12 48	12 54	1	T	7 05	4.6	7 16	4.9	12 52	1 10
2	W	7 40	4.5	7 50	5.1	1 30	1 38	2	F	7 41	4.9	7 52	5.0	1 30	1 53
3	T	8 17	4.7	8 25	5.2	2 10	2 21	3	S	8 12	5.2	8 26	5.0	2 08	2 36
4	F	8 50	4.9	8 58	5.1	2 47	3 03	4	S	8 41	5.3	8 59	4.9	2 45	3 18
5	S	9 20	5.0	9 28	5.1	3 23	3 43	5	M	9 10	5.5	9 34	4.8	3 21	4 00
6	S	9 47	5.1	10 00	4.9	3 57	4 21	6	T	9 42	5.5	10 14	4.6	3 57	4 41
7	M	10 16	5.1	10 35	4.7	4 28	4 59	7	W	10 22	5.5	11 02	4.5	4 33	5 25
8	T	10 50	5.2	11 17	4.5	4 58	5 38	8	T	11 10	5.4	...	...	5 13	6 14
9	W	11 34	5.2	...	...	5 30	6 24	9	F	12 03	4.3	12 11	5.2	5 58	7 15
10	T	12 10	4.3	12 25	5.1	6 09	7 24	10	S	1 08	4.2	1 17	5.1	6 59	8 27
11	F	1 10	4.2	1 25	5.1	7 00	8 42	11	S	2 15	4.2	2 26	5.0	8 22	9 36
12	S	2 15	4.1	2 31	5.1	8 23	9 55	12	M	3 21	4.3	3 35	5.0	9 42	10 37
13	S	3 24	4.2	3 42	5.1	9 50	10 57	13	T	4 28	4.6	4 45	5.1	10 48	11 31
14	M	4 37	4.3	4 57	5.2	10 59	11 53	14	W	5 32	4.9	5 49	5.2	11 47	...
15	T	5 46	4.7	6 05	5.5	11 59	...	15	T	6 28	5.3	6 45	5.3	12 21	12 42
16	W	6 46	5.1	7 03	5.7	12 44	12 56	16	F	7 18	5.7	7 35	5.3	1 08	1 34
17	T	7 37	5.5	7 53	5.8	1 34	1 50	17	S	8 03	5.9	8 21	5.3	1 55	2 25
18	F	8 25	5.8	8 40	5.7	2 22	2 42	18	S	8 46	6.0	9 05	5.1	2 40	3 13
19	S	9 10	6.0	9 26	5.6	3 08	3 32	19	M	9 28	5.9	9 50	4.9	3 24	4 00
20	S	9 56	5.9	10 13	5.3	3 52	4 20	20	T	10 11	5.7	10 38	4.6	4 06	4 44
21	M	10 42	5.8	11 04	4.9	4 35	5 07	21	W	10 56	5.4	11 30	4.3	4 47	5 29
22	T	11 31	5.5	11 57	4.6	5 17	5 54	22	T	11 45	5.0	...	...	5 28	6 14
23	W	...	...	12 22	5.2	5 59	6 44	23	F	12 25	4.1	12 37	4.8	6 10	7 05
24	T	12 52	4.3	1 13	4.9	6 46	7 41	24	S	1 19	3.9	1 29	4.5	7 00	8 02
25	F	1 47	4.1	2 06	4.7	7 42	8 44	25	S	2 12	3.8	2 21	4.4	8 03	9 02
26	S	2 42	3.9	3 00	4.5	8 49	9 46	26	M	3 05	3.8	3 13	4.3	9 13	9 58
27	S	3 40	3.9	3 57	4.5	9 53	10 41	27	T	3 58	3.9	4 07	4.3	10 13	10 46
28	M	4 38	3.9	4 55	4.5	10 50	11 29	28	W	4 51	4.1	5 01	4.3	11 06	11 29
29	T	5 35	4.1	5 49	4.6	11 39	...	29	T	5 40	4.3	5 51	4.4	11 54	...
30	W	6 24	4.4	6 36	4.8	12 11	12 25	30	F	6 22	4.6	6 36	4.5	12 09	12 39
								31	S	6 59	5.0	7 17	4.6	12 48	1 24

Dates when Ht. of **Low** Water is below Mean Lower Low with Ht. of lowest given for each period and Date of lowest in ( ):

17th - 21st: -0.6' (18th - 19th)   15th - 19th: -0.4' (16th - 18th)

Average Rise and Fall 4.6 ft.

When a high tide exceeds av. ht., the *following* low tide will be lower than av.

2009 HIGH & LOW WATER
THE BATTERY, NY HARBOR
40°42'N, 74°00.9'W

Standard Time starts Nov. 1 at 2 a.m.   Standard Time

NOVEMBER

Day of Month	Day of Week	HIGH a.m.	Ht.	HIGH p.m.	Ht.	LOW a.m.	LOW p.m.
1	S	*6 33	5.3	*6 55	4.7	1 28	*1 10
2	M	7 06	5.5	7 33	4.7	1 08	1 55
3	T	7 41	5.6	8 14	4.6	1 50	2 40
4	W	8 21	5.7	9 00	4.5	2 33	3 26
5	T	9 06	5.6	9 55	4.4	3 17	4 13
6	F	10 01	5.4	10 59	4.3	4 04	5 05
7	S	11 06	5.3	...	...	4 56	6 02
8	S	12 06	4.3	12 13	5.1	5 59	7 07
9	M	1 10	4.4	1 19	4.9	7 14	8 12
10	T	2 10	4.5	2 22	4.8	8 28	9 12
11	W	3 12	4.7	3 26	4.7	9 34	10 06
12	T	4 12	5.0	4 28	4.7	10 32	10 55
13	F	5 08	5.3	5 25	4.7	11 26	11 42
14	S	5 57	5.5	6 16	4.7	...	12 17
15	S	6 42	5.6	7 03	4.7	12 29	1 07
16	M	7 24	5.7	7 47	4.6	1 14	1 54
17	T	8 05	5.5	8 31	4.4	1 58	2 39
18	W	8 45	5.4	9 17	4.3	2 41	3 23
19	T	9 28	5.1	10 05	4.1	3 22	4 05
20	F	10 13	4.9	10 56	3.9	4 02	4 47
21	S	11 02	4.6	11 48	3.8	4 41	5 30
22	S	11 51	4.4	...	...	5 23	6 17
23	M	12 37	3.7	12 38	4.2	6 13	7 09
24	T	1 23	3.7	1 24	4.1	7 20	8 03
25	W	2 09	3.8	2 11	4.0	8 28	8 54
26	T	2 55	4.0	3 01	3.9	9 27	9 40
27	F	3 42	4.2	3 56	3.9	10 19	10 23
28	S	4 29	4.5	4 50	4.0	11 08	11 06
29	S	5 13	4.8	5 41	4.1	11 56	11 50
30	M	5 56	5.2	6 27	4.3	...	12 45

DECEMBER

Day of Month	Day of Week	HIGH a.m.	Ht.	HIGH p.m.	Ht.	LOW a.m.	LOW p.m.
1	T	6 39	5.4	7 13	4.4	12 37	1 34
2	W	7 22	5.6	8 00	4.5	1 26	2 22
3	T	8 09	5.7	8 51	4.5	2 15	3 11
4	F	9 00	5.6	9 48	4.5	3 06	4 00
5	S	9 57	5.5	10 51	4.5	3 57	4 50
6	S	11 00	5.2	11 53	4.5	4 50	5 44
7	M	...	...	12 03	5.0	5 50	6 42
8	T	12 53	4.6	1 02	4.8	6 58	7 43
9	W	1 51	4.6	2 02	4.5	8 09	8 42
10	T	2 48	4.7	3 02	4.3	9 15	9 37
11	F	3 46	4.9	4 04	4.1	10 14	10 29
12	S	4 43	5.0	5 05	4.1	11 08	11 17
13	S	5 36	5.1	5 59	4.1	11 59	...
14	M	6 23	5.2	6 47	4.1	12 05	12 48
15	T	7 06	5.2	7 32	4.1	12 51	1 35
16	W	7 47	5.2	8 14	4.1	1 37	2 19
17	T	8 26	5.1	8 57	4.0	2 20	3 01
18	F	9 06	4.9	9 41	4.0	3 01	3 41
19	S	9 47	4.7	10 26	3.9	3 40	4 19
20	S	10 29	4.5	11 11	3.8	4 17	4 56
21	M	11 10	4.3	11 53	3.7	4 53	5 32
22	T	11 51	4.1	...	...	5 31	6 10
23	W	12 33	3.8	12 31	3.9	6 20	6 52
24	T	1 10	3.8	1 13	3.8	7 30	7 44
25	F	1 50	4.0	2 01	3.6	8 41	8 40
26	S	2 35	4.1	2 58	3.6	9 42	9 35
27	S	3 28	4.4	4 03	3.6	10 37	10 27
28	M	4 27	4.7	5 07	3.8	11 29	11 20
29	T	5 25	5.0	6 05	4.0	...	12 22
30	W	6 19	5.3	6 57	4.3	12 14	1 14
31	T	7 10	5.5	7 48	4.5	1 08	2 04

*Standard Time starts

Dates when Ht. of **Low** Water is below Mean Lower Low with Ht. of lowest given for each period and Date of lowest in ( ):

4th: -0.2'
12th - 17th: -0.3' (13th, 15th - 16th)

1st - 8th: -0.6' (3rd - 4th)
10th - 11th: -0.2'
14th - 18th: -0.3' (15th - 16th)
29th - 31st: -0.8' (31st)

Average Rise and Fall 4.6 ft.

When a high tide exceeds av. ht., the *following* low tide will be lower than av.

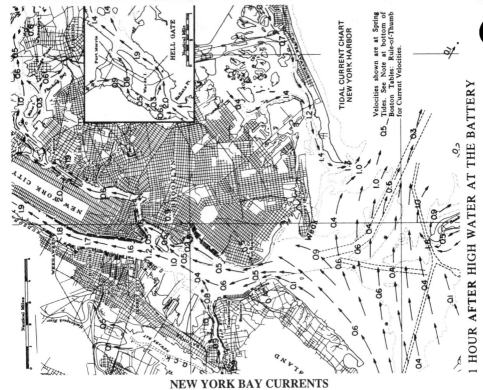

NEW YORK BAY CURRENTS
Ward Is and Randall Is. are cojoined- no water between them.
Bronx Kill N. of Randall Is. not navigable

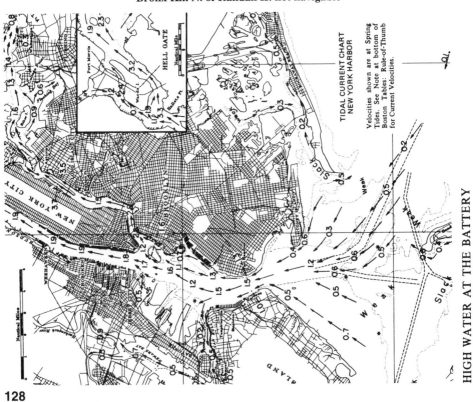

HIGH WATER AT THE BATTERY

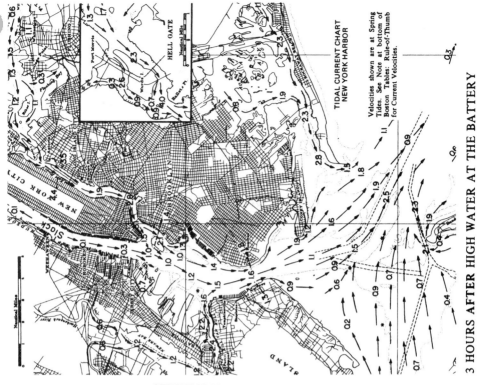

3 HOURS AFTER HIGH WATER AT THE BATTERY

NEW YORK BAY CURRENTS
Ward Is and Randall Is. are cojoined- no water between them.
Bronx Kill N. of Randall Is. not navigable

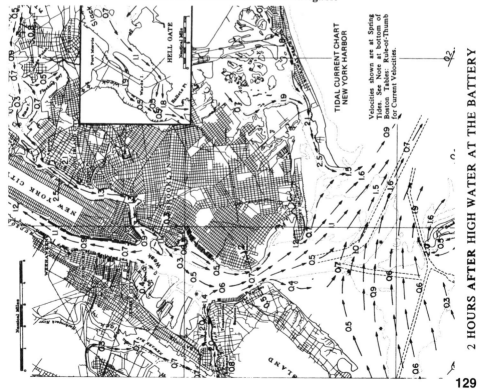

2 HOURS AFTER HIGH WATER AT THE BATTERY

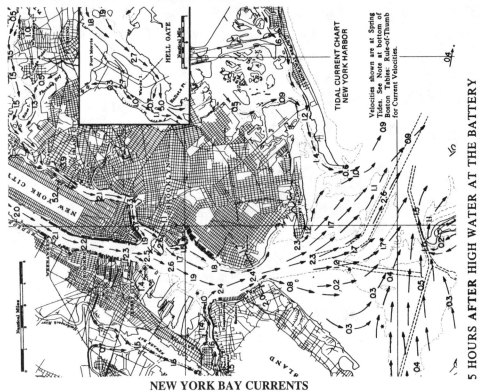

NEW YORK BAY CURRENTS
Ward Is and Randall Is. are cojoined- no water between them.
Bronx Kill N. of Randall Is. not navigable

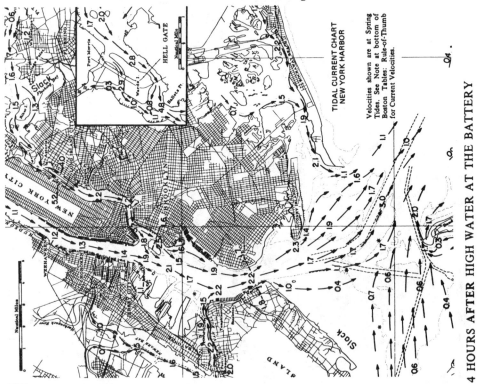

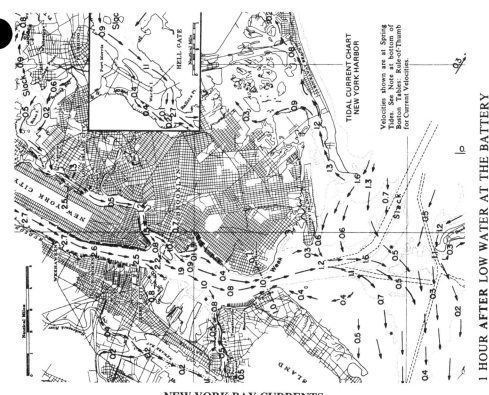

Within the upper chart:

HELL GATE

TIDAL CURRENT CHART
NEW YORK HARBOR

Velocities shown are at Spring
Tides. See Note at bottom of
Boston Tables: Rule-of-Thumb
for Current Velocities.

1 HOUR AFTER LOW WATER AT THE BATTERY

NEW YORK BAY CURRENTS
Ward Is and Randall Is. are cojoined- no water between them.
Bronx Kill N. of Randall Is. not navigable

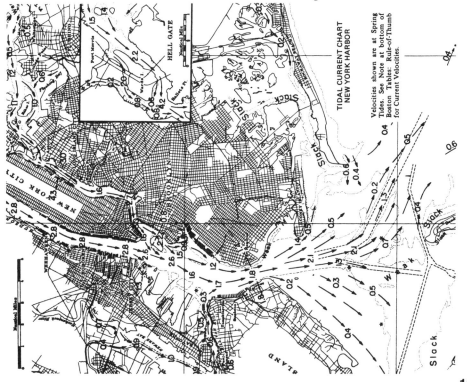

HELL GATE

TIDAL CURRENT CHART
NEW YORK HARBOR

Velocities shown are at Spring
Tides. See Note at bottom of
Boston Tables: Rule-of-Thumb
for Current Velocities.

LOW WATER AT THE BATTERY

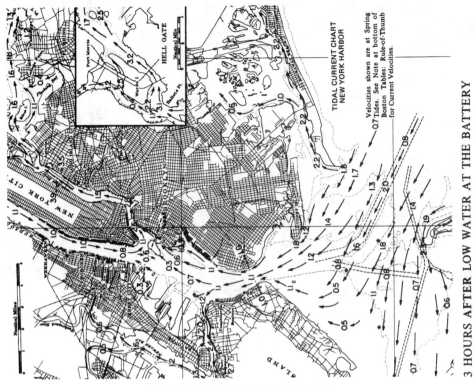

TIDAL CURRENT CHART
NEW YORK HARBOR

Velocities shown are at Spring
Tides. See Note at bottom of
Boston Tables: Rule-of-Thumb
for Current Velocities.

HELL GATE

3 HOURS AFTER LOW WATER AT THE BATTERY

NEW YORK BAY CURRENTS
Ward Is and Randall Is. are cojoined- no water between them.
Bronx Kill N. of Randall Is. not navigable

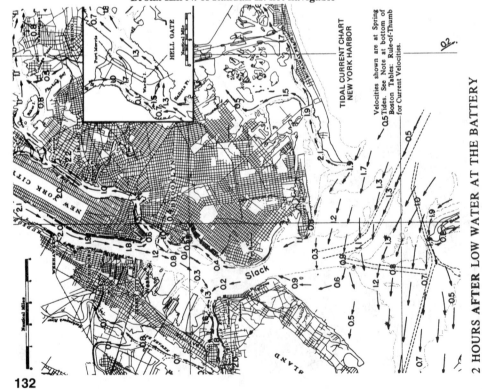

TIDAL CURRENT CHART
NEW YORK HARBOR

Velocities shown are at Spring
Tides. See Note at bottom of
Boston Tables: Rule-of-Thumb
for Current Velocities.

HELL GATE

Slack

2 HOURS AFTER LOW WATER AT THE BATTERY

132

NEW YORK BAY CURRENTS
Ward Is and Randall Is. are cojoined- no water between them.
Bronx Kill N. of Randall Is. not navigable

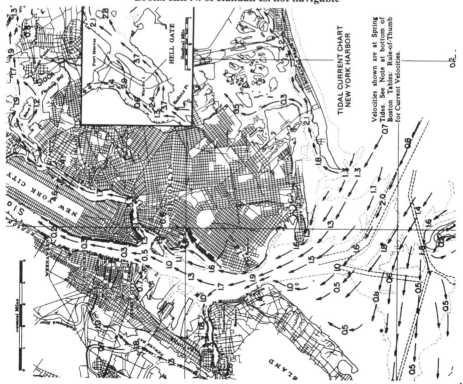

2009 HIGH & LOW WATER
SANDY HOOK, NJ
40°28'N, 74°00.6'W

Standard Time   **Standard Time**

DAY OF MONTH	DAY OF WEEK	JANUARY HIGH a.m.	Ht.	HIGH p.m.	Ht.	LOW a.m.	LOW p.m.	DAY OF MONTH	DAY OF WEEK	FEBRUARY HIGH a.m.	Ht.	HIGH p.m.	Ht.	LOW a.m.	LOW p.m.
1	T	10 28	4.5	11 04	4.0	4 21	4 52	1	S	11 35	4.2	...	...	5 29	5 35
2	F	11 12	4.3	11 48	4.1	5 01	5 28	2	M	12 03	4.6	12 31	4.0	6 32	6 30
3	S	...	...	12 01	4.2	5 51	6 12	3	T	12 59	4.7	1 33	3.9	7 53	7 47
4	S	12 37	4.3	12 55	4.0	7 01	7 10	4	W	2 02	4.8	2 42	3.8	9 09	9 04
5	M	1 29	4.5	1 54	3.9	8 20	8 18	5	T	3 12	4.9	3 55	4.0	10 14	10 11
6	T	2 28	4.8	3 01	3.9	9 29	9 24	6	F	4 23	5.1	5 03	4.3	11 13	11 12
7	W	3 33	5.0	4 10	4.0	10 31	10 25	7	S	5 27	5.4	6 02	4.7	...	12 08
8	T	4 39	5.3	5 16	4.3	11 28	11 24	8	S	6 24	5.7	6 55	5.0	12 10	1 00
9	F	5 41	5.6	6 16	4.6	...	12 24	9	M	7 15	5.9	7 46	5.3	1 05	1 49
10	S	6 36	5.9	7 10	4.8	12 22	1 19	10	T	8 02	5.8	8 33	5.4	1 57	2 35
11	S	7 29	6.0	8 02	5.0	1 18	2 10	11	W	8 49	5.6	9 21	5.4	2 47	3 17
12	M	8 20	6.0	8 55	5.1	2 12	2 59	12	T	9 36	5.3	10 10	5.3	3 34	3 58
13	T	9 11	5.8	9 48	5.1	3 04	3 45	13	F	10 24	4.9	10 58	5.1	4 19	4 38
14	W	10 02	5.4	10 41	5.0	3 54	4 30	14	S	11 13	4.5	11 46	4.8	5 06	5 18
15	T	10 54	5.0	11 34	4.9	4 44	5 15	15	S	...	...	12 02	4.1	5 55	6 02
16	F	11 45	4.6	...	...	5 36	6 02	16	M	12 33	4.6	12 53	3.8	6 53	6 56
17	S	12 24	4.8	12 36	4.2	6 33	6 53	17	T	1 23	4.3	1 46	3.5	7 59	8 01
18	S	1 14	4.6	1 27	3.9	7 37	7 49	18	W	2 16	4.2	2 44	3.4	9 03	9 06
19	M	2 04	4.5	2 21	3.6	8 42	8 47	19	T	3 14	4.1	3 46	3.5	9 59	10 02
20	T	2 57	4.4	3 19	3.5	9 40	9 41	20	F	4 14	4.2	4 46	3.6	10 49	10 53
21	W	3 53	4.4	4 20	3.5	10 32	10 31	21	S	5 09	4.4	5 37	3.9	11 34	11 40
22	T	4 49	4.4	5 16	3.6	11 20	11 19	22	S	5 56	4.7	6 21	4.2	...	12 17
23	F	5 39	4.6	6 05	3.8	...	12 05	23	M	6 36	4.9	6 59	4.4	12 25	12 57
24	S	6 23	4.8	6 48	4.0	12 04	12 48	24	T	7 13	5.0	7 34	4.6	1 08	1 36
25	S	7 03	4.9	7 27	4.1	12 49	1 30	25	W	7 48	5.0	8 07	4.8	1 50	2 12
26	M	7 40	5.0	8 03	4.2	1 32	2 08	26	T	8 23	5.0	8 41	4.9	2 30	2 46
27	T	8 14	4.9	8 38	4.3	2 12	2 44	27	F	8 59	4.8	9 16	5.0	3 09	3 20
28	W	8 48	4.9	9 12	4.3	2 50	3 18	28	S	9 39	4.7	9 57	5.0	3 48	3 53
29	T	9 23	4.7	9 47	4.4	3 27	3 49								
30	F	10 01	4.6	10 27	4.5	4 04	4 21								
31	S	10 45	4.4	11 12	4.6	4 43	4 54								

Dates when Ht. of **Low** Water is below Mean Lower Low with Ht. of lowest given for each period and Date of lowest in ( ):

7th - 16th: -1.2' (12th)   6th - 13th: -1.2' (9th-10th)
26th - 30th: -0.3' (27th - 29th)   24th - 28th: -0.4' (26th)

Average Rise and Fall 4.6 ft.

When a high tide exceeds av. ht., the *following* low tide will be lower than av.

2009 HIGH & LOW WATER
SANDY HOOK, NJ
40°28'N, 74°00.6'W

Daylight Time starts March 8 at 2 a.m.

Daylight Saving Time

DAY OF MONTH	DAY OF WEEK	MARCH HIGH a.m.	Ht.	HIGH p.m.	Ht.	LOW a.m.	LOW p.m.	DAY OF MONTH	DAY OF WEEK	APRIL HIGH a.m.	Ht.	HIGH p.m.	Ht.	LOW a.m.	LOW p.m.
1	S	10 26	4.4	10 45	5.0	4 29	4 30	1	W	12 28	5.2	1 16	4.3	7 11	7 07
2	M	11 20	4.2	11 40	5.0	5 17	5 13	2	T	1 32	5.1	2 19	4.3	8 24	8 27
3	T	...	...	12 20	4.1	6 18	6 11	3	F	2 37	5.0	3 24	4.4	9 37	9 44
4	W	12 41	4.9	1 24	4.0	7 37	7 32	4	S	3 44	5.0	4 29	4.6	10 39	10 51
5	T	1 47	4.9	2 32	4.0	8 54	8 53	5	S	4 50	5.0	5 32	4.9	11 33	11 49
6	F	2 58	4.9	3 43	4.2	9 59	10 02	6	M	5 51	5.1	6 27	5.3	...	12 22
7	S	4 08	5.1	4 49	4.6	10 55	11 02	7	T	6 45	5.2	7 16	5.6	12 42	1 08
8	S	*6 12	5.3	*6 47	5.0	...	*12 48	8	W	7 33	5.3	8 00	5.8	1 33	1 52
9	M	7 08	5.5	7 39	5.4	12 58	1 36	9	T	8 19	5.2	8 43	5.9	2 21	2 35
10	T	7 56	5.6	8 24	5.6	1 50	2 23	10	F	9 01	5.1	9 22	5.8	3 07	3 15
11	W	8 41	5.6	9 09	5.7	2 40	3 06	11	S	9 43	4.8	10 03	5.6	3 50	3 54
12	T	9 25	5.4	9 52	5.7	3 27	3 47	12	S	10 27	4.6	10 44	5.3	4 30	4 31
13	F	10 09	5.1	10 36	5.5	4 12	4 25	13	M	11 14	4.3	11 28	5.0	5 10	5 07
14	S	10 54	4.7	11 20	5.2	4 55	5 02	14	T	...	...	12 03	4.0	5 51	5 44
15	S	11 42	4.4	...	...	5 36	5 39	15	W	12 15	4.7	12 53	3.9	6 35	6 27
16	M	12 05	4.9	12 31	4.0	6 20	6 18	16	T	1 04	4.5	1 44	3.8	7 27	7 25
17	T	12 53	4.6	1 22	3.8	7 10	7 05	17	F	1 55	4.3	2 35	3.8	8 30	8 40
18	W	1 42	4.4	2 14	3.6	8 11	8 12	18	S	2 47	4.2	3 28	3.9	9 32	9 50
19	T	2 35	4.2	3 10	3.6	9 19	9 26	19	S	3 42	4.3	4 22	4.1	10 24	10 47
20	F	3 32	4.1	4 09	3.6	10 20	10 29	20	M	4 38	4.3	5 14	4.4	11 10	11 38
21	S	4 31	4.2	5 08	3.8	11 11	11 23	21	T	5 32	4.5	6 02	4.8	11 53	...
22	S	5 29	4.4	6 01	4.1	11 56	...	22	W	6 21	4.7	6 45	5.2	12 26	12 35
23	M	6 18	4.6	6 45	4.5	12 11	12 38	23	T	7 07	4.8	7 26	5.6	1 13	1 17
24	T	7 02	4.8	7 25	4.8	12 57	1 18	24	F	7 51	4.9	8 06	5.8	2 00	2 00
25	W	7 42	5.0	8 01	5.1	1 42	1 57	25	S	8 35	5.0	8 47	6.0	2 48	2 45
26	T	8 20	5.0	8 36	5.4	2 26	2 36	26	S	9 21	4.9	9 32	6.0	3 35	3 30
27	F	8 58	5.0	9 12	5.5	3 09	3 14	27	M	10 12	4.8	10 22	5.9	4 22	4 17
28	S	9 38	4.9	9 51	5.6	3 51	3 53	28	T	11 08	4.7	11 19	5.7	5 11	5 06
29	S	10 24	4.7	10 36	5.5	4 34	4 32	29	W	...	...	12 09	4.6	6 03	6 01
30	M	11 15	4.6	11 28	5.4	5 20	5 15	30	T	12 21	5.5	1 10	4.6	7 02	7 04
31	T	...	...	12 14	4.4	6 10	6 04								

*Daylight Saving Time starts

Dates when Ht. of **Low** Water is below Mean Lower Low with Ht. of lowest given for each period and Date of lowest in ( ):

7th - 14th: -0.8' (10th - 11th)   6th - 11th: -0.4' (7th - 10th)
26th - 29th: -0.3' (27th - 28th)   25th - 28th: -0.4' (26th - 27th)

Average Rise and Fall 4.6 ft.

When a high tide exceeds av. ht., the *following* low tide will be lower than av.

2009 HIGH & LOW WATER
SANDY HOOK, NJ
40°28'N, 74°00.6'W

Daylight Saving Time   Daylight Saving Time

DAY OF MONTH	DAY OF WEEK	MAY HIGH a.m.	Ht.	MAY HIGH p.m.	Ht.	MAY LOW a.m.	MAY LOW p.m.	DAY OF MONTH	DAY OF WEEK	JUNE HIGH a.m.	Ht.	JUNE HIGH p.m.	Ht.	JUNE LOW a.m.	JUNE LOW p.m.
1	F	1 23	5.3	2 11	4.7	8 08	8 18	1	M	3 01	4.9	3 46	5.2	9 41	10 15
2	S	2 24	5.1	3 11	4.8	9 14	9 31	2	T	3 58	4.6	4 41	5.3	10 33	11 12
3	S	3 25	5.0	4 11	5.0	10 13	10 35	3	W	4 56	4.5	5 34	5.4	11 21	...
4	M	4 26	4.9	5 09	5.2	11 05	11 32	4	T	5 53	4.4	6 24	5.5	12 03	12 06
5	T	5 25	4.8	6 03	5.5	11 52	...	5	F	6 45	4.4	7 09	5.6	12 52	12 51
6	W	6 20	4.8	6 51	5.7	12 24	12 37	6	S	7 33	4.4	7 51	5.6	1 38	1 35
7	T	7 09	4.8	7 35	5.8	1 13	1 21	7	S	8 17	4.4	8 31	5.5	2 23	2 19
8	F	7 55	4.8	8 15	5.8	2 00	2 03	8	M	8 59	4.4	9 10	5.4	3 06	3 01
9	S	8 39	4.7	8 56	5.7	2 45	2 45	9	T	9 42	4.3	9 51	5.3	3 46	3 42
10	S	9 20	4.6	9 34	5.5	3 28	3 25	10	W	10 24	4.2	10 30	5.1	4 25	4 21
11	M	10 04	4.4	10 14	5.3	4 08	4 04	11	T	11 08	4.2	11 11	4.9	5 02	4 58
12	T	10 49	4.2	10 56	5.0	4 47	4 42	12	F	11 52	4.1	11 52	4.7	5 38	5 36
13	W	11 36	4.1	11 41	4.8	5 26	5 19	13	S	...	...	12 35	4.2	6 14	6 17
14	T	...	...	12 24	4.0	6 06	5 59	14	S	12 35	4.6	1 17	4.3	6 52	7 08
15	F	12 28	4.6	1 12	3.9	6 49	6 48	15	M	1 19	4.5	1 59	4.4	7 36	8 14
16	S	1 15	4.5	1 58	4.0	7 39	7 51	16	T	2 05	4.4	2 44	4.7	8 28	9 24
17	S	2 02	4.4	2 44	4.1	8 35	9 03	17	W	2 57	4.3	3 34	4.9	9 25	10 27
18	M	2 51	4.3	3 32	4.4	9 30	10 07	18	T	3 56	4.2	4 29	5.2	10 21	11 25
19	T	3 44	4.3	4 22	4.7	10 19	11 02	19	F	5 00	4.3	5 28	5.5	11 15	...
20	W	4 41	4.4	5 14	5.0	11 06	11 54	20	S	6 04	4.5	6 26	5.9	12 20	12 10
21	T	5 39	4.5	6 04	5.4	11 52	...	21	S	7 03	4.7	7 21	6.2	1 15	1 06
22	F	6 33	4.6	6 53	5.8	12 45	12 40	22	M	7 58	4.9	8 14	6.3	2 09	2 03
23	S	7 25	4.8	7 40	6.1	1 36	1 29	23	T	8 52	5.1	9 06	6.4	3 02	2 59
24	S	8 15	4.9	8 28	6.2	2 28	2 21	24	W	9 46	5.2	10 00	6.2	3 54	3 54
25	M	9 06	5.0	9 18	6.2	3 19	3 13	25	T	10 43	5.2	10 56	6.0	4 43	4 47
26	T	10 00	4.9	10 12	6.1	4 09	4 05	26	F	11 41	5.3	11 53	5.7	5 32	5 41
27	W	10 58	4.9	11 11	5.9	5 00	4 58	27	S	...	...	12 37	5.3	6 21	6 37
28	T	11 59	4.9	...	...	5 51	5 54	28	S	12 48	5.4	1 32	5.3	7 13	7 39
29	F	12 11	5.7	12 58	5.0	6 45	6 54	29	M	1 41	5.0	2 24	5.3	8 08	8 46
30	S	1 09	5.4	1 55	5.0	7 44	8 02	30	T	2 34	4.7	3 16	5.3	9 04	9 50
31	S	2 05	5.1	2 50	5.1	8 44	9 11								

Dates when Ht. of **Low** Water is below Mean Lower Low with Ht. of lowest given for each period and Date of lowest in ( ):

24th - 29th: -0.5' (25th - 27th)   22nd - 28th: -0.7' (24th - 25th)

Average Rise and Fall 4.6 ft.

When a high tide exceeds av. ht., the *following* low tide will be lower than av.

2009 HIGH & LOW WATER
SANDY HOOK, NJ
40°28'N, 74°00.6'W

Daylight Saving Time   **Daylight Saving Time**

DAY OF MONTH	DAY OF WEEK	JULY HIGH a.m.	Ht.	p.m.	Ht.	LOW a.m.	p.m.	DAY OF MONTH	DAY OF WEEK	AUGUST HIGH a.m.	Ht.	p.m.	Ht.	LOW a.m.	p.m.
1	W	3 28	4.4	4 09	5.2	9 59	10 48	1	S	4 58	3.9	5 27	5.0	11 12	...
2	T	4 26	4.2	5 03	5.2	10 50	11 40	2	S	5 56	4.0	6 19	5.1	12 03	12 01
3	F	5 25	4.1	5 56	5.2	11 38	...	3	M	6 48	4.2	7 05	5.2	12 48	12 47
4	S	6 21	4.1	6 44	5.3	12 28	12 24	4	T	7 33	4.4	7 47	5.3	1 32	1 32
5	S	7 11	4.2	7 29	5.4	1 15	1 10	5	W	8 13	4.5	8 25	5.4	2 13	2 16
6	M	7 56	4.3	8 10	5.4	1 59	1 55	6	T	8 51	4.7	9 00	5.3	2 52	2 58
7	T	8 38	4.4	8 49	5.4	2 42	2 39	7	F	9 26	4.7	9 34	5.2	3 29	3 37
8	W	9 18	4.4	9 26	5.3	3 22	3 20	8	S	10 00	4.8	10 08	5.1	4 02	4 14
9	T	9 58	4.4	10 04	5.2	4 00	4 00	9	S	10 35	4.8	10 44	4.9	4 34	4 51
10	F	10 36	4.4	10 39	5.0	4 34	4 37	10	M	11 09	4.9	11 23	4.7	5 03	5 28
11	S	11 15	4.4	11 16	4.8	5 07	5 13	11	T	11 49	5.0	...	...	5 34	6 09
12	S	11 53	4.5	11 56	4.7	5 38	5 50	12	W	12 09	4.5	12 36	5.0	6 09	7 02
13	M	...	...	12 32	4.6	6 10	6 34	13	T	1 03	4.3	1 29	5.1	6 55	8 16
14	T	12 40	4.5	1 15	4.7	6 46	7 31	14	F	2 02	4.2	2 28	5.2	8 01	9 35
15	W	1 28	4.3	2 01	4.9	7 33	8 45	15	S	3 07	4.2	3 33	5.3	9 23	10 43
16	T	2 23	4.2	2 54	5.1	8 35	9 57	16	S	4 17	4.3	4 44	5.5	10 35	11 42
17	F	3 24	4.2	3 55	5.3	9 44	11 01	17	M	5 27	4.6	5 51	5.8	11 39	...
18	S	4 33	4.2	5 01	5.5	10 49	11 59	18	T	6 30	5.0	6 51	6.1	12 37	12 38
19	S	5 42	4.4	6 06	5.8	11 51	...	19	W	7 26	5.4	7 44	6.3	1 30	1 35
20	M	6 45	4.7	7 05	6.1	12 56	12 50	20	T	8 18	5.7	8 34	6.3	2 20	2 30
21	T	7 42	5.1	8 00	6.3	1 51	1 49	21	F	9 07	5.9	9 22	6.1	3 08	3 22
22	W	8 36	5.3	8 52	6.4	2 43	2 45	22	S	9 56	6.0	10 11	5.8	3 53	4 12
23	T	9 29	5.5	9 43	6.3	3 33	3 39	23	S	10 46	5.9	11 01	5.4	4 36	5 00
24	F	10 22	5.6	10 35	6.0	4 20	4 31	24	M	11 36	5.7	11 52	5.0	5 18	5 49
25	S	11 16	5.6	11 28	5.6	5 06	5 22	25	T	...	...	12 27	5.5	6 01	6 40
26	S	...	...	12 10	5.6	5 51	6 15	26	W	12 45	4.6	1 17	5.2	6 46	7 38
27	M	12 21	5.2	1 02	5.5	6 38	7 11	27	T	1 38	4.3	2 08	5.0	7 40	8 43
28	T	1 14	4.8	1 52	5.3	7 28	8 14	28	F	2 32	4.0	3 01	4.8	8 44	9 48
29	W	2 06	4.5	2 43	5.2	8 23	9 19	29	S	3 29	3.9	3 57	4.7	9 49	10 44
30	T	3 00	4.2	3 35	5.0	9 23	10 20	30	S	4 29	3.9	4 55	4.8	10 46	11 33
31	F	3 57	4.0	4 30	4.9	10 19	11 14	31	M	5 28	4.1	5 49	4.9	11 36	...

Dates when Ht. of **Low** Water is below Mean Lower Low with Ht. of lowest given for each period and Date of lowest in ( ):

22st - 26th: -0.8' (23rd - 24th)   19th - 24th: - 0.7' (21st - 22nd)

Average Rise and Fall 4.6 ft.

When a high tide exceeds av. ht., the *following* low tide will be lower than av.

2009 HIGH & LOW WATER
SANDY HOOK, NJ
40°28'N, 74°00.6'W

Daylight Saving Time   Daylight Saving Time

DAY OF MONTH	DAY OF WEEK	SEPTEMBER HIGH a.m.	Ht.	HIGH p.m.	Ht.	LOW a.m.	LOW p.m.	DAY OF MONTH	DAY OF WEEK	OCTOBER HIGH a.m.	Ht.	HIGH p.m.	Ht.	LOW a.m.	LOW p.m.
1	T	6 20	4.3	6 37	5.1	12 17	12 22	1	T	6 29	4.7	6 43	5.0	12 20	12 39
2	W	7 04	4.6	7 18	5.2	12 59	1 07	2	F	7 08	5.0	7 23	5.1	12 59	1 22
3	T	7 43	4.8	7 56	5.3	1 38	1 50	3	S	7 44	5.3	8 00	5.1	1 37	2 06
4	F	8 19	5.0	8 31	5.3	2 16	2 32	4	S	8 17	5.5	8 36	5.1	2 14	2 48
5	S	8 52	5.1	9 05	5.2	2 52	3 13	5	M	8 51	5.6	9 14	5.0	2 51	3 30
6	S	9 24	5.2	9 39	5.1	3 27	3 52	6	T	9 26	5.7	9 55	4.8	3 28	4 12
7	M	9 56	5.3	10 16	4.9	3 59	4 30	7	W	10 07	5.6	10 43	4.6	4 07	4 55
8	T	10 32	5.3	10 58	4.7	4 32	5 09	8	T	10 55	5.5	11 40	4.4	4 47	5 43
9	W	11 17	5.3	11 51	4.5	5 06	5 53	9	F	11 55	5.4	...	...	5 33	6 39
10	T	...	...	12 08	5.3	5 45	6 47	10	S	12 44	4.3	12 59	5.3	6 30	7 49
11	F	12 49	4.3	1 08	5.2	6 35	8 00	11	S	1 48	4.3	2 04	5.2	7 45	9 03
12	S	1 53	4.2	2 13	5.2	7 48	9 19	12	M	2 52	4.4	3 09	5.2	9 07	10 07
13	S	2 59	4.3	3 21	5.3	9 14	10 27	13	T	3 56	4.7	4 14	5.2	10 17	11 03
14	M	4 07	4.5	4 30	5.4	10 27	11 24	14	W	4 59	5.0	5 16	5.3	11 18	11 53
15	T	5 14	4.8	5 36	5.6	11 29	...	15	T	5 56	5.4	6 13	5.4	...	12 13
16	W	6 15	5.3	6 34	5.9	12 17	12 26	16	F	6 48	5.8	7 04	5.5	12 39	1 05
17	T	7 08	5.7	7 25	6.0	1 06	1 21	17	S	7 35	6.1	7 51	5.4	1 25	1 55
18	F	7 57	6.0	8 13	6.0	1 53	2 13	18	S	8 19	6.2	8 36	5.3	2 09	2 44
19	S	8 43	6.2	8 59	5.8	2 39	3 03	19	M	9 01	6.1	9 21	5.0	2 52	3 30
20	S	9 28	6.2	9 45	5.5	3 22	3 51	20	T	9 43	5.9	10 07	4.7	3 34	4 14
21	M	10 14	6.0	10 33	5.1	4 04	4 37	21	W	10 27	5.5	10 55	4.4	4 14	4 56
22	T	11 01	5.7	11 23	4.7	4 45	5 22	22	T	11 14	5.2	11 47	4.2	4 54	5 39
23	W	11 50	5.4	...	...	5 25	6 09	23	F	...	...	12 04	4.9	5 34	6 25
24	T	12 16	4.4	12 41	5.1	6 07	7 01	24	S	12 41	4.0	12 56	4.6	6 20	7 18
25	F	1 10	4.1	1 32	4.8	6 57	8 02	25	S	1 34	3.9	1 47	4.5	7 17	8 19
26	S	2 04	4.0	2 25	4.6	8 02	9 08	26	M	2 26	3.9	2 39	4.4	8 28	9 19
27	S	3 00	3.9	3 20	4.6	9 13	10 07	27	T	3 18	3.9	3 31	4.4	9 36	10 11
28	M	3 56	4.0	4 16	4.6	10 15	10 56	28	W	4 10	4.1	4 23	4.4	10 32	10 55
29	T	4 53	4.1	5 10	4.7	11 07	11 40	29	T	5 00	4.4	5 15	4.5	11 22	11 36
30	W	5 44	4.4	6 00	4.9	11 54	...	30	F	5 46	4.7	6 03	4.6	...	12 08
								31	S	6 29	5.1	6 48	4.7	12 16	12 54

Dates when Ht. of **Low** Water is below Mean Lower Low with Ht. of lowest given for each period and Date of lowest in ( ):

17th - 21st: -0.6' (18th - 19th)   14th - 19th: -0.5' (17th)

Average Rise and Fall 4.6 ft.

When a high tide exceeds av. ht., the *following* low tide will be lower than av.

2009 HIGH & LOW WATER
SANDY HOOK, NJ
40°28'N, 74°00.6'W

Standard Time starts Nov. 1 at 2 a.m.   Standard Time

DAY OF MONTH	DAY OF WEEK	NOVEMBER HIGH a.m.	Ht.	HIGH p.m.	Ht.	LOW a.m.	LOW p.m.	DAY OF MONTH	DAY OF WEEK	DECEMBER HIGH a.m.	Ht.	HIGH p.m.	Ht.	LOW a.m.	LOW p.m.
1	S	*6 08	5.4	*6 30	4.8	12 56	*12 39	1	T	6 19	5.6	6 50	4.6	12 06	1 04
2	M	6 45	5.7	7 11	4.8	12 37	1 25	2	W	7 04	5.8	7 39	4.6	12 56	1 54
3	T	7 24	5.8	7 54	4.8	1 20	2 11	3	T	7 52	5.9	8 29	4.6	1 46	2 43
4	W	8 05	5.9	8 41	4.7	2 04	2 57	4	F	8 42	5.8	9 24	4.6	2 37	3 32
5	T	8 52	5.8	9 34	4.5	2 49	3 44	5	S	9 37	5.7	10 23	4.6	3 29	4 22
6	F	9 45	5.6	10 34	4.4	3 37	4 34	6	S	10 36	5.5	11 23	4.6	4 22	5 13
7	S	10 46	5.4	11 37	4.4	4 28	5 30	7	M	11 36	5.2	...	...	5 19	6 09
8	S	11 50	5.3	...	...	5 28	6 33	8	T	12 22	4.7	12 34	5.0	6 24	7 10
9	M	12 40	4.5	12 53	5.1	6 39	7 40	9	W	1 20	4.8	1 32	4.7	7 35	8 10
10	T	1 39	4.6	1 53	5.0	7 55	8 42	10	T	2 16	4.9	2 28	4.4	8 44	9 06
11	W	2 39	4.9	2 53	4.9	9 04	9 37	11	F	3 12	5.0	3 28	4.3	9 45	9 58
12	T	3 38	5.1	3 53	4.8	10 03	10 26	12	S	4 09	5.1	4 27	4.2	10 40	10 46
13	F	4 34	5.4	4 51	4.8	10 58	11 12	13	S	5 02	5.3	5 23	4.2	11 31	11 33
14	S	5 26	5.7	5 43	4.8	11 49	11 57	14	M	5 51	5.3	6 13	4.2	...	12 19
15	S	6 12	5.8	6 31	4.8	...	12 37	15	T	6 35	5.4	6 59	4.3	12 18	1 05
16	M	6 55	5.8	7 16	4.7	12 42	1 24	16	W	7 17	5.3	7 42	4.2	1 03	1 49
17	T	7 37	5.7	8 00	4.6	1 25	2 09	17	T	7 57	5.2	8 24	4.2	1 47	2 30
18	W	8 18	5.5	8 44	4.4	2 08	2 52	18	F	8 37	5.1	9 06	4.1	2 28	3 10
19	T	8 59	5.3	9 30	4.2	2 49	3 33	19	S	9 17	4.9	9 49	3.9	3 07	3 47
20	F	9 43	5.0	10 18	4.0	3 29	4 13	20	S	9 58	4.7	10 33	3.9	3 45	4 23
21	S	10 30	4.7	11 09	3.8	4 08	4 53	21	M	10 39	4.5	11 17	3.8	4 22	4 58
22	S	11 18	4.5	11 58	3.8	4 49	5 37	22	T	11 21	4.3	11 59	3.9	5 01	5 34
23	M	...	...	12 05	4.3	5 35	6 25	23	W	...	...	12 04	4.1	5 47	6 15
24	T	12 46	3.8	12 52	4.2	6 35	7 19	24	T	12 41	4.0	12 49	3.9	6 47	7 04
25	W	1 32	3.9	1 40	4.1	7 45	8 14	25	F	1 24	4.1	1 38	3.8	7 59	8 01
26	T	2 19	4.1	2 30	4.1	8 50	9 03	26	S	2 12	4.3	2 34	3.7	9 05	8 59
27	F	3 08	4.3	3 24	4.1	9 45	9 49	27	S	3 06	4.5	3 37	3.8	10 04	9 55
28	S	3 57	4.6	4 19	4.1	10 36	10 34	28	M	4 04	4.8	4 40	3.9	10 58	10 49
29	S	4 46	5.0	5 12	4.3	11 25	11 19	29	T	5 03	5.2	5 39	4.2	11 51	11 43
30	M	5 33	5.3	6 02	4.4	...	12 14	30	W	5 58	5.5	6 33	4.4	...	12 44
								31	T	6 50	5.8	7 24	4.7	12 38	1 36

*Standard Time starts

Dates when Ht. of **Low** Water is below Mean Lower Low with Ht. of lowest given for each period and Date of lowest in ( ):

3rd - 4th: -0.2'
12th - 17th: -0.3' (13th - 16th)

1st - 8th: -0.6' (3rd - 4th)
14th - 17th: -0.2' (29th - 31st)
29th - 31st: -0.7' (31st)

Average Rise and Fall 4.6 ft.

When a high tide exceeds av. ht., the *following* low tide will be lower than av.

2009 CURRENT TABLE
DELAWARE BAY ENTRANCE
38°46.85'N, 75°02.58'W

Standard Time   Standard Time

JANUARY   FEBRUARY

DAY OF MONTH	DAY OF WEEK	NORTHWEST Flood Starts a.m.	**p.m.**	Kts.	SOUTHEAST Ebb Starts a.m.	**p.m.**	Kts.	DAY OF MONTH	DAY OF WEEK	NORTHWEST Flood Starts a.m.	**p.m.**	Kts.	SOUTHEAST Ebb Starts a.m.	**p.m.**	Kts.
1	T	6 28	**7 10**	a1.3	12 36	**12 47**	p1.3	1	S	7 55	**7 57**	p1.5	1 32	**1 44**	1.2
2	F	7 17	**7 48**	p1.3	1 17	**1 27**	p1.2	2	M	8 55	**8 48**	p1.5	2 22	**2 37**	a1.2
3	S	8 11	**8 30**	p1.3	2 02	**2 11**	p1.2	3	T	10 01	**9 47**	p1.6	3 20	**3 39**	a1.3
4	S	9 13	**9 19**	p1.4	2 53	**3 04**	p1.2	4	W	11 10	**10 50**	p1.6	4 24	**4 46**	a1.3
5	M	10 20	**10 12**	p1.5	3 49	**4 04**	a1.2	5	T	...	**12 17**	1.2	5 29	**5 54**	a1.4
6	T	11 27	**11 10**	p1.6	4 47	**5 07**	a1.3	6	F	12 01	**1 22**	a1.7	6 36	**7 01**	a1.6
7	W	...	**12 32**	1.2	5 47	**6 10**	a1.5	7	S	1 04	**2 22**	a1.8	7 41	**8 05**	a1.7
8	T	12 11	**1 37**	a1.8	6 49	**7 15**	a1.6	8	S	2 08	**3 17**	a1.8	8 40	**9 03**	a1.8
9	F	1 15	**2 38**	a1.9	7 51	**8 18**	a1.8	9	M	3 08	**4 08**	a1.9	9 35	**9 57**	a1.8
10	S	2 17	**3 33**	a2.0	8 51	**9 17**	a1.9	10	T	4 02	**4 54**	a1.9	10 25	**10 48**	a1.9
11	S	3 16	**4 27**	a2.0	9 48	**10 12**	a1.9	11	W	4 56	**5 39**	a1.9	11 12	**11 37**	a1.8
12	M	4 13	**5 18**	a2.0	10 41	**11 07**	a1.9	12	T	5 48	**6 21**	a1.7	11 57	**...**	1.6
13	T	5 09	**6 08**	a1.9	11 34	**...**	1.9	13	F	6 39	**7 00**	1.5	12 23	**12 40**	a1.5
14	W	6 05	**6 56**	a1.8	12 01	**12 23**	p1.7	14	S	7 31	**7 38**	p1.4	1 08	**1 22**	a1.3
15	T	7 00	**7 41**	a1.6	12 52	**1 11**	p1.5	15	S	8 25	**8 15**	p1.3	1 54	**2 05**	a1.2
16	F	7 58	**8 26**	1.4	1 44	**1 59**	1.3	16	M	9 24	**8 57**	p1.2	2 43	**2 52**	a1.1
17	S	8 58	**9 11**	p1.3	2 37	**2 48**	a1.2	17	T	10 27	**9 45**	p1.1	3 37	**3 46**	a1.0
18	S	10 04	**9 56**	p1.2	3 32	**3 39**	a1.1	18	W	11 28	**10 39**	p1.1	4 34	**4 43**	a0.9
19	M	11 08	**10 41**	p1.2	4 27	**4 32**	a1.1	19	T	-A-	**12 25**	0.7	5 31	**5 40**	a1.0
20	T	11 59	**11 28**	p1.2	5 20	**5 25**	a1.0	20	F	...	**1 17**	0.8	6 27	**6 37**	a1.0
21	W	...	**1 03**	0.8	6 11	**6 19**	a1.1	21	S	12 30	**2 01**	a1.2	7 19	**7 32**	a1.2
22	T	12 16	**1 54**	a1.3	7 02	**7 12**	a1.2	22	S	1 24	**2 39**	a1.3	8 07	**8 21**	a1.3
23	F	1 05	**2 38**	a1.3	7 51	**8 04**	a1.2	23	M	2 14	**3 14**	a1.4	8 49	**9 06**	a1.4
24	S	1 54	**3 17**	a1.4	8 36	**8 51**	a1.3	24	T	3 01	**3 46**	a1.5	9 29	**9 46**	a1.4
25	S	2 39	**3 52**	a1.4	9 18	**9 34**	a1.4	25	W	3 45	**4 18**	a1.5	10 06	**10 24**	a1.5
26	M	3 22	**4 25**	a1.5	9 57	**10 15**	a1.4	26	T	4 29	**4 50**	1.5	10 41	**11 02**	a1.5
27	T	4 03	**4 57**	a1.5	10 33	**10 53**	a1.5	27	F	5 13	**5 25**	p1.6	11 18	**11 41**	1.4
28	W	4 45	**5 28**	a1.5	11 09	**11 31**	a1.5	28	S	6 00	**6 02**	p1.6	11 56	**...**	1.4
29	T	5 28	**6 00**	1.4	11 44	**...**	1.4								
30	F	6 13	**6 35**	1.4	12 09	**12 20**	p1.4								
31	S	7 01	**7 13**	p1.5	12 49	**12 59**	p1.3								

A also at 11:34 p.m. 1.1

The Kts. (knots) columns show the **maximum** predicted velocities of the stronger one of the Flood Currents and the stronger one of the Ebb Currents for each day. The letter "a" means the velocity shown should occur **after** the **a.m.** Current Change. The letter "p" means the velocity shown should occur **after** the **p.m.** Current Change (even if next morning). No "a" or "p" means a.m. and p.m. velocities are the same for that day.

Av. Max. Vel.: Fl. 1.8 Kts., Ebb 1.9 Kts.

Max. Flood & Max. Ebb 3 hrs. 5 min. after Flood Starts & Ebb Starts, ±15 min.

See pp. 22-29 for Current Change at other points.

Daylight Time starts March 8 at 2 a.m.   **Daylight Saving Time**

		MARCH					APRIL		
		CURRENT TURNS TO					CURRENT TURNS TO		
		NORTHWEST Flood Starts	SOUTHEAST Ebb Starts				NORTHWEST Flood Starts	SOUTHEAST Ebb Starts	
DAY OF MONTH	DAY OF WEEK	a.m. **p.m.** Kts.	a.m. **p.m.** Kts.		DAY OF MONTH	DAY OF WEEK	a.m. **p.m.** Kts.	a.m. **p.m.** Kts.	
1	S	6 49 **6 45** p1.6	12 22 **12 38** a1.4		1	W	9 31 **9 17** p1.5	2 45 **3 13** a1.4	
2	M	7 41 **7 32** p1.6	1 08 **1 25** a1.4		2	T	10 37 **10 26** p1.4	3 49 **4 20** a1.3	
3	T	8 41 **8 28** p1.5	2 00 **2 21** a1.3		3	F	11 45 **11 39** p1.4	5 00 **5 32** a1.3	
4	W	9 48 **9 31** p1.5	3 01 **3 26** a1.3		4	S	... **12 49** 1.3	6 11 **6 41** a1.4	
5	T	10 57 **10 41** p1.5	4 09 **4 36** a1.3		5	S	12 48 **1 48** a1.5	7 16 **7 44** a1.5	
6	F	11 59 **11 51** p1.6	5 19 **5 46** a1.4		6	M	1 54 **2 41** a1.6	8 15 **8 42** a1.6	
7	S	... **1 07** 1.3	6 27 **6 53** a1.5		7	T	2 54 **3 29** 1.6	9 08 **9 32** 1.6	
8	S	12 59 ***3 04** a1.7	*8 31 ***8 54** a1.7		8	W	3 48 **4 11** 1.7	9 55 **10 18** 1.6	
9	M	3 02 **3 56** a1.8	9 27 **9 49** a1.8		9	T	4 40 **4 52** p1.7	10 38 **10 59** 1.6	
10	T	3 58 **4 42** a1.8	10 17 **10 39** a1.8		10	F	5 26 **5 28** p1.7	11 19 **11 39** p1.6	
11	W	4 51 **5 25** a1.8	11 03 **11 25** a1.8		11	S	6 10 **6 02** p1.6	11 59 ... 1.3	
12	T	5 41 **6 05** 1.7	11 47 ... 1.7		12	S	6 54 **6 35** p1.5	12 18 **12 37** a1.5	
13	F	6 29 **6 42** p1.7	12 08 **12 28** a1.6		13	M	7 36 **7 08** p1.4	12 57 **1 16** a1.3	
14	S	7 17 **7 17** p1.5	12 50 **1 08** a1.5		14	T	8 17 **7 45** p1.2	1 37 **1 56** a1.2	
15	S	8 02 **7 50** p1.4	1 31 **1 47** a1.3		15	W	9 00 **8 27** p1.1	2 20 **2 39** a1.1	
16	M	8 50 **8 25** p1.2	2 13 **2 28** a1.2		16	T	9 49 **9 16** p1.0	3 08 **3 28** a1.0	
17	T	9 40 **9 06** p1.1	2 58 **3 12** a1.0		17	F	10 42 **10 14** p1.0	4 03 **4 26** a0.9	
18	W	10 37 **9 55** p1.0	3 49 **4 04** a0.9		18	S	11 36 **11 18** p1.0	5 02 **5 26** a0.9	
19	T	11 38 **10 53** p1.0	4 48 **5 03** a0.9		19	S	... **12 24** 0.9	5 58 **6 23** a1.0	
20	F	... **12 34** 0.7	5 48 **6 03** a0.9		20	M	12 21 **1 08** a1.1	6 51 **7 17** 1.1	
21	S	12 01 **1 24** a1.1	6 45 **7 01** a1.0		21	T	1 21 **1 51** 1.2	7 42 **8 07** p1.3	
22	S	12 56 **2 09** a1.1	7 39 **7 57** a1.1		22	W	2 19 **2 32** p1.4	8 31 **8 54** p1.4	
23	M	1 53 **2 49** a1.2	8 29 **8 47** 1.2		23	T	3 14 **3 14** p1.6	9 17 **9 38** p1.6	
24	T	2 48 **3 25** 1.3	9 13 **9 31** 1.3		24	F	4 05 **3 57** p1.8	10 02 **10 22** p1.7	
25	W	3 38 **4 00** 1.4	9 55 **10 13** p1.5		25	S	4 54 **4 40** p1.9	10 47 **11 07** p1.8	
26	T	4 26 **4 36** p1.6	10 34 **10 53** p1.6		26	S	5 43 **5 27** p1.9	11 32 **11 53** p1.7	
27	F	5 12 **5 13** p1.7	11 13 **11 33** p1.6		27	M	6 34 **6 17** p1.9	... **12 20** 1.4	
28	S	5 59 **5 52** p1.8	11 54 ... 1.4		28	T	7 27 **7 09** p1.8	12 43 **1 11** a1.7	
29	S	6 47 **6 36** p1.8	12 15 **12 37** a1.6		29	W	8 21 **8 08** p1.6	1 38 **2 07** a1.6	
30	M	7 38 **7 23** p1.7	1 00 **1 23** a1.6		30	T	9 20 **9 11** p1.5	2 36 **3 08** a1.5	
31	T	8 31 **8 17** p1.6	1 49 **2 14** a1.5						

*Daylight Saving Time starts

The Kts. (knots) columns show the **maximum** predicted velocities of the stronger one of the Flood Currents and the stronger one of the Ebb Currents for each day. The letter "a" means the velocity shown should occur **after** the **a.m.** Current Change. The letter "p" means the velocity shown should occur **after** the **p.m.** Current Change (even if next morning). No "a" or "p" means a.m. and p.m. velocities are the same for that day.

Av. Max. Vel.: Fl. 1.8 Kts., Ebb 1.9 Kts.

Max. Flood & Max. Ebb 3 hrs. 5 min. after Flood Starts & Ebb Starts, ±15 min.

See pp. 22-29 for Current Change at other points.

2009 CURRENT TABLE
DELAWARE BAY ENTRANCE
38°46.85'N, 75°02.58'W

Daylight Saving Time   Daylight Saving Time

		MAY							JUNE				
DAY OF MONTH	DAY OF WEEK	CURRENT TURNS TO					DAY OF MONTH	DAY OF WEEK	CURRENT TURNS TO				
		NORTHWEST Flood Starts			SOUTHEAST Ebb Starts				NORTHWEST Flood Starts			SOUTHEAST Ebb Starts	
		a.m.	p.m.	Kts.	a.m.	p.m.	Kts.			a.m.	p.m.	Kts.	a.m.	p.m.	Kts.
1	F	10 24	10 21	p1.4	3 40	4 16	a1.4	1	M	11 58	...	1.4	5 31	6 13	1.3
2	S	11 28	11 34	p1.4	4 49	5 26	a1.4	2	T	12 30	12 48	p1.5	6 28	7 09	1.3
3	S	...	12 27	1.3	5 56	6 32	a1.4	3	W	1 33	1 36	p1.5	7 23	8 00	p1.4
4	M	12 42	1 21	p1.5	6 56	7 31	1.4	4	T	2 31	2 21	p1.5	8 14	8 47	p1.4
5	T	1 46	2 11	p1.6	7 52	8 24	1.5	5	F	3 23	3 03	p1.6	9 02	9 29	p1.5
6	W	2 44	2 57	p1.6	8 43	9 11	p1.6	6	S	4 10	3 43	p1.5	9 46	10 09	p1.5
7	T	3 37	3 38	p1.7	9 30	9 53	p1.6	7	S	4 52	4 20	p1.5	10 27	10 48	p1.4
8	F	4 25	4 17	p1.7	10 12	10 33	p1.6	8	M	5 33	4 57	p1.5	11 08	11 27	p1.4
9	S	5 11	4 52	p1.6	10 52	11 11	p1.5	9	T	6 12	5 34	p1.5	11 47	...	1.1
10	S	5 52	5 26	p1.6	11 31	11 49	p1.4	10	W	6 47	6 10	p1.4	12 06	12 27	a1.4
11	M	6 32	6 00	p1.5	...	12 10	1.1	11	T	7 21	6 50	p1.3	12 44	1 07	a1.3
12	T	7 10	6 36	p1.4	12 28	12 50	a1.3	12	F	7 55	7 32	p1.2	1 23	1 48	a1.3
13	W	7 48	7 13	p1.3	1 08	1 30	a1.2	13	S	8 30	8 19	p1.1	2 02	2 30	a1.2
14	T	8 27	7 57	p1.2	1 49	2 12	a1.1	14	S	9 08	9 11	p1.1	2 43	3 16	a1.1
15	F	9 08	8 44	p1.1	2 33	2 58	a1.1	15	M	9 48	10 11	a1.1	3 28	4 07	a1.1
16	S	9 52	9 40	p1.0	3 21	3 51	a1.0	16	T	10 32	11 16	a1.2	4 18	5 00	1.1
17	S	10 39	10 42	p1.0	4 14	4 47	a1.0	17	W	11 20	...	1.3	5 12	5 53	p1.2
18	M	11 25	11 47	1.0	5 08	5 42	1.0	18	T	12 20	12 11	p1.5	6 09	6 47	p1.4
19	T	...	12 11	1.2	6 01	6 34	1.1	19	F	1 23	1 05	p1.6	7 07	7 42	p1.5
20	W	12 50	12 57	p1.4	6 53	7 25	p1.3	20	S	2 26	2 01	p1.8	8 06	8 39	p1.7
21	T	1 51	1 44	p1.6	7 47	8 16	p1.5	21	S	3 24	2 59	p1.9	9 05	9 35	p1.8
22	F	2 49	2 33	p1.7	8 40	9 06	p1.7	22	M	4 20	3 58	p2.0	10 02	10 31	p1.9
23	S	3 44	3 23	p1.9	9 31	9 56	p1.8	23	T	5 13	4 55	p2.0	10 58	11 26	p1.9
24	S	4 37	4 15	p1.9	10 22	10 46	p1.8	24	W	6 07	5 52	p2.0	11 53	...	1.5
25	M	5 29	5 08	p2.0	11 13	11 38	p1.8	25	T	6 59	6 50	p1.9	12 21	12 49	a1.9
26	T	6 21	6 02	p1.9	...	12 06	1.4	26	F	7 51	7 50	p1.7	1 16	1 45	a1.8
27	W	7 15	7 00	p1.8	12 32	1 01	a1.8	27	S	8 43	8 51	1.5	2 09	2 43	a1.7
28	T	8 09	8 00	p1.7	1 28	1 59	a1.7	28	S	9 35	9 56	a1.5	3 03	3 43	a1.5
29	F	9 06	9 03	p1.5	2 26	2 59	a1.6	29	M	10 28	11 05	a1.4	3 59	4 45	a1.4
30	S	10 04	10 12	1.4	3 26	4 04	a1.5	30	T	11 20	...	1.4	4 57	5 44	1.2
31	S	11 02	11 23	a1.4	4 29	5 11	a1.4								

The Kts. (knots) columns show the **maximum** predicted velocities of the stronger one of the Flood Currents and the stronger one of the Ebb Currents for each day. The letter "a" means the velocity shown should occur **after** the **a.m.** Current Change. The letter "p" means the velocity shown should occur **after** the **p.m.** Current Change (even if next morning). No "a" or "p" means a.m. and p.m. velocities are the same for that day.
Av. Max. Vel.: Fl. 1.8 Kts., Ebb 1.9 Kts.
Max. Flood & Max. Ebb 3 hrs. 5 min. after Flood Starts & Ebb Starts, ±15 min.

See pp. 22-29 for Current Change at other points.

2009 CURRENT TABLE
DELAWARE BAY ENTRANCE
38°46.85'N, 75°02.58'W

Daylight Saving Time						Daylight Saving Time					

		JULY							AUGUST				

DAY OF MONTH	DAY OF WEEK	NORTHWEST Flood Starts			SOUTHEAST Ebb Starts			DAY OF MONTH	DAY OF WEEK	NORTHWEST Flood Starts			SOUTHEAST Ebb Starts		
		a.m.	**p.m.**	Kts.	a.m.	**p.m.**	Kts.			a.m.	**p.m.**	Kts.	a.m.	**p.m.**	Kts.
1	W	12 12	**12 09**	p1.4	5 53	**6 39**	p1.2	1	S	1 48	**1 05**	p1.2	7 03	**7 49**	p1.1
2	T	1 14	**12 57**	p1.4	6 48	**7 31**	p1.3	2	S	2 40	**1 55**	p1.3	7 58	**8 39**	p1.2
3	F	2 12	**1 43**	p1.4	7 40	**8 20**	p1.3	3	M	3 26	**2 44**	p1.4	8 50	**9 24**	p1.3
4	S	3 05	**2 29**	p1.4	8 31	**9 05**	p1.3	4	T	4 06	**3 30**	p1.4	9 37	**10 05**	p1.4
5	S	3 51	**3 12**	p1.4	9 19	**9 48**	p1.4	5	W	4 41	**4 12**	p1.5	10 20	**10 43**	p1.4
6	M	4 33	**3 54**	p1.5	10 02	**10 28**	p1.4	6	T	5 14	**4 52**	p1.5	11 01	**11 20**	p1.5
7	T	5 11	**4 33**	p1.5	10 44	**11 07**	p1.4	7	F	5 44	**5 32**	p1.5	11 40	**11 55**	p1.5
8	W	5 46	**5 12**	p1.5	11 26	**11 44**	p1.4	8	S	6 13	**6 13**	p1.4	...	**12 17**	1.2
9	T	6 20	**5 51**	p1.4	...	**12 05**	1.1	9	S	6 43	**6 57**	1.3	12 29	**12 52**	a1.4
10	F	6 50	**6 30**	p1.4	12 21	**12 44**	a1.4	10	M	7 14	**7 40**	a1.4	1 03	**1 29**	a1.3
11	S	7 20	**7 12**	p1.3	12 57	**1 21**	a1.4	11	T	7 49	**8 29**	a1.4	1 39	**2 08**	a1.3
12	S	7 52	**7 58**	1.2	1 32	**2 00**	a1.3	12	W	8 29	**9 23**	a1.5	2 19	**2 53**	1.2
13	M	8 27	**8 48**	a1.2	2 09	**2 41**	a1.2	13	T	9 16	**10 26**	a1.5	3 07	**3 46**	p1.2
14	T	9 05	**9 45**	a1.3	2 49	**3 27**	a1.2	14	F	10 11	**11 33**	a1.5	4 04	**4 47**	p1.2
15	W	9 49	**10 48**	a1.4	3 37	**4 19**	p1.2	15	S	11 13	**...**	1.5	5 09	**5 53**	p1.3
16	T	10 40	**11 54**	a1.5	4 32	**5 16**	p1.2	16	S	12 40	**12 20**	p1.6	6 17	**7 00**	p1.4
17	F	11 37	**...**	1.5	5 33	**6 15**	p1.3	17	M	1 46	**1 29**	p1.7	7 25	**8 07**	p1.6
18	S	1 00	**12 37**	p1.7	6 37	**7 17**	p1.5	18	T	2 48	**2 35**	p1.8	8 31	**9 09**	p1.8
19	S	2 04	**1 40**	p1.8	7 41	**8 20**	p1.6	19	W	3 44	**3 38**	p1.9	9 32	**10 05**	p1.9
20	M	3 06	**2 44**	p1.9	8 45	**9 21**	p1.8	20	T	4 36	**4 36**	p2.0	10 28	**10 57**	p1.9
21	T	4 02	**3 46**	p2.0	9 46	**10 19**	p1.9	21	F	5 24	**5 31**	p2.0	11 20	**11 47**	p1.9
22	W	4 57	**4 45**	p2.0	10 42	**11 13**	p1.9	22	S	6 10	**6 25**	1.8	...	**12 11**	1.7
23	T	5 48	**5 42**	p2.0	11 38	**...**	1.7	23	S	6 55	**7 19**	a1.8	12 34	**12 59**	a1.7
24	F	6 38	**6 39**	p1.9	12 06	**12 32**	a1.9	24	M	7 37	**8 12**	a1.7	1 19	**1 47**	1.5
25	S	7 26	**7 36**	1.7	12 57	**1 25**	a1.8	25	T	8 18	**9 07**	a1.5	2 04	**2 34**	1.3
26	S	8 12	**8 33**	a1.6	1 46	**2 18**	a1.6	26	W	8 59	**10 07**	a1.4	2 49	**3 24**	p1.2
27	M	8 59	**9 34**	a1.5	2 35	**3 11**	a1.4	27	T	9 43	**11 11**	a1.2	3 37	**4 19**	p1.0
28	T	9 45	**10 39**	a1.4	3 25	**4 07**	1.2	28	F	10 33	**...**	1.1	4 31	**5 18**	p1.0
29	W	10 34	**11 46**	a1.3	4 18	**5 05**	1.1	29	S	12 15	**-A-**	0.7	5 28	**6 17**	p1.0
30	T	11 24	**...**	1.3	5 13	**6 02**	p1.1	30	S	1 13	**12 25**	p1.1	6 26	**7 13**	p1.0
31	F	12 49	**12 14**	p1.2	6 09	**6 57**	p1.1	31	M	2 05	**1 21**	p1.2	7 23	**8 06**	p1.1

A also at 11:28 a.m. 1.1

The Kts. (knots) columns show the **maximum** predicted velocities of the stronger one of the Flood Currents and the stronger one of the Ebb Currents for each day. The letter "a" means the velocity shown should occur **after** the **a.m.** Current Change. The letter "p" means the velocity shown should occur **after** the **p.m.** Current Change (even if next morning). No "a" or "p" means a.m. and p.m. velocities are the same for that day.
Av. Max. Vel.: Fl. 1.8 Kts., Ebb 1.9 Kts.
Max. Flood & Max. Ebb 3 hrs. 5 min. after Flood Starts & Ebb Starts, ±15 min.

See pp. 22-29 for Current Change at other points.

Daylight Saving Time   Daylight Saving Time

SEPTEMBER | OCTOBER

D A Y O F M O N T H	D A Y O F W E E K	CURRENT TURNS TO						D A Y O F M O N T H	D A Y O F W E E K	CURRENT TURNS TO					
		NORTHWEST Flood Starts			SOUTHEAST Ebb Starts					NORTHWEST Flood Starts			SOUTHEAST Ebb Starts		
		a.m.	p.m.	Kts.	a.m.	p.m.	Kts.			a.m.	p.m.	Kts.	a.m.	p.m.	Kts.
1	T	2 49	2 14	p1.3	8 18	8 53	p1.2	1	T	2 34	2 36	p1.3	8 32	8 57	p1.3
2	W	3 28	3 04	p1.4	9 08	9 35	p1.3	2	F	3 09	3 25	p1.3	9 17	9 39	1.3
3	T	4 01	3 50	p1.4	9 52	10 14	p1.4	3	S	3 43	4 11	1.4	9 58	10 18	1.4
4	F	4 32	4 32	p1.4	10 32	10 51	p1.4	4	S	4 17	4 56	a1.5	10 37	10 56	a1.5
5	S	5 02	5 14	1.4	11 10	11 26	1.4	5	M	4 52	5 40	a1.6	11 14	11 35	a1.6
6	S	5 32	5 57	a1.5	11 46	...	1.4	6	T	5 29	6 24	a1.7	11 53	...	1.6
7	M	6 04	6 40	a1.5	12 01	12 22	1.4	7	W	6 10	7 11	a1.7	12 15	12 35	p1.5
8	T	6 39	7 25	a1.6	12 37	12 59	p1.4	8	T	6 55	8 01	a1.7	12 58	1 20	p1.5
9	W	7 19	8 14	a1.6	1 16	1 41	1.3	9	F	7 46	8 58	a1.6	1 45	2 12	p1.4
10	T	8 02	9 08	a1.6	1 58	2 28	p1.3	10	S	8 41	9 58	a1.5	2 39	3 11	p1.3
11	F	8 53	10 10	a1.5	2 48	3 23	p1.2	11	S	9 47	11 04	a1.4	3 42	4 18	p1.3
12	S	9 54	11 18	a1.5	3 48	4 29	p1.2	12	M	10 58	...	1.4	4 52	5 29	p1.3
13	S	11 02	...	1.5	4 57	5 39	p1.3	13	T	12 09	12 10	p1.5	6 02	6 36	p1.4
14	M	12 25	12 14	p1.5	6 08	6 48	p1.4	14	W	1 09	1 19	p1.5	7 08	7 38	p1.5
15	T	1 29	1 23	p1.6	7 15	7 54	p1.6	15	T	2 04	2 23	p1.6	8 08	8 35	p1.6
16	W	2 28	2 29	p1.8	8 20	8 54	p1.7	16	F	2 56	3 21	1.7	9 03	9 27	1.6
17	T	3 22	3 30	p1.8	9 18	9 48	p1.8	17	S	3 42	4 15	a1.8	9 52	10 15	a1.7
18	F	4 11	4 26	p1.9	10 11	10 37	p1.8	18	S	4 26	5 06	a1.8	10 37	10 59	a1.7
19	S	4 57	5 19	1.8	11 00	11 23	a1.8	19	M	5 07	5 54	a1.8	11 20	11 42	a1.7
20	S	5 39	6 10	a1.8	11 46	...	1.7	20	T	5 45	6 40	a1.7	...	12 01	1.6
21	M	6 20	7 00	a1.8	12 08	12 30	1.6	21	W	6 21	7 26	a1.6	12 23	12 42	p1.4
22	T	6 59	7 49	a1.7	12 51	1 14	p1.5	22	T	6 58	8 09	a1.4	1 04	1 24	p1.3
23	W	7 37	8 38	a1.5	1 33	1 58	p1.3	23	F	7 36	8 54	a1.3	1 45	2 08	p1.1
24	T	8 14	9 30	a1.3	2 16	2 43	p1.1	24	S	8 17	9 41	a1.2	2 28	2 56	p1.0
25	F	8 56	10 28	a1.2	3 00	3 35	p1.0	25	S	9 04	10 32	a1.1	3 16	3 48	p1.0
26	S	9 44	11 29	a1.1	3 51	4 33	p0.9	26	M	10 00	11 23	a1.0	4 11	4 45	p0.9
27	S	10 42	...	1.0	4 49	5 33	p0.9	27	T	11 03	...	1.0	5 11	5 41	p1.0
28	M	12 25	-A-	0.7	5 49	6 30	p1.0	28	W	12 09	12 06	p1.0	6 08	6 33	p1.0
29	T	1 13	12 44	p1.1	6 47	7 23	p1.1	29	T	12 51	1 06	p1.1	7 01	7 23	p1.1
30	W	1 56	1 41	p1.2	7 42	8 12	p1.2	30	F	1 31	2 04	1.2	7 51	8 12	1.2
								31	S	2 12	2 58	a1.3	8 37	8 58	a1.4

A also at 11:44 a.m. 1.0

The Kts. (knots) columns show the **maximum** predicted velocities of the stronger one of the Flood Currents and the stronger one of the Ebb Currents for each day. The letter "a" means the velocity shown should occur **after** the **a.m.** Current Change. The letter "p" means the velocity shown should occur **after** the **p.m.** Current Change (even if next morning). No "a" or "p" means a.m. and p.m. velocities are the same for that day.

Av. Max. Vel.: Fl. 1.8 Kts., Ebb 1.9 Kts.

Max. Flood & Max. Ebb 3 hrs. 5 min. after Flood Starts & Ebb Starts, ±15 min.

See pp. 22-29 for Current Change at other points.

2009 CURRENT TABLE
DELAWARE BAY ENTRANCE
38°46.85'N, 75°02.58'W

Standard Time starts Nov. 1 at 2 a.m. — Standard Time

NOVEMBER

Day of Month	Day of Week	NORTHWEST Flood Starts			SOUTHEAST Ebb Starts		
		a.m.	p.m.	Kts.	a.m.	p.m.	Kts.
1	S	1 52	*2 48	a1.5	*8 21	*8 43	p1.6
2	M	2 34	3 35	a1.6	9 03	9 27	a1.6
3	T	3 17	4 22	a1.8	9 45	10 10	a1.7
4	W	4 01	5 10	a1.8	10 30	10 56	a1.7
5	T	4 49	5 59	a1.8	11 17	11 44	a1.6
6	F	5 40	6 51	a1.8	...	12 08	1.6
7	S	6 36	7 46	a1.7	12 37	1 02	p1.5
8	S	7 36	8 45	a1.5	1 34	2 02	p1.4
9	M	8 43	9 48	a1.5	2 38	3 07	p1.4
10	T	9 54	10 48	1.4	3 47	4 14	p1.4
11	W	11 06	11 44	p1.5	4 54	5 17	p1.4
12	T	...	12 12	1.4	5 56	6 16	1.4
13	F	12 37	1 15	a1.6	6 52	7 12	a1.5
14	S	1 26	2 12	a1.7	7 44	8 04	a1.6
15	S	2 12	3 04	a1.7	8 31	8 52	a1.6
16	M	2 55	3 52	a1.7	9 14	9 36	a1.6
17	T	3 35	4 38	a1.7	9 56	10 18	a1.6
18	W	4 13	5 21	a1.6	10 36	10 58	a1.5
19	T	4 50	6 02	a1.5	11 16	11 38	a1.4
20	F	5 27	6 41	a1.4	11 57	...	1.3
21	S	6 05	7 20	a1.3	12 19	12 38	p1.2
22	S	6 46	7 58	a1.2	1 01	1 21	p1.1
23	M	7 32	8 39	a1.1	1 46	2 07	p1.1
24	T	8 24	9 21	a1.0	2 36	2 56	p1.0
25	W	9 24	10 05	1.0	3 31	3 48	p1.0
26	T	10 28	10 49	p1.1	4 25	4 40	p1.0
27	F	11 30	11 33	p1.3	5 16	5 32	1.1
28	S	...	12 30	1.1	6 06	6 24	a1.2
29	S	12 19	1 29	a1.4	6 55	7 17	a1.4
30	M	1 08	2 23	a1.6	7 44	8 08	a1.5

DECEMBER

Day of Month	Day of Week	NORTHWEST Flood Starts			SOUTHEAST Ebb Starts		
		a.m.	p.m.	Kts.	a.m.	p.m.	Kts.
1	T	1 58	3 14	a1.7	8 33	8 59	a1.7
2	W	2 49	4 04	a1.8	9 22	9 48	a1.7
3	T	3 40	4 55	a1.9	10 12	10 39	a1.8
4	F	4 34	5 46	a1.9	11 04	11 32	a1.8
5	S	5 30	6 38	a1.8	11 58	...	1.7
6	S	6 29	7 31	a1.7	12 28	12 53	p1.6
7	M	7 30	8 26	a1.6	1 26	1 50	p1.5
8	T	8 36	9 23	a1.5	2 28	2 50	p1.5
9	W	9 47	10 21	p1.5	3 34	3 52	p1.5
10	T	10 56	11 14	p1.5	4 38	4 53	1.3
11	F	...	12 02	1.2	5 37	5 51	a1.4
12	S	12 05	1 04	a1.5	6 32	6 47	a1.4
13	S	12 55	2 00	a1.6	7 23	7 40	a1.5
14	M	1 42	2 51	a1.6	8 11	8 29	a1.5
15	T	2 26	3 38	a1.6	8 54	9 13	a1.5
16	W	3 08	4 20	a1.6	9 35	9 54	a1.5
17	T	3 47	5 01	a1.5	10 15	10 35	a1.5
18	F	4 24	5 38	a1.5	10 54	11 15	a1.4
19	S	5 01	6 12	a1.4	11 33	11 55	a1.4
20	S	5 40	6 45	a1.4	...	12 11	1.3
21	M	6 20	7 17	a1.3	12 35	12 49	p1.2
22	T	7 05	7 51	a1.2	1 17	1 29	p1.2
23	W	7 54	8 28	1.1	2 00	2 11	p1.1
24	T	8 50	9 09	p1.2	2 48	2 57	p1.1
25	F	9 52	9 55	p1.3	3 39	3 49	1.0
26	S	10 56	10 43	p1.4	4 30	4 43	a1.1
27	S	11 58	11 36	p1.5	5 23	5 40	a1.2
28	M	...	12 59	1.1	6 17	6 39	a1.4
29	T	12 31	1 58	a1.7	7 13	7 38	a1.5
30	W	1 30	2 53	a1.8	8 10	8 35	a1.7
31	T	2 29	3 47	a1.9	9 05	9 29	a1.8

*Standard Time starts

The Kts. (knots) columns show the **maximum** predicted velocities of the stronger one of the Flood Currents and the stronger one of the Ebb Currents for each day. The letter "a" means the velocity shown should occur **after** the **a.m.** Current Change. The letter "p" means the velocity shown should occur **after** the **p.m.** Current Change (even if next morning). No "a" or "p" means a.m. and p.m. velocities are the same for that day.
Av. Max. Vel.: Fl. 1.8 Kts., Ebb 1.9 Kts.
Max. Flood & Max. Ebb 3 hrs. 5 min. after Flood Starts & Ebb Starts, ±15 min.

See pp. 22-29 for Current Change at other points.

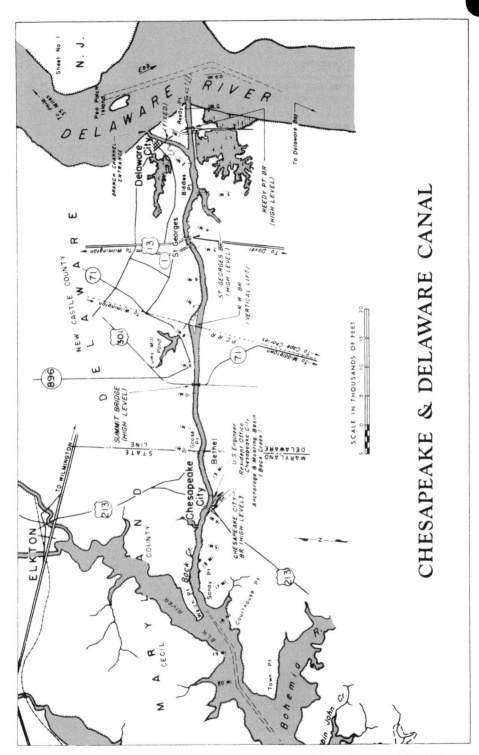

CHESAPEAKE & DELAWARE CANAL

See Chesapeake & Delaware Canal Current Tables, Pages 148-153

CHESAPEAKE & DELAWARE CANAL REGULATIONS

(Traffic Dispatcher is located at Chesapeake City.)

Philadelphia District Engineer issues notices periodically showing available channel depths and navigation conditions.

Projected Channel dimensions are 35 ft. deep and 450 ft. wide. (The branch to Delaware City is 8 ft. deep and 50 ft. wide.) The distance from the Delaware River Ship Channel to the Elk River is 19.1 miles.

1. Traffic controls, located at Reedy Point and Old Town Point Wharf flash green when Canal is open, flash red when it is closed.
2. Vessel identification and monitoring are performed by TV cameras at Reedy Point and Old Town Point Wharf.
3. Vessels, Tugs and Tows are required to have radiotelephones as applied to the following:
   a. Power vessels of 300 gross tons and upward.
   b. Vessels of 100 gross tons and upward carrying 1 or more passengers for hire.
   c. Every towing vessel of 26 feet or over.
4. Vessels listed in 3. will not enter the Canal until radio communication is made with the dispatcher and clearance is received. Ships' captains will tell the dispatcher the estimated time of passing Reedy Point or Town Point. Communication is to be established on Channel 13 (156.65 MHz). Dispatcher also monitors Channel 16 (156.8 MHz) to respond to emergencies.
5. A westbound vessel must be able to pass Reedy Is. or Pea Patch Is. within 120 min. of receiving clearance; an eastbound vessel must be able to pass Arnold Point within 120 min. If passage is not made within these 120 min., a new clearance must be solicited. Vessels must also report to the dispatcher the time of passing the outer end of the jetties at Reedy Point and Old Town Point Wharf.
6. Vessels exceeding 800 feet are required to have operable bow thrusters.
7. Maximum combined extreme breadth of vessels meeting and overtaking each other is 190 feet.
8. Vessels of all types are required to travel at a safe speed to avoid damage by suction or wash to wharves, landings, other boats, etc. Operators of yachts, motorboats, etc. are cautioned that large, deep-draft ocean-going and commercial vessels ply the Canal and to moor or anchor well away from the main ship channel.
9. Vessels proceeding *with* the current shall have the right-of-way but all small pleasure craft shall relinquish the right-of-way to deeper draft vessels which have a limited maneuvering ability.
10. Vessels under sail will not be permitted in the Canal.
11. Vessels difficult to handle must use the Canal during daylight hours and must have tug assistance. They should clear Reedy Point Bridge (going east) or Chesapeake City Bridge (going west) before dark.

Anchorage and wharfage facilities for small vessels only are at Chesapeake City and permission to use them for more than 24 hours must be obtained from the dispatcher.

The 5 Highway Bridges are high level and fixed and the 1 vertical lift RR bridge has a clearance when closed of 45 ft. at MHW.

Normal tide range is 5.4 ft. at Delaware R. end of the Canal and 2.6 ft. at Chesapeake City. Local mean low water at Courthouse Pt. is 2.5 ft. and decreases gradually eastward to 0.6 ft. at Delaware R. (See pp. 18 and 19 for times of High Water in this area.)
Note: A violent northeast storm may raise tide 4 to 5 ft. above normal in the Canal; a westerly storm may cause low tide to fall slightly below normal at Chesapeake City and as much as 4.0 ft. below normal at Reedy Point.

2009 CURRENT TABLE
CHESAPEAKE & DELAWARE CANAL

39°31.89'N, 75°49.65'W at Chesapeake City

JANUARY

Standard Time

DAY OF MONTH	DAY OF WEEK	EAST Flood Starts a.m.	p.m.	Kts.	WEST Ebb Starts a.m.	p.m.	Kts.
1	T	7 05	7 17	p2.4	1 09	12 39	p2.1
2	F	7 55	7 57	p2.3	1 41	1 37	1.9
3	S	8 50	8 40	p2.2	2 15	2 41	a2.0
4	S	9 52	9 27	p2.1	2 54	3 58	a2.1
5	M	10 59	10 20	p2.0	3 38	5 21	a2.2
6	T	11 59	11 18	p2.0	4 27	6 38	a2.4
7	W	...	1 09	1.9	5 20	7 48	a2.5
8	T	12 19	2 10	p2.1	6 17	8 49	a2.7
9	F	1 25	3 09	p2.3	7 19	9 41	a2.8
10	S	2 29	4 03	p2.5	8 23	10 27	a2.9
11	S	3 30	4 54	p2.5	9 26	11 11	a2.9
12	M	4 29	5 44	2.5	10 28	11 54	a2.9
13	T	5 27	6 30	a2.6	11 30	...	2.7
14	W	6 25	7 13	a2.5	12 35	12 30	p2.5
15	T	7 21	7 53	a2.4	1 16	1 30	p2.2
16	F	8 18	8 31	a2.1	1 55	2 29	a2.1
17	S	9 16	9 09	a1.9	2 35	3 32	a2.1
18	S	10 17	9 49	1.6	3 14	4 37	a2.0
19	M	11 14	10 32	1.5	3 53	5 42	a2.0
20	T	11 59	11 18	a1.5	4 31	6 42	a1.9
21	W	...	12 53	1.6	5 08	7 37	a1.9
22	T	12 07	1 35	p1.7	5 47	8 25	a2.0
23	F	12 58	2 15	p1.9	6 31	9 05	a2.1
24	S	1 51	2 54	p2.0	7 20	9 40	a2.2
25	S	2 41	3 33	p2.1	8 12	10 13	a2.3
26	M	3 29	4 12	p2.3	9 05	10 44	a2.4
27	T	4 16	4 51	p2.4	9 56	11 14	a2.4
28	W	5 03	5 30	p2.5	10 47	11 44	a2.3
29	T	5 50	6 09	p2.5	11 40	...	2.2
30	F	6 38	6 48	p2.4	12 16	12 34	a2.1
31	S	7 28	7 27	p2.3	12 50	1 32	a2.2

FEBRUARY

Standard Time

DAY OF MONTH	DAY OF WEEK	EAST Flood Starts a.m.	p.m.	Kts.	WEST Ebb Starts a.m.	p.m.	Kts.
1	S	8 22	8 09	p2.2	1 27	2 37	a2.3
2	M	9 25	8 57	p2.0	2 09	3 54	a2.3
3	T	10 36	9 56	p1.9	2 57	5 14	a2.4
4	W	11 46	11 04	p1.9	3 55	6 27	a2.4
5	T	...	12 53	2.0	5 00	7 29	a2.5
6	F	12 12	1 56	p2.1	6 10	8 23	a2.6
7	S	1 19	2 54	p2.3	7 20	9 10	a2.7
8	S	2 23	3 46	2.4	8 28	9 52	a2.7
9	M	3 23	4 36	a2.6	9 31	10 33	a2.7
10	T	4 18	5 20	a2.7	10 30	11 13	a2.6
11	W	5 14	6 02	a2.7	11 28	11 53	a2.5
12	T	6 08	6 41	a2.6	...	12 24	2.2
13	F	7 01	7 17	a2.4	12 32	1 19	a2.3
14	S	7 53	7 51	a2.1	1 08	2 14	a2.2
15	S	8 43	8 26	a1.8	1 43	3 12	a2.1
16	M	9 35	9 05	a1.6	2 15	4 13	a2.0
17	T	10 26	9 53	a1.5	2 48	5 13	a1.9
18	W	11 15	10 48	a1.5	3 26	6 08	a1.8
19	T	11 59	11 44	a1.5	4 13	6 56	a1.8
20	F	...	12 45	1.7	5 08	7 39	a1.9
21	S	12 38	1 30	p1.8	6 07	8 16	a2.0
22	S	1 30	2 14	p2.0	7 07	8 49	a2.2
23	M	2 20	2 57	p2.1	8 05	9 20	a2.2
24	T	3 08	3 38	p2.2	9 01	9 50	a2.2
25	W	3 54	4 18	p2.3	9 53	10 20	2.2
26	T	4 39	4 58	p2.4	10 44	10 52	p2.3
27	F	5 26	5 37	2.4	11 37	11 26	p2.5
28	S	6 15	6 18	2.3	...	12 33	1.8

The Kts. (knots) columns show the **maximum** predicted velocities of the stronger one of the Flood Currents and the stronger one of the Ebb Currents for each day.

The letter "a" means the velocity shown should occur **after** the a.m. Current Change. The letter "p" means the velocity shown should occur **after** the p.m. Current Change (even if next morning). No "a" or "p" means a.m. and p.m. velocities are the same for that day.

Av. Max. Vel.: Fl. 2.0 Kts., Ebb 1.9 Kts.

Max. Flood 3 hrs. 10 min. after Flood Starts ±45 min.

Max. Ebb 2 hrs. 45 min. after Ebb Starts ±45 min

See pp. 22-29 for Current Change at other points.

Note: *from NOS. These predictions should be considered questionable. Caution is advised.*

Daylight Time starts March 8 at 2 a.m.   Daylight Saving Time

MARCH

DAY OF MONTH	DAY OF WEEK	EAST Flood Starts a.m.	p.m.	Kts.	WEST Ebb Starts a.m.	p.m.	Kts.
1	S	7 06	7 00	2.2	12 04	1 33	a2.5
2	M	8 01	7 46	a2.1	12 46	2 39	a2.5
3	T	9 04	8 40	a2.0	1 33	3 51	a2.5
4	W	10 16	9 47	1.9	2 29	5 03	a2.4
5	T	11 28	11 01	1.9	3 37	6 05	a2.4
6	F	...	12 34	2.0	4 55	6 59	a2.3
7	S	12 10	1 36	2.1	6 12	7 48	a2.4
8	S	1 14	*3 31	a2.3	*8 24	*9 32	a2.5
9	M	3 16	4 22	a2.5	9 31	10 13	a2.5
10	T	4 12	5 06	a2.7	10 31	10 52	a2.4
11	W	5 07	5 47	a2.7	11 27	11 31	p2.4
12	T	5 59	6 27	a2.7	...	12 22	2.1
13	F	6 50	7 04	a2.5	12 08	1 15	a2.4
14	S	7 38	7 39	a2.3	12 43	2 07	a2.4
15	S	8 23	8 14	a2.1	1 15	2 59	a2.3
16	M	9 04	8 51	a1.9	1 45	3 51	a2.1
17	T	9 45	9 34	a1.7	2 14	4 45	a2.0
18	W	10 27	10 26	a1.6	2 48	5 38	a1.9
19	T	11 14	11 26	a1.6	3 33	6 25	a1.8
20	F	...	12 04	1.7	4 31	7 07	a1.8
21	S	12 24	12 54	p1.8	5 41	7 45	a1.7
22	S	1 18	1 42	p1.9	6 51	8 19	a1.8
23	M	2 09	2 29	p2.0	7 57	8 52	1.9
24	T	2 59	3 15	p2.1	8 58	9 24	p2.1
25	W	3 47	3 59	p2.2	9 55	9 57	p2.3
26	T	4 33	4 41	2.3	10 49	10 30	p2.5
27	F	5 20	5 23	a2.4	11 42	11 05	p2.6
28	S	6 08	6 06	a2.5	-A-	12 38	1.6
29	S	6 58	6 51	a2.5	...	1 37	1.5
30	M	7 51	7 39	a2.4	12 26	2 36	a2.8
31	T	8 47	8 33	a2.3	1 15	3 38	a2.7

APRIL

DAY OF MONTH	DAY OF WEEK	EAST Flood Starts a.m.	p.m.	Kts.	WEST Ebb Starts a.m.	p.m.	Kts.
1	W	9 49	9 34	a2.1	2 10	4 40	a2.6
2	T	10 58	10 45	a2.1	3 14	5 40	a2.4
3	F	...	12 07	2.0	4 32	6 34	a2.3
4	S	12 01	1 10	2.0	5 56	7 22	a2.2
5	S	1 04	2 07	2.1	7 14	8 08	a2.2
6	M	2 06	2 59	a2.4	8 24	8 51	2.2
7	T	3 06	3 46	a2.5	9 28	9 32	p2.4
8	W	4 02	4 29	a2.6	10 26	10 11	p2.5
9	T	4 55	5 09	a2.6	11 20	10 47	p2.5
10	F	5 43	5 47	a2.5	-B-	12 13	1.6
11	S	6 29	6 24	a2.4	...	1 04	1.4
12	S	7 11	7 02	a2.3	12 01	1 54	a2.4
13	M	7 49	7 42	a2.1	12 20	2 41	a2.3
14	T	8 23	8 23	a2.0	12 50	3 26	a2.2
15	W	8 57	9 10	a1.9	1 25	4 11	a2.1
16	T	9 35	10 02	a1.9	2 07	4 54	a2.0
17	F	10 20	11 01	a1.9	3 00	5 35	a1.8
18	S	11 11	11 59	a1.9	4 04	6 12	a1.7
19	S	...	12 04	1.9	5 21	6 48	1.7
20	M	12 54	12 54	p2.0	6 37	7 22	p1.9
21	T	1 46	1 42	p2.1	7 45	7 56	p2.2
22	W	2 37	2 29	p2.1	8 49	8 31	p2.4
23	T	3 28	3 16	p2.2	9 49	9 08	p2.6
24	F	4 17	4 03	a2.3	10 46	9 47	p2.8
25	S	5 05	4 49	a2.4	11 43	10 28	p2.9
26	S	5 55	5 38	a2.5	-C-	12 40	1.3
27	M	6 47	6 30	a2.5	...	1 37	1.3
28	T	7 42	7 26	a2.5	12 02	2 32	a2.9
29	W	8 37	8 25	a2.4	12 58	3 25	a2.8
30	T	9 36	9 29	a2.3	2 01	4 17	a2.6

*Daylight Saving Time starts

A also at 11:44 p.m. 2.7   **B** also at 11:21 p.m. 2.5   **C** also at 11:12 p.m. 2.9

The Kts. (knots) columns show the **maximum** predicted velocities of the stronger one of the Flood Currents and the stronger one of the Ebb Currents for each day.

The letter "a" means the velocity shown should occur **after** the **a.m.** Current Change. The letter "p" means the velocity shown should occur **after** the **p.m.** Current Change (even if next morning). No "a" or "p" means a.m. and p.m. velocities are the same for that day.

Av. Max. Vel.: Fl. 2.0 Kts., Ebb 1.9 Kts.

Max. Flood 3 hrs. 10 min. after Flood Starts ±45 min.

Max. Ebb 2 hrs. 45 min. after Ebb Starts ±45 min

See pp. 22-29 for Current Change at other points.

Note: *from NOS. These predictions should be considered questionable. Caution is advised.*   **149**

2009 CURRENT TABLE
CHESAPEAKE & DELAWARE CANAL
39°31.89'N, 75°49.65'W at Chesapeake City

Daylight Saving Time | Daylight Saving Time

		MAY								JUNE					
		CURRENT TURNS TO								CURRENT TURNS TO					
		EAST Flood Starts			WEST Ebb Starts					EAST Flood Starts			WEST Ebb Starts		
DAY OF MONTH	DAY OF WEEK	a.m.	p.m.	Kts.	a.m.	p.m.	Kts.	DAY OF MONTH	DAY OF WEEK	a.m.	p.m.	Kts.	a.m.	p.m.	Kts.
1	F	10 38	10 38	a2.2	3 13	5 08	a2.4	1	M	11 59	...	2.0	5 45	6 03	p2.1
2	S	11 40	11 48	2.1	4 33	5 57	a2.2	2	T	12 40	12 47	a2.0	6 56	6 48	p2.2
3	S	...	12 37	2.0	5 54	6 43	a2.1	3	W	1 42	1 33	a2.1	8 03	7 31	p2.4
4	M	12 54	1 30	a2.1	7 08	7 28	p2.2	4	T	2 40	2 17	a2.1	9 06	8 11	p2.4
5	T	1 56	2 18	a2.3	8 16	8 10	p2.4	5	F	3 33	3 02	a2.1	10 03	8 48	p2.4
6	W	2 55	3 04	a2.4	9 19	8 51	p2.5	6	S	4 18	3 46	a2.1	10 54	9 22	p2.4
7	T	3 49	3 46	a2.4	10 16	9 30	p2.5	7	S	4 58	4 30	a2.2	11 40	9 55	p2.4
8	F	4 39	4 26	a2.4	11 10	10 04	p2.5	8	M	5 33	5 14	a2.2	-D-	12 23	1.1
9	S	5 25	5 07	a2.3	11 59	10 36	p2.5	9	T	6 07	5 59	a2.2	-E-	1 02	1.2
10	S	6 04	5 47	a2.3	-A-	12 49	1.2	10	W	6 39	6 43	a2.3	...	1 37	1.3
11	M	6 40	6 29	a2.2	-B-	1 34	1.1	11	T	7 13	7 29	a2.3	12 01	2 09	a2.4
12	T	7 13	7 12	a2.2	...	2 15	1.2	12	F	7 48	8 15	a2.3	12 39	2 39	a2.3
13	W	7 45	7 57	a2.2	12 11	2 53	a2.3	13	S	8 25	9 04	a2.3	1 32	3 09	a2.1
14	T	8 19	8 44	a2.2	12 54	3 28	a2.2	14	S	9 05	9 57	a2.3	2 29	3 41	a1.9
15	F	8 57	9 35	a2.1	1 44	4 03	a2.1	15	M	9 47	10 57	a2.2	3 31	4 16	p2.0
16	S	9 39	10 31	a2.1	2 41	4 38	a1.9	16	T	10 33	11 58	a2.2	4 43	4 55	p2.1
17	S	10 26	11 30	a2.1	3 46	5 14	a1.8	17	W	11 23	...	2.1	6 03	5 37	p2.3
18	M	11 16	...	2.1	5 02	5 50	p1.9	18	T	12 58	12 14	p2.0	7 20	6 21	p2.4
19	T	12 27	12 07	p2.1	6 19	6 27	p2.1	19	F	1 57	1 07	p2.0	8 32	7 08	p2.6
20	W	1 23	12 56	p2.1	7 31	7 05	p2.4	20	S	2 55	2 04	2.0	9 38	7 59	p2.8
21	T	2 17	1 44	p2.1	8 40	7 45	p2.6	21	S	3 51	3 06	a2.2	10 35	8 54	p2.9
22	F	3 11	2 35	2.1	9 44	8 28	p2.8	22	M	4 45	4 07	a2.4	11 27	9 53	p3.0
23	S	4 03	3 28	a2.2	10 44	9 14	p2.9	23	T	5 38	5 07	a2.5	-F-	12 15	1.2
24	S	4 55	4 23	a2.4	11 40	10 03	p3.0	24	W	6 31	6 06	a2.6	...	1 00	1.4
25	M	5 47	5 18	a2.5	-C-	12 34	1.2	25	T	7 22	7 06	a2.6	12 01	1 43	a2.9
26	T	6 40	6 17	a2.6	...	1 26	1.3	26	F	8 11	8 05	a2.5	1 00	2 25	a2.8
27	W	7 34	7 17	a2.6	12 01	2 14	a3.0	27	S	8 58	9 05	a2.4	2 05	3 07	a2.6
28	T	8 27	8 17	a2.5	12 55	2 59	a2.8	28	S	9 43	10 09	a2.2	3 11	3 50	a2.2
29	F	9 20	9 19	a2.4	2 03	3 44	a2.6	29	M	10 29	11 16	a2.1	4 20	4 35	p2.1
30	S	10 13	10 25	a2.2	3 14	4 30	a2.4	30	T	11 16	...	1.9	5 31	5 21	p2.2
31	S	11 07	11 34	2.1	4 29	5 17	a2.1								

A also at 11:05 p.m. 2.4 **B** also at 11:36 p.m. 2.4 **C** also at 10:55 p.m. 3.0
D also at 10:30 p.m. 2.4 **E** also at 11:08 p.m. 2.4 **F** also at 10:53 p.m. 3.0

The Kts. (knots) columns show the **maximum** predicted velocities of the stronger one of the Flood Currents and the stronger one of the Ebb Currents for each day.
The letter "a" means the velocity shown should occur **after** the **a.m.** Current Change. The letter "p" means the velocity shown should occur **after** the **p.m.** Current Change (even if next morning). No "a" or "p" means a.m. and p.m. velocities are the same for that day.
Av. Max. Vel.: Fl. 2.0 Kts., Ebb 1.9 Kts.
Max. Flood 3 hrs. 10 min. after Flood Starts ±45 min.
Max. Ebb 2 hrs. 45 min. after Ebb Starts ±45 min

See pp. 22-29 for Current Change at other points.

Note: *from NOS. These predictions should be considered questionable. Caution is advised.*

2009 CURRENT TABLE
CHESAPEAKE & DELAWARE CANAL
39°31.89'N, 75°49.65'W at Chesapeake City

Daylight Saving Time   **Daylight Saving Time**

JULY

Day of Month	Day of Week	CURRENT TURNS TO EAST Flood Starts a.m.	**p.m.**	Kts.	WEST Ebb Starts a.m.	**p.m.**	Kts.
1	W	12 22	12 02	1.8	6 40	6 07	p2.1
2	T	1 23	12 49	a1.8	7 46	6 49	p2.2
3	F	2 18	1 36	a1.9	8 48	7 29	p2.2
4	S	3 07	2 24	a1.9	9 42	8 08	p2.2
5	S	3 48	3 14	a1.9	10 28	8 46	p2.2
6	M	4 25	4 01	a2.0	11 08	9 26	p2.3
7	T	4 58	4 45	a2.1	11 44	10 09	p2.3
8	W	5 31	5 29	a2.2	-A-	12 16	1.4
9	T	6 07	6 14	a2.3	-B-	12 47	1.6
10	F	6 41	6 59	a2.4	...	1 16	1.7
11	S	7 18	7 45	a2.4	12 32	1 45	a2.3
12	S	7 55	8 33	a2.4	1 24	2 14	a2.1
13	M	8 33	9 25	a2.3	2 19	2 47	p2.1
14	T	9 12	10 23	a2.3	3 20	3 23	p2.2
15	W	9 56	11 28	a2.1	4 31	4 05	p2.3
16	T	10 46	...	2.0	5 53	4 53	p2.3
17	F	12 34	-C-	1.7	7 10	5 46	p2.4
18	S	1 38	12 46	p1.9	8 21	6 44	p2.6
19	S	2 40	1 50	2.0	9 22	7 45	p2.7
20	M	3 39	2 56	a2.2	10 14	8 51	p2.8
21	T	4 34	3 59	2.3	11 00	9 56	p2.9
22	W	5 26	4 58	p2.5	11 42	10 59	p2.9
23	T	6 15	5 55	p2.6	...	12 24	1.8
24	F	7 03	6 53	p2.7	12 01	1 05	a2.8
25	S	7 48	7 51	p2.6	1 02	1 46	a2.6
26	S	8 30	8 49	2.3	2 03	2 26	a2.4
27	M	9 10	9 49	2.1	3 04	3 08	p2.2
28	T	9 51	10 52	a1.9	4 07	3 50	p2.1
29	W	10 34	11 54	1.7	5 14	4 34	p2.1
30	T	11 22	...	1.6	6 21	5 19	p2.0
31	F	12 52	12 13	a1.6	7 23	6 04	p1.9

AUGUST

Day of Month	Day of Week	CURRENT TURNS TO EAST Flood Starts a.m.	**p.m.**	Kts.	WEST Ebb Starts a.m.	**p.m.**	Kts.
1	S	1 42	1 05	a1.7	8 19	6 48	p1.9
2	S	2 27	1 57	a1.7	9 08	7 32	p2.0
3	M	3 07	2 47	a1.8	9 48	8 19	p2.1
4	T	3 44	3 34	a1.9	10 23	9 09	p2.1
5	W	4 19	4 19	a2.0	10 54	9 58	p2.2
6	T	4 55	5 02	a2.2	11 23	10 47	p2.2
7	F	5 32	5 46	a2.3	11 51	11 36	p2.2
8	S	6 09	6 31	a2.4	...	12 19	2.1
9	S	6 47	7 18	a2.4	12 26	12 49	p2.2
10	M	7 24	8 05	a2.4	1 19	1 22	p2.3
11	T	8 02	8 56	a2.3	2 14	1 58	p2.4
12	W	8 42	9 53	a2.2	3 15	2 38	p2.4
13	T	9 27	11 00	a2.0	4 26	3 24	p2.4
14	F	10 22	11 59	a1.9	5 45	4 18	p2.4
15	S	11 29	...	1.8	6 57	5 23	p2.4
16	S	1 18	12 39	p1.9	8 00	6 33	p2.4
17	M	2 22	1 47	p2.1	8 54	7 45	p2.6
18	T	3 22	2 51	p2.3	9 41	8 56	p2.7
19	W	4 16	3 51	p2.5	10 23	10 02	p2.7
20	T	5 05	4 49	p2.7	11 03	11 03	p2.6
21	F	5 52	5 45	p2.8	11 43	...	2.2
22	S	6 36	6 41	p2.8	12 02	12 23	a2.5
23	S	7 18	7 36	p2.6	1 00	1 03	p2.4
24	M	7 57	8 31	p2.3	1 58	1 44	p2.4
25	T	8 36	9 24	2.0	2 55	2 23	p2.2
26	W	9 15	10 19	p1.8	3 53	3 02	p2.1
27	T	9 59	11 13	p1.6	4 55	3 41	p1.9
28	F	10 50	11 59	p1.5	5 55	4 24	p1.8
29	S	11 47	...	1.2	6 51	5 13	p1.7
30	S	12 51	12 42	a1.5	7 39	6 08	p1.7
31	M	1 34	1 34	a1.6	8 20	7 05	p1.8

A also at 10:54 p.m. 2.3   **B** also at 11:41 p.m. 2.3   **C** also at 11:44 a.m. 1.9

The Kts. (knots) columns show the **maximum** predicted velocities of the stronger one of the Flood Currents and the stronger one of the Ebb Currents for each day.
The letter "a" means the velocity shown should occur **after** the **a.m.** Current Change. The letter "p" means the velocity shown should occur **after** the **p.m.** Current Change (even if next morning). No "a" or "p" means a.m. and p.m. velocities are the same for that day.
Av. Max. Vel.: Fl. 2.0 Kts., Ebb 1.9 Kts.
Max. Flood 3 hrs. 10 min. after Flood Starts ±45 min.
Max. Ebb 2 hrs. 45 min. after Ebb Starts ±45 min
See pp. 22-29 for Current Change at other points.

Note: *from NOS. These predictions should be considered questionable. Caution is advised.*   **151**

2009 CURRENT TABLE
CHESAPEAKE & DELAWARE CANAL
39°31.89'N, 75°49.65'W at Chesapeake City

Daylight Saving Time   Daylight Saving Time

SEPTEMBER

DAY OF MONTH	DAY OF WEEK	EAST Flood Starts a.m.	EAST Flood Starts p.m.	Kts.	WEST Ebb Starts a.m.	WEST Ebb Starts p.m.	Kts.
1	T	2 15	2 22	a1.7	8 56	8 01	p1.9
2	W	2 56	3 08	a1.9	9 28	8 57	p2.0
3	T	3 36	3 53	a2.0	9 57	9 50	p2.0
4	F	4 16	4 37	2.1	10 26	10 41	2.0
5	S	4 54	5 21	p2.3	10 55	11 31	a2.2
6	S	5 33	6 07	p2.4	11 25	...	2.4
7	M	6 12	6 54	2.4	12 23	12 01	p2.5
8	T	6 52	7 42	2.3	1 17	12 35	p2.6
9	W	7 35	8 35	a2.2	2 14	1 16	p2.6
10	T	8 19	9 30	2.0	3 16	2 02	p2.5
11	F	9 10	10 37	1.9	4 23	2 54	p2.4
12	S	10 13	11 48	p1.9	5 32	3 57	p2.4
13	S	11 26	...	1.8	6 34	5 13	p2.3
14	M	12 57	12 37	1.9	7 28	6 34	p2.3
15	T	2 00	1 42	p2.2	8 16	7 50	p2.4
16	W	2 57	2 44	p2.5	9 00	9 00	p2.4
17	T	3 50	3 43	p2.7	9 42	10 03	p2.4
18	F	4 37	4 39	p2.8	10 23	11 02	2.3
19	S	5 21	5 34	p2.8	11 02	11 59	a2.5
20	S	6 03	6 28	p2.7	11 42	...	2.5
21	M	6 44	7 20	p2.5	12 55	12 21	p2.5
22	T	7 24	8 09	p2.3	1 50	12 59	p2.4
23	W	8 04	8 54	p2.0	2 44	1 35	p2.2
24	T	8 46	9 37	p1.8	3 37	2 10	p2.0
25	F	9 32	10 20	p1.7	4 30	2 47	p1.8
26	S	10 24	11 04	p1.6	5 21	3 30	p1.7
27	S	11 22	11 50	p1.6	6 08	4 26	p1.6
28	M	...	12 17	1.2	6 49	5 34	p1.6
29	T	12 35	1 07	a1.7	7 24	6 42	p1.6
30	W	1 20	1 55	a1.8	7 57	7 46	p1.7

OCTOBER

DAY OF MONTH	DAY OF WEEK	EAST Flood Starts a.m.	EAST Flood Starts p.m.	Kts.	WEST Ebb Starts a.m.	WEST Ebb Starts p.m.	Kts.
1	T	2 04	2 42	a1.9	8 28	8 46	a1.9
2	F	2 48	3 29	p2.1	8 59	9 42	a2.2
3	S	3 31	4 15	p2.2	9 30	10 35	a2.4
4	S	4 13	5 01	p2.3	10 03	11 28	a2.6
5	M	4 55	5 48	p2.4	10 38	...	2.7
6	T	5 38	6 35	p2.5	12 22	-A-	1.5
7	W	6 23	7 25	p2.4	1 19	12 01	p2.8
8	T	7 11	8 18	p2.3	2 16	12 44	p2.8
9	F	8 05	9 16	p2.2	3 13	1 37	p2.6
10	S	9 02	10 18	p2.1	4 11	2 37	p2.5
11	S	10 09	11 27	p2.0	5 09	3 49	p2.3
12	M	11 22	...	1.9	6 02	5 13	p2.2
13	T	12 31	12 31	p2.1	6 50	6 36	p2.1
14	W	1 31	1 35	p2.3	7 36	7 50	p2.1
15	T	2 25	2 36	p2.5	8 19	8 58	a2.2
16	F	3 15	3 35	p2.6	9 02	10 00	a2.4
17	S	4 01	4 31	p2.7	9 44	10 57	a2.6
18	S	4 45	5 23	p2.6	10 24	11 53	a2.6
19	M	5 26	6 13	p2.6	11 02	...	2.6
20	T	6 09	7 00	p2.4	12 47	-B-	1.5
21	W	6 52	7 42	p2.2	1 39	12 14	p2.4
22	T	7 36	8 19	p2.1	2 28	12 48	p2.2
23	F	8 20	8 53	p2.0	3 13	1 23	p2.0
24	S	9 06	9 27	p1.9	3 56	2 04	p1.9
25	S	9 56	10 05	p1.8	4 37	2 52	p1.7
26	M	10 50	10 50	p1.8	5 16	3 53	p1.6
27	T	11 45	11 38	p1.9	5 51	5 06	p1.6
28	W	...	12 37	1.4	6 25	6 20	a1.7
29	T	12 26	1 28	a1.9	6 57	7 28	a1.9
30	F	1 13	2 18	a2.0	7 29	8 33	a2.2
31	S	1 58	3 08	a2.1	8 03	9 33	a2.4

A also at 11:16 a.m. 2.8   **B** also at 11:39 a.m. 2.5

The Kts. (knots) columns show the **maximum** predicted velocities of the stronger one of the Flood Currents and the stronger one of the Ebb Currents for each day.

The letter "a" means the velocity shown should occur **after** the **a.m.** Current Change. The letter "p" means the velocity shown should occur **after** the **p.m.** Current Change (even if next morning). No "a" or "p" means a.m. and p.m. velocities are the same for that day.

Av. Max. Vel.: Fl. 2.0 Kts., Ebb 1.9 Kts.
Max. Flood 3 hrs. 10 min. after Flood Starts ±45 min.
Max. Ebb 2 hrs. 45 min. after Ebb Starts ±45 min

See pp. 22-29 for Current Change at other points.

   Note: *from NOS. These predictions should be considered questionable. Caution is advised.*

2009 CURRENT TABLE
CHESAPEAKE & DELAWARE CANAL

39°31.89'N, 75°49.65'W at Chesapeake City

Standard Time starts Nov. 1 at 2 a.m.　　　　　　　Standard Time

NOVEMBER

Day of Month	Day of Week	EAST Flood Starts a.m.	p.m.	Kts.	WEST Ebb Starts a.m.	p.m.	Kts.
1	S	1 45	*2 57	p2.2	*7 40	*9 30	a2.6
2	M	2 32	3 45	p2.3	8 19	10 26	a2.8
3	T	3 20	4 33	p2.4	9 01	11 21	a2.9
4	W	4 09	5 23	p2.5	9 45	...	2.9
5	T	5 01	6 14	p2.5	12 16	-A-	1.2
6	F	5 56	7 07	p2.4	1 09	-B-	1.2
7	S	6 55	8 03	p2.3	1 58	12 27	p2.7
8	S	7 56	9 01	p2.2	2 47	1 34	p2.5
9	M	9 02	10 03	p2.1	3 36	2 50	p2.3
10	T	10 11	11 01	2.1	4 24	4 13	p2.1
11	W	11 20	11 56	a2.1	5 11	5 32	1.9
12	T	...	12 25	2.2	5 56	6 44	a2.1
13	F	12 47	1 27	p2.4	6 41	7 51	a2.4
14	S	1 35	2 26	p2.5	7 24	8 53	a2.6
15	S	2 22	3 20	p2.5	8 07	9 50	a2.6
16	M	3 07	4 09	p2.4	8 48	10 43	a2.6
17	T	3 52	4 54	p2.4	9 26	11 33	a2.5
18	W	4 36	5 34	p2.3	10 02	...	2.4
19	T	5 22	6 09	p2.2	12 20	-C-	1.2
20	F	6 08	6 41	p2.2	1 01	-D-	1.2
21	S	6 52	7 12	p2.1	1 39	12 01	p2.1
22	S	7 36	7 45	p2.1	2 13	12 38	p2.0
23	M	8 22	8 21	p2.1	2 45	1 30	p1.8
24	T	9 13	9 02	p2.1	3 17	2 30	p1.7
25	W	10 08	9 48	p2.1	3 49	3 42	a1.8
26	T	11 05	10 36	p2.0	4 23	4 58	a1.9
27	F	11 59	11 24	p2.0	4 59	6 11	a2.1
28	S	...	12 54	1.8	5 37	7 20	a2.4
29	S	12 12	1 48	a2.1	6 17	8 26	a2.6
30	M	1 03	2 41	2.1	6 59	9 25	a2.7

DECEMBER

Day of Month	Day of Week	EAST Flood Starts a.m.	p.m.	Kts.	WEST Ebb Starts a.m.	p.m.	Kts.
1	T	1 56	3 32	p2.3	7 45	10 20	a2.9
2	W	2 53	4 22	p2.4	8 35	11 12	a2.9
3	T	3 49	5 13	p2.5	9 27	11 59	a3.0
4	F	4 46	6 05	p2.6	10 22	...	3.0
5	S	5 45	6 57	p2.5	12 47	-E-	1.3
6	S	6 45	7 48	p2.4	1 31	12 28	p2.7
7	M	7 45	8 40	p2.3	2 14	1 37	p2.5
8	T	8 49	9 33	2.2	2 57	2 50	p2.2
9	W	9 58	10 26	2.1	3 43	4 07	2.0
10	T	11 07	11 17	a2.1	4 31	5 23	a2.1
11	F	...	12 13	2.1	5 18	6 33	a2.3
12	S	12 06	1 15	p2.2	6 04	7 41	a2.4
13	S	12 54	2 13	p2.2	6 50	8 42	a2.5
14	M	1 44	3 04	p2.2	7 34	9 36	a2.5
15	T	2 33	3 48	p2.2	8 16	10 24	a2.4
16	W	3 22	4 27	p2.2	8 55	11 07	a2.4
17	T	4 08	5 02	p2.2	9 32	11 46	a2.3
18	F	4 53	5 34	p2.2	10 09	...	2.3
19	S	5 36	6 04	p2.2	12 21	-F-	1.3
20	S	6 19	6 36	p2.3	12 53	-G-	1.1
21	M	7 01	7 09	p2.3	1 21	12 23	p2.1
22	T	7 46	7 45	p2.2	1 48	1 16	p1.9
23	W	8 35	8 23	p2.2	2 17	2 14	a1.9
24	T	9 31	9 05	p2.2	2 48	3 22	a2.0
25	F	10 31	9 52	p2.1	3 25	4 40	a2.2
26	S	11 32	10 43	p2.0	4 07	5 57	a2.3
27	S	-H-	12 31	1.7	4 52	7 09	a2.4
28	M	...	1 29	1.9	5 39	8 16	a2.6
29	T	12 33	2 25	p2.1	6 30	9 13	a2.7
30	W	1 34	3 19	p2.3	7 25	10 03	a2.8
31	T	2 37	4 11	p2.4	8 24	10 49	a2.9

*Standard Time starts

A also at 10:33 a.m. 2.9　**B** also at 11:26 a.m. 2.9　**C** also at 10:36 a.m. 2.3
D also at 11:11 a.m. 2.2　**E** also at 11:23 a.m. 2.9　**F** also at 10:49 a.m. 2.2
G also at 11:34 a.m. 2.2　**H** also at 11:36 p.m. 2.0

The Kts. (knots) columns show the **maximum** predicted velocities of the stronger one of the Flood Currents and the stronger one of the Ebb Currents for each day.

The letter "a" means the velocity shown should occur **after** the **a.m.** Current Change. The letter "p" means the velocity shown should occur **after** the **p.m.** Current Change (even if next morning). No "a" or "p" means a.m. and p.m. velocities are the same for that day.

Av. Max. Vel.: Fl. 2.0 Kts., Ebb 1.9 Kts.

Max. Flood 3 hrs. 10 min. after Flood Starts ±45 min.

Max. Ebb 2 hrs. 45 min. after Ebb Starts ±45 min

See pp. 22-29 for Current Change at other points.

Note: *from NOS. These predictions should be considered questionable. Caution is advised.*

Upper Chesapeake Bay Currents

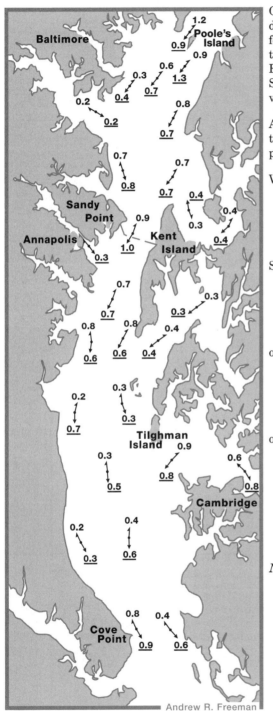

Andrew R. Freeman

On this Current Diagram, the arrows denote maximum velocities. Refer to the four areas listed below for the specific times. Double-headed arrows are for Ebb and the velocities are <u>underlined</u>. Single-headed arrows are for Flood and velocities and <u>not</u> underlined.

All times below are in hours and relate to the time of High Water at Baltimore, pp. 156-159.

West of Pooles Island:
 Flood begins 3 1/2 before
 Flood max. 1 1/2 before (1.2 kts.)
 Ebb begins 2 1/2 after
 Ebb max. 4 1/2 after (0.9 kts.)

Sandy Point:
 Flood begins 3 1/2 before
 Flood max. 1 1/2 before (0.9 kts.)
 Ebb begins 1 1/2 after
 Ebb max. 4 1/2 after (1.0 kts.)

off Tilghman Island:
 Flood begins 5 1/2 before
 Flood max. 3 1/2 before (0.3 kts.)
 Ebb begins 1/2 after
 Ebb max. 3 1/2 after (0.7 kts.)

off Cove Point:
 Flood begins 6 1/2 before
 Flood max. 4 1/2 before (0.9 kts.)
 Ebb begins 1/2 before
 Ebb max. 1 1/2 after (0.8 kts.)

Note:
 From the beginning of the Flood Current at Cove Pt. until the Ebb Current begins off Baltimore, a northbound vessel will have over 8 hours of Fair Tide. A vessel bound southward from Sandy Pt. can expect only 4 hours of Fair Tide.

RELATIONSHIP OF HIGH WATER AND EBB CURRENT

Many people wonder why the times of High Water and the start of Ebb Current at the mouths of bays and inlets are not simultaneous. After doing a rough survey of tide and current data, involving principally the New Jersey Coast and south shore of Long Island, we have made diagrams of what we believe happens hour by hour. Picture the rising Tide, borne by the Flood Current, as a long wave. The wave enters the inlet and the crest reaches its maximum height in or at the inlet. But, the body of water inside the inlet - in the bay - has yet to be filled and the Flood Current continues to pour water through the inlet for a good period *after* the crest has already passed the inlet. The Ebb Current will not start until the level of the water in the ocean is lower than the water in the bay. This does not necessarily apply to the mouths of small bays with wide entrances. The narrowness of the inlet and the size of the bay are the controlling factors. Picture a siphon effect.

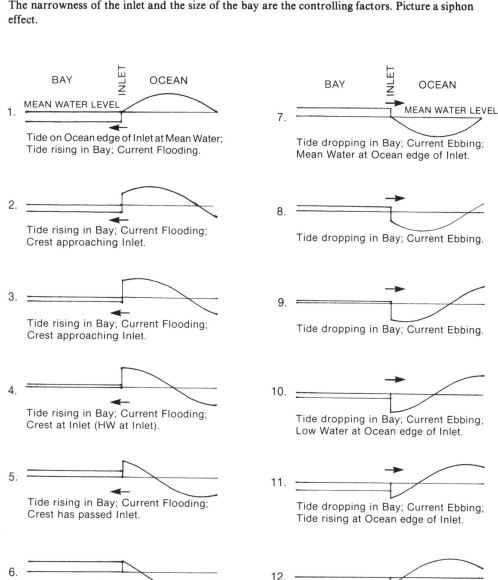

1. Tide on Ocean edge of Inlet at Mean Water; Tide rising in Bay; Current Flooding.

2. Tide rising in Bay; Current Flooding; Crest approaching Inlet.

3. Tide rising in Bay; Current Flooding; Crest approaching Inlet.

4. Tide rising in Bay; Current Flooding; Crest at Inlet (HW at Inlet).

5. Tide rising in Bay; Current Flooding; Crest has passed Inlet.

6. High Water in Bay; Ebb Current about to start.

7. Tide dropping in Bay; Current Ebbing; Mean Water at Ocean edge of Inlet.

8. Tide dropping in Bay; Current Ebbing.

9. Tide dropping in Bay; Current Ebbing.

10. Tide dropping in Bay; Current Ebbing; Low Water at Ocean edge of Inlet.

11. Tide dropping in Bay; Current Ebbing; Tide rising at Ocean edge of Inlet.

12. Low Water in Bay; Flood Current about to start.

155

2009 HIGH WATER
BALTIMORE, MD

At Ft. McHenry 39°16'N, 76°34.7'W

Standard Time   Standard Time   Daylight Time starts
March 8 at 2 a.m.

DAY OF MONTH	DAY OF WEEK	JANUARY a.m.	Ht.	p.m.	Ht.	DAY OF WEEK	FEBRUARY a.m.	Ht.	p.m.	Ht.	DAY OF WEEK	MARCH a.m.	Ht.	p.m.	Ht.	DAY OF MONTH
1	T	9 33	0.7	9 53	1.1	S	10 43	1.0	10 47	0.7	S	9 26	1.3	9 41	0.8	1
2	F	10 22	0.8	10 32	1.0	M	11 38	1.1	11 38	0.6	M	10 16	1.3	10 30	0.8	2
3	S	11 14	0.9	11 16	0.9	T	...	...	12 38	1.2	T	11 12	1.3	11 26	0.7	3
4	S	...	...	12 10	1.0	W	12 37	0.6	1 42	1.2	W	...	...	12 14	1.3	4
5	M	12 04	0.8	1 09	1.1	T	1 41	0.5	2 50	1.2	T	12 28	0.7	1 23	1.3	5
6	T	12 58	0.6	2 09	1.2	F	2 46	0.6	3 56	1.3	F	1 34	0.8	2 35	1.3	6
7	W	1 57	0.6	3 10	1.3	S	3 49	0.7	4 58	1.3	S	2 39	0.9	3 43	1.3	7
8	T	3 00	0.5	4 10	1.4	S	4 49	0.8	5 54	1.3	S	*4 40	1.0	*5 43	1.3	8
9	F	4 03	0.5	5 10	1.4	M	5 46	0.9	6 46	1.3	M	5 39	1.1	6 36	1.3	9
10	S	5 01	0.6	6 06	1.4	T	6 39	1.0	7 31	1.2	T	6 33	1.2	7 21	1.2	10
11	S	5 59	0.7	7 00	1.4	W	7 33	1.0	8 13	1.1	W	7 25	1.3	8 02	1.2	11
12	M	6 55	0.8	7 51	1.3	T	8 25	1.1	8 54	1.0	T	8 15	1.4	8 42	1.1	12
13	T	7 50	0.8	8 39	1.2	F	9 18	1.1	9 35	0.9	F	9 03	1.4	9 22	1.0	13
14	W	8 46	0.9	9 25	1.1	S	10 12	1.1	10 17	0.8	S	9 50	1.4	10 02	0.9	14
15	T	9 44	0.9	10 10	1.0	S	11 06	1.1	11 01	0.7	S	10 36	1.4	10 45	0.9	15
16	F	10 43	0.9	10 54	0.9	M	11 59	1.1	11 50	0.6	M	11 24	1.3	11 31	0.8	16
17	S	11 45	1.0	11 39	0.7	T	...	...	1 04	1.0	T	...	...	12 15	1.3	17
18	S	...	...	12 48	1.0	W	12 45	0.6	2 08	1.0	W	12 21	0.8	1 11	1.2	18
19	M	12 26	0.6	1 51	1.0	T	1 43	0.6	3 11	1.0	T	1 17	0.8	2 13	1.1	19
20	T	1 18	0.5	2 53	1.0	F	2 42	0.6	4 07	1.0	F	2 15	0.9	3 16	1.1	20
21	W	2 13	0.5	3 50	1.0	S	3 37	0.7	4 53	1.0	S	3 13	0.9	4 14	1.1	21
22	T	3 09	0.5	4 40	1.1	S	4 27	0.7	5 32	1.0	S	4 08	1.0	5 02	1.1	22
23	F	4 03	0.5	5 25	1.1	M	5 12	0.8	6 06	1.1	M	4 58	1.1	5 43	1.1	23
24	S	4 52	0.5	6 04	1.1	T	5 55	0.9	6 37	1.1	T	5 45	1.2	6 20	1.1	24
25	S	5 37	0.6	6 38	1.1	W	6 36	0.9	7 09	1.1	W	6 28	1.3	6 56	1.1	25
26	M	6 20	0.6	7 11	1.1	T	7 16	1.0	7 42	1.0	T	7 09	1.4	7 32	1.1	26
27	T	7 00	0.7	7 42	1.1	F	7 57	1.1	8 18	1.0	F	7 50	1.5	8 11	1.1	27
28	W	7 41	0.7	8 13	1.0	S	8 40	1.2	8 57	0.9	S	8 31	1.6	8 52	1.0	28
29	T	8 22	0.8	8 46	1.0						S	9 15	1.7	9 38	1.0	29
30	F	9 06	0.8	9 22	0.9						M	10 02	1.7	10 27	0.9	30
31	S	9 53	0.9	10 02	0.8						T	10 54	1.7	11 22	0.9	31

*Daylight Saving Time starts

Dates when Ht. of **Low** Water is below Mean Low with Ht. of lowest given for each period and Date of lowest in ( ):

3rd - 31st: -0.6' (10th - 12th)   1st - 25th: -0.5' (7th - 10th)   1st - 11th: -0.3' (8th - 10th)

Average Rise and Fall 1.1 ft.

When a high tide exceeds av. ht., the *following* low tide will be lower than av.

2009 HIGH WATER
BALTIMORE, MD
At Ft. McHenry 39°16'N, 76°34.7'W

		Daylight Saving Time APRIL					Daylight Saving Time MAY					Daylight Saving Time JUNE				
D.o.M	D.o.W	a.m.	Ht.	p.m.	Ht.	D.o.W	a.m.	Ht.	p.m.	Ht.	D.o.W	a.m.	Ht.	p.m.	Ht.	D.o.M
1	W	11 52	1.6	...	...	F	12 16	1.2	12 42	1.6	M	2 13	1.6	2 14	1.4	1
2	T	12 21	1.0	12 57	1.5	S	1 19	1.3	1 48	1.5	T	3 16	1.8	3 08	1.3	2
3	F	1 25	1.0	2 07	1.4	S	2 23	1.4	2 50	1.4	W	4 14	1.9	4 01	1.2	3
4	S	2 29	1.1	3 18	1.4	M	3 26	1.6	3 48	1.4	T	5 07	1.9	4 53	1.1	4
5	S	3 33	1.2	4 22	1.3	T	4 25	1.7	4 41	1.3	F	5 54	2.0	5 43	1.1	5
6	M	4 33	1.4	5 17	1.3	W	5 20	1.8	5 29	1.2	S	6 37	2.0	6 32	1.1	6
7	T	5 29	1.5	6 05	1.3	T	6 09	1.9	6 15	1.2	S	7 17	2.0	7 20	1.1	7
8	W	6 21	1.6	6 49	1.2	F	6 54	1.9	6 59	1.1	M	7 54	1.9	8 05	1.1	8
9	T	7 11	1.7	7 31	1.2	S	7 37	1.9	7 44	1.1	T	8 32	1.9	8 50	1.2	9
10	F	7 56	1.7	8 11	1.1	S	8 15	1.9	8 27	1.1	W	9 07	1.8	9 33	1.2	10
11	S	8 39	1.7	8 52	1.0	M	8 53	1.9	9 11	1.1	T	9 44	1.7	10 17	1.2	11
12	S	9 20	1.7	9 34	1.0	T	9 31	1.8	9 56	1.1	F	10 21	1.7	11 03	1.3	12
13	M	10 02	1.6	10 19	1.0	W	10 11	1.7	10 42	1.2	S	10 59	1.6	11 51	1.4	13
14	T	10 44	1.6	11 06	1.0	T	10 53	1.6	11 30	1.2	S	11 39	1.5	...	...	14
15	W	11 30	1.5	11 56	1.0	F	11 38	1.5	...	...	M	12 41	1.4	12 22	1.5	15
16	T	...	...	12 21	1.4	S	12 21	1.3	12 24	1.5	T	1 33	1.5	1 08	1.4	16
17	F	12 50	1.1	1 16	1.3	S	1 15	1.3	1 13	1.4	W	2 25	1.7	1 59	1.3	17
18	S	1 46	1.1	2 13	1.3	M	2 09	1.4	2 02	1.3	T	3 16	1.8	2 55	1.1	18
19	S	2 42	1.2	3 07	1.2	T	3 02	1.5	2 52	1.3	F	4 07	2.0	3 55	1.1	19
20	M	3 37	1.3	3 57	1.2	W	3 53	1.6	3 44	1.2	S	4 58	2.1	4 57	1.0	20
21	T	4 27	1.4	4 43	1.2	T	4 40	1.8	4 36	1.1	S	5 49	2.2	5 58	1.1	21
22	W	5 14	1.5	5 27	1.2	F	5 27	1.9	5 29	1.1	M	6 42	2.2	6 57	1.1	22
23	T	5 58	1.7	6 11	1.1	S	6 13	2.1	6 23	1.1	T	7 35	2.2	7 54	1.2	23
24	F	6 40	1.8	6 56	1.1	S	7 00	2.1	7 18	1.1	W	8 28	2.1	8 51	1.3	24
25	S	7 23	1.9	7 42	1.1	M	7 49	2.2	8 12	1.1	T	9 20	2.0	9 48	1.4	25
26	S	8 08	2.0	8 31	1.1	T	8 39	2.1	9 08	1.2	F	10 12	1.9	10 48	1.5	26
27	M	8 55	2.0	9 23	1.1	W	9 33	2.0	10 05	1.3	S	11 03	1.8	11 49	1.6	27
28	T	9 45	1.9	10 17	1.1	T	10 28	1.9	11 04	1.3	S	11 53	1.6	...	...	28
29	W	10 40	1.9	11 15	1.1	F	11 25	1.8	...	...	M	12 52	1.7	12 43	1.5	29
30	T	11 39	1.7	...	...	S	12 05	1.4	12 22	1.7	T	1 55	1.8	1 34	1.3	30
31						S	1 09	1.5	1 19	1.5						31

Dates when Ht. of **Low** Water is below Mean Low with Ht. of lowest given for each period and Date of lowest in ( ):

Average Rise and Fall 1.1 ft.

When a high tide exceeds av. ht., the *following* low tide will be lower than av.

2009 HIGH WATER
BALTIMORE, MD
At Ft. McHenry 39°16'N, 76°34.7'W

		Daylight Saving Time					Daylight Saving Time					Daylight Saving Time				
DAY OF MONTH	DAY OF WEEK	JULY				DAY OF WEEK	AUGUST				DAY OF WEEK	SEPTEMBER				DAY OF MONTH
		a.m.	Ht.	p.m.	Ht.		a.m.	Ht.	p.m.	Ht.		a.m.	Ht.	p.m.	Ht.	
1	W	2 57	1.9	2 28	1.2	S	4 24	1.9	3 54	1.1	T	5 28	1.8	5 21	1.3	1
2	T	3 56	1.9	3 24	1.1	S	5 14	1.9	4 53	1.1	W	6 06	1.8	6 09	1.4	2
3	F	4 49	2.0	4 21	1.0	M	5 59	1.9	5 46	1.2	T	6 39	1.8	6 53	1.5	3
4	S	5 36	2.0	5 17	1.0	T	6 38	1.8	6 34	1.2	F	7 09	1.7	7 34	1.5	4
5	S	6 20	2.0	6 09	1.1	W	7 13	1.8	7 19	1.3	S	7 39	1.7	8 13	1.6	5
6	M	6 59	1.9	6 57	1.1	T	7 45	1.8	8 01	1.4	S	8 11	1.7	8 52	1.7	6
7	T	7 36	1.9	7 43	1.2	F	8 15	1.8	8 41	1.4	M	8 44	1.6	9 31	1.8	7
8	W	8 11	1.8	8 25	1.2	S	8 45	1.7	9 21	1.5	T	9 21	1.5	10 13	1.9	8
9	T	8 45	1.8	9 09	1.3	S	9 17	1.7	10 03	1.6	W	10 04	1.4	10 59	2.0	9
10	F	9 16	1.8	9 50	1.3	M	9 50	1.6	10 45	1.7	T	10 51	1.3	11 50	2.0	10
11	S	9 49	1.7	10 33	1.4	T	10 27	1.5	11 30	1.8	F	11 46	1.2	...	...	11
12	S	10 23	1.6	11 18	1.5	W	11 10	1.4	...	...	S	12 47	2.0	12 49	1.2	12
13	M	11 00	1.6	11 59	1.6	T	12 19	1.9	12 01	1.3	S	1 51	2.0	1 57	1.2	13
14	T	11 41	1.4	...	...	F	1 13	2.0	12 59	1.2	M	2 58	2.0	3 06	1.3	14
15	W	12 55	1.7	12 27	1.3	S	2 12	2.0	2 05	1.1	T	4 02	1.9	4 12	1.4	15
16	T	1 47	1.8	1 21	1.2	S	3 14	2.0	3 14	1.2	W	5 01	1.9	5 13	1.5	16
17	F	2 41	1.9	2 22	1.1	M	4 18	2.1	4 21	1.2	T	5 53	1.9	6 11	1.7	17
18	S	3 37	2.0	3 28	1.1	T	5 18	2.1	5 24	1.4	F	6 40	1.8	7 06	1.8	18
19	S	4 35	2.1	4 35	1.1	W	6 14	2.0	6 23	1.5	S	7 24	1.7	7 59	1.9	19
20	M	5 32	2.1	5 38	1.2	T	7 05	2.0	7 20	1.6	S	8 06	1.6	8 49	2.0	20
21	T	6 28	2.1	6 38	1.3	F	7 52	1.9	8 15	1.7	M	8 47	1.5	9 39	2.0	21
22	W	7 22	2.1	7 36	1.4	S	8 36	1.8	9 10	1.8	T	9 30	1.4	10 28	2.0	22
23	T	8 13	2.0	8 33	1.5	S	9 19	1.7	10 05	1.9	W	10 14	1.3	11 18	1.9	23
24	F	9 02	2.0	9 30	1.6	M	10 01	1.6	10 59	1.9	T	11 02	1.2	...	...	24
25	S	9 48	1.8	10 28	1.7	T	10 45	1.4	11 55	1.9	F	12 11	1.9	12 01	1.2	25
26	S	10 34	1.7	11 27	1.8	W	11 32	1.3	...	...	S	1 07	1.8	12 54	1.2	26
27	M	11 19	1.5	...	...	T	12 52	1.9	12 24	1.2	S	2 06	1.7	1 57	1.2	27
28	T	12 27	1.8	12 06	1.4	F	1 51	1.9	1 22	1.2	M	3 05	1.7	3 00	1.2	28
29	W	1 28	1.9	12 56	1.2	S	2 52	1.8	2 25	1.2	T	3 56	1.6	3 59	1.3	29
30	T	2 29	1.9	1 52	1.1	S	3 51	1.8	3 29	1.2	W	4 40	1.6	4 53	1.4	30
31	F	3 29	1.9	2 53	1.1	M	4 44	1.8	4 28	1.2						31

Dates when Ht. of **Low** Water is below Mean Low with Ht. of lowest given for each period and Date of lowest in ( ):

Average Rise and Fall 1.1 ft.

When a high tide exceeds av. ht., the *following* low tide will be lower than av.

2009 HIGH WATER
BALTIMORE, MD
At Ft. McHenry 39°16'N, 76°34.7'W

Daylight Saving Time — Standard Time starts Nov. 1 at 2 a.m. — **Standard Time**

DAY OF MONTH	DAY OF WEEK	OCTOBER a.m.	Ht.	p.m.	Ht.	DAY OF WEEK	NOVEMBER a.m.	Ht.	p.m.	Ht.	DAY OF WEEK	DECEMBER a.m.	Ht.	p.m.	Ht.	DAY OF MONTH
1	T	5 18	1.6	5 42	1.5	S	*4 41	1.2	*5 34	1.7	T	4 50	0.8	5 47	1.6	1
2	F	5 52	1.6	6 25	1.6	M	5 23	1.2	6 14	1.8	W	5 42	0.8	6 34	1.7	2
3	S	6 26	1.6	7 05	1.7	T	6 07	1.1	6 55	1.9	T	6 34	0.8	7 24	1.7	3
4	S	7 00	1.5	7 43	1.8	W	6 53	1.0	7 39	1.9	F	7 27	0.8	8 15	1.6	4
5	M	7 37	1.4	8 22	1.9	T	7 42	1.0	8 27	1.9	S	8 22	0.8	9 08	1.6	5
6	T	8 16	1.3	9 02	1.9	F	8 35	1.0	9 19	1.8	S	9 19	0.9	10 02	1.5	6
7	W	8 59	1.3	9 46	2.0	S	9 31	1.0	10 16	1.7	M	10 20	0.9	10 57	1.3	7
8	T	9 47	1.2	10 35	2.0	S	10 32	1.0	11 16	1.6	T	11 25	1.0	11 51	1.2	8
9	F	10 41	1.2	11 30	2.0	M	11 38	1.1	...	...	W	...	...	12 34	1.1	9
10	S	11 39	1.2	...	...	T	12 16	1.5	12 45	1.2	T	12 44	1.1	1 41	1.2	10
11	S	12 30	1.9	12 44	1.2	W	1 16	1.5	1 53	1.3	F	1 37	1.0	2 46	1.3	11
12	M	1 35	1.8	1 52	1.2	T	2 11	1.4	2 58	1.4	S	2 29	0.8	3 46	1.4	12
13	T	2 40	1.8	3 00	1.4	F	3 03	1.3	3 57	1.5	S	3 21	0.7	4 40	1.4	13
14	W	3 41	1.7	4 05	1.5	S	3 52	1.2	4 51	1.6	M	4 13	0.7	5 28	1.4	14
15	T	4 36	1.6	5 05	1.6	S	4 39	1.1	5 41	1.7	T	5 02	0.7	6 12	1.4	15
16	F	5 25	1.6	6 01	1.7	M	5 25	1.0	6 26	1.7	W	5 49	0.6	6 54	1.4	16
17	S	6 10	1.5	6 53	1.9	T	6 10	0.9	7 09	1.7	T	6 34	0.7	7 33	1.3	17
18	S	6 53	1.4	7 42	1.9	W	6 55	0.9	7 51	1.7	F	7 18	0.7	8 11	1.3	18
19	M	7 36	1.3	8 28	1.9	T	7 39	0.9	8 32	1.6	S	8 01	0.7	8 48	1.2	19
20	T	8 19	1.2	9 13	1.9	F	8 24	0.9	9 14	1.5	S	8 44	0.7	9 24	1.2	20
21	W	9 02	1.2	9 57	1.9	S	9 10	0.9	9 56	1.4	M	9 30	0.7	9 59	1.1	21
22	T	9 48	1.1	10 43	1.8	S	9 59	0.9	10 39	1.3	T	10 19	0.7	10 36	1.0	22
23	F	10 36	1.1	11 31	1.7	M	10 52	0.9	11 23	1.3	W	11 11	0.8	11 14	1.0	23
24	S	11 28	1.1	...	...	T	11 49	0.9	...	...	T	...	...	12 06	0.8	24
25	S	12 22	1.6	12 24	1.1	W	12 06	1.2	12 48	1.0	F	12 01	0.9	1 01	0.9	25
26	M	1 15	1.5	1 25	1.1	T	12 50	1.1	1 47	1.1	S	12 41	0.8	1 55	1.0	26
27	T	2 06	1.5	2 27	1.2	F	1 35	1.1	2 41	1.2	S	1 32	0.7	2 48	1.2	27
28	W	2 54	1.4	3 26	1.2	S	2 21	1.0	3 31	1.3	M	2 28	0.6	3 41	1.3	28
29	T	3 38	1.4	4 21	1.3	S	3 09	0.9	4 17	1.4	T	3 27	0.6	4 34	1.4	29
30	F	4 20	1.3	5 09	1.4	M	3 59	0.8	5 02	1.5	W	4 25	0.6	5 27	1.4	30
31	S	5 00	1.3	5 53	1.5						T	5 21	0.6	6 19	1.4	31

*Standard Time starts

Dates when Ht. of **Low** Water is below Mean Low with Ht. of lowest given for each period and Date of lowest in ( ):

30th: -0.2'

1st - 6th: -0.3' (3rd - 4th)
11th - 20th: -0.3' (12th - 17tth)
25th - 31st: -0.5' (31st)

Average Rise and Fall 1.1 ft.

When a high tide exceeds av. ht., the *following* low tide will be lower than av.

2009 HIGH WATER
MIAMI HARBOR ENTRANCE, FL
25°45.8'N, 80°07.8'W

		Standard Time		Standard Time					Daylight Time starts Mar. 8 at 2 a.m.			

DAY OF MONTH	DAY OF WEEK	JANUARY				DAY OF WEEK	FEBRUARY				DAY OF WEEK	MARCH				DAY OF MONTH
		a.m.	Ht.	p.m.	Ht.		a.m.	Ht.	p.m.	Ht.		a.m.	Ht.	p.m.	Ht.	
1	T	11 31	2.2	11 53	2.1	S	12 16	2.1	12 20	2.0	S	11 14	2.2	11 57	2.2	1
2	F	...	...	12 09	2.1	M	1 10	2.0	1 10	1.9	M	11 59	2.1	...	...	2
3	S	12 42	2.1	12 52	2.1	T	2 12	2.0	2 11	1.9	T	12 52	2.2	12 53	2.0	3
4	S	1 37	2.1	1 42	2.0	W	3 22	2.0	3 23	1.9	W	1 55	2.1	1 58	2.0	4
5	M	2 39	2.1	2 39	2.0	T	4 33	2.1	4 37	2.0	T	3 05	2.1	3 14	2.0	5
6	T	3 45	2.2	3 45	2.1	F	5 38	2.2	5 46	2.2	F	4 16	2.1	4 29	2.1	6
7	W	4 52	2.3	4 52	2.1	S	6 36	2.4	6 47	2.4	S	5 21	2.3	5 37	2.3	7
8	T	5 54	2.4	5 56	2.3	S	7 28	2.5	7 42	2.5	S	*7 17	2.4	*7 36	2.5	8
9	F	6 52	2.6	6 58	2.4	M	8 17	2.6	8 35	2.6	M	8 08	2.5	8 30	2.6	9
10	S	7 45	2.7	7 53	2.6	T	9 01	2.7	9 23	2.6	T	8 53	2.6	9 17	2.7	10
11	S	8 35	2.8	8 47	2.6	W	9 45	2.6	10 10	2.6	W	9 36	2.6	10 03	2.7	11
12	M	9 23	2.8	9 40	2.7	T	10 28	2.5	10 56	2.5	T	10 18	2.6	10 47	2.7	12
13	T	10 10	2.7	10 31	2.6	F	11 09	2.4	11 43	2.3	F	10 57	2.5	11 29	2.5	13
14	W	10 57	2.6	11 23	2.5	S	11 51	2.2	...	...	S	11 36	2.4	...	...	14
15	T	11 43	2.5	...	...	S	12 30	2.1	12 35	2.0	S	12 11	2.4	12 16	2.2	15
16	F	12 14	2.3	12 29	2.3	M	1 20	1.9	1 22	1.8	M	12 53	2.2	12 56	2.0	16
17	S	1 08	2.2	1 18	2.1	T	2 16	1.8	2 16	1.7	T	1 39	2.0	1 41	1.9	17
18	S	2 04	2.0	2 09	1.9	W	3 18	1.7	3 18	1.7	W	2 31	1.9	2 33	1.8	18
19	M	3 03	1.9	3 04	1.8	T	4 23	1.7	4 23	1.7	T	3 29	1.8	3 35	1.7	19
20	T	4 05	1.9	4 03	1.8	F	5 21	1.8	5 23	1.8	F	4 33	1.8	4 43	1.8	20
21	W	5 04	1.9	5 01	1.8	S	6 10	1.9	6 16	1.9	S	5 34	1.9	5 47	1.8	21
22	T	5 57	1.9	5 54	1.8	S	6 53	2.0	7 02	2.0	S	6 27	2.0	6 43	2.0	22
23	F	6 43	2.0	6 42	1.9	M	7 32	2.2	7 44	2.1	M	7 13	2.1	7 32	2.2	23
24	S	7 24	2.1	7 26	2.0	T	8 10	2.2	8 25	2.2	T	7 54	2.2	8 16	2.3	24
25	S	8 03	2.2	8 07	2.1	W	8 46	2.3	9 05	2.3	W	8 34	2.3	8 59	2.5	25
26	M	8 40	2.2	8 48	2.1	T	9 21	2.3	9 45	2.4	T	9 13	2.4	9 41	2.6	26
27	T	9 15	2.3	9 27	2.1	F	9 57	2.3	10 26	2.4	F	9 51	2.4	10 23	2.6	27
28	W	9 51	2.2	10 06	2.1	S	10 34	2.2	11 09	2.3	S	10 31	2.4	11 07	2.6	28
29	T	10 25	2.2	10 47	2.1						S	11 12	2.4	11 53	2.6	29
30	F	11 01	2.1	11 29	2.1						M	11 57	2.4	...	...	30
31	S	11 38	2.1	...	...						T	12 44	2.5	12 48	2.3	31

*Daylight Saving Time starts

Dates when Ht. of **Low** Water is below Mean Low with Ht. of lowest given for each period and Date of lowest in ( ):

6th - 17th: -0.8' (11th)	1st - 14th: -0.7' (8th - 10th)	1st - 15th: -0.6' (11th - 12th)
24th - 31st: -0.3' (27th)	24th - 28th: -0.4' (27th - 28th)	26th - 31st: -0.4' (27th - 29th)

Average Rise and Fall 2.5 ft.

When a high tide exceeds av. ht., the *following* low tide will be lower than av.

2009 HIGH WATER
MIAMI HARBOR ENTRANCE, FL
25°45.8'N, 80°07.8'W

DAY OF MONTH	DAY OF WEEK	APRIL a.m.	Ht.	p.m.	Ht.	DAY OF WEEK	MAY a.m.	Ht.	p.m.	Ht.	DAY OF WEEK	JUNE a.m.	Ht.	p.m.	Ht.	DAY OF MONTH
1	W	1 40	2.4	1 47	2.2	F	2 27	2.5	2 49	2.3	M	3 59	2.3	4 43	2.3	1
2	T	2 42	2.3	2 55	2.2	S	3 28	2.4	3 58	2.3	T	4 56	2.3	5 43	2.4	2
3	F	3 49	2.3	4 08	2.2	S	4 30	2.4	5 05	2.4	W	5 51	2.2	6 39	2.4	3
4	S	4 56	2.3	5 20	2.3	M	5 29	2.4	6 07	2.5	T	6 43	2.2	7 29	2.4	4
5	S	5 57	2.4	6 25	2.4	T	6 23	2.4	7 02	2.6	F	7 30	2.2	8 15	2.4	5
6	M	6 52	2.5	7 21	2.6	W	7 13	2.4	7 52	2.6	S	8 15	2.2	8 57	2.4	6
7	T	7 42	2.5	8 12	2.7	T	7 59	2.4	8 37	2.6	S	8 57	2.2	9 37	2.4	7
8	W	8 27	2.6	8 58	2.8	F	8 42	2.4	9 19	2.6	M	9 37	2.2	10 16	2.4	8
9	T	9 10	2.6	9 42	2.7	S	9 23	2.4	10 00	2.6	T	10 18	2.2	10 55	2.3	9
10	F	9 49	2.5	10 22	2.7	S	10 01	2.3	10 38	2.5	W	10 57	2.1	11 32	2.3	10
11	S	10 28	2.5	11 02	2.6	M	10 40	2.3	11 17	2.4	T	11 38	2.1	...	...	11
12	S	11 06	2.4	11 42	2.4	T	11 19	2.2	11 56	2.3	F	12 10	2.2	12 20	2.0	12
13	M	11 44	2.2	...	...	W	11 59	2.1	...	...	S	12 49	2.2	1 05	2.0	13
14	T	12 22	2.3	12 24	2.1	T	12 37	2.2	12 43	2.0	S	1 29	2.1	1 53	2.0	14
15	W	1 05	2.2	1 08	2.0	F	1 20	2.1	1 31	2.0	M	2 11	2.1	2 45	2.0	15
16	T	1 52	2.0	1 59	1.9	S	2 05	2.1	2 24	1.9	T	2 56	2.0	3 42	2.0	16
17	F	2 45	2.0	2 58	1.9	S	2 54	2.0	3 22	1.9	W	3 47	2.0	4 42	2.1	17
18	S	3 42	1.9	4 02	1.9	M	3 45	2.0	4 22	2.0	T	4 43	2.1	5 43	2.2	18
19	S	4 39	2.0	5 06	2.0	T	4 38	2.0	5 22	2.1	F	5 43	2.1	6 43	2.4	19
20	M	5 34	2.0	6 04	2.1	W	5 32	2.1	6 18	2.3	S	6 43	2.2	7 40	2.5	20
21	T	6 24	2.1	6 56	2.3	T	6 24	2.2	7 12	2.5	S	7 41	2.4	8 35	2.7	21
22	W	7 10	2.2	7 44	2.5	F	7 16	2.3	8 04	2.6	M	8 39	2.5	9 27	2.8	22
23	T	7 55	2.4	8 31	2.6	S	8 07	2.4	8 54	2.7	T	9 34	2.6	10 18	2.8	23
24	F	8 39	2.4	9 17	2.7	S	8 58	2.5	9 45	2.8	W	10 30	2.6	11 09	2.8	24
25	S	9 23	2.5	10 03	2.8	M	9 49	2.5	10 35	2.8	T	11 25	2.6	11 58	2.8	25
26	S	10 08	2.5	10 50	2.8	T	10 42	2.6	11 26	2.8	F	...	...	12 20	2.6	26
27	M	10 56	2.5	11 40	2.7	W	11 37	2.6	...	...	S	12 48	2.7	1 16	2.5	27
28	T	11 47	2.5	...	...	T	12 18	2.7	12 34	2.5	S	1 39	2.5	2 14	2.4	28
29	W	12 32	2.6	12 42	2.4	F	1 11	2.6	1 33	2.5	M	2 31	2.4	3 13	2.3	29
30	T	1 28	2.5	1 43	2.4	S	2 06	2.5	2 35	2.4	T	3 25	2.3	4 14	2.3	30
31						S	3 02	2.4	3 39	2.4						31

Dates when Ht. of **Low** Water is below Mean Low with Ht. of lowest given for each period and Date of lowest in ( ):

6th - 12th: -0.4' (8th - 10th)	5th - 10th: -0.3' (7th - 8th)	3rd - 7th: -0.2'
23rd - 29th: -0.5' (26th - 27th)	22nd - 28th: -0.6' (24th - 25th)	19th - 29th: -0.7' (23rd)

Average Rise and Fall 2.5 ft.

When a high tide exceeds av. ht., the *following* low tide will be lower than av.

161

2009 HIGH WATER
MIAMI HARBOR ENTRANCE, FL
25°45.8'N, 80°07.8'W

		Daylight Saving Time					Daylight Saving Time					Daylight Saving Time				
D A Y O F M O N T H	D A Y O F W E E K	JULY				D A Y O F W E E K	AUGUST				D A Y O F W E E K	SEPTEMBER				D A Y O F M O N T H
		a.m.	Ht.	p.m.	Ht.		a.m.	Ht.	p.m.	Ht.		a.m.	Ht.	p.m.	Ht.	
1	W	4 20	2.2	5 15	2.2	S	5 41	2.1	6 37	2.2	T	6 58	2.4	7 34	2.5	1
2	T	5 17	2.1	6 12	2.2	S	6 36	2.1	7 25	2.2	W	7 44	2.5	8 13	2.6	2
3	F	6 12	2.1	7 05	2.2	M	7 25	2.2	8 08	2.3	T	8 26	2.6	8 50	2.7	3
4	S	7 03	2.1	7 52	2.2	T	8 10	2.2	8 47	2.4	F	9 07	2.7	9 26	2.7	4
5	S	7 50	2.1	8 35	2.3	W	8 53	2.3	9 24	2.5	S	9 46	2.8	10 01	2.8	5
6	M	8 34	2.1	9 14	2.3	T	9 33	2.4	10 00	2.5	S	10 25	2.8	10 37	2.7	6
7	T	9 15	2.1	9 52	2.3	F	10 12	2.4	10 35	2.5	M	11 05	2.8	11 13	2.7	7
8	W	9 56	2.2	10 29	2.3	S	10 51	2.4	11 09	2.5	T	11 47	2.8	11 51	2.6	8
9	T	10 37	2.2	11 06	2.3	S	11 32	2.4	11 45	2.4	W	...	...	12 34	2.7	9
10	F	11 16	2.2	11 41	2.3	M	...	...	12 11	2.4	T	12 34	2.6	1 25	2.7	10
11	S	11 56	2.1	...	...	T	12 20	2.4	12 55	2.4	F	1 25	2.5	2 25	2.6	11
12	S	12 17	2.2	12 38	2.1	W	12 59	2.3	1 45	2.4	S	2 27	2.5	3 31	2.6	12
13	M	12 53	2.2	1 22	2.1	T	1 44	2.3	2 42	2.3	S	3 39	2.5	4 40	2.7	13
14	T	1 32	2.1	2 12	2.1	F	2 39	2.3	3 47	2.3	M	4 53	2.6	5 44	2.8	14
15	W	2 15	2.1	3 07	2.1	S	3 45	2.3	4 56	2.4	T	6 02	2.8	6 42	2.9	15
16	T	3 06	2.1	4 09	2.2	S	4 58	2.3	6 02	2.5	W	7 03	3.0	7 34	3.1	16
17	F	4 06	2.1	5 15	2.2	M	6 09	2.5	7 02	2.7	T	7 58	3.2	8 22	3.2	17
18	S	5 13	2.2	6 20	2.4	T	7 13	2.7	7 56	2.9	F	8 49	3.3	9 08	3.2	18
19	S	6 21	2.3	7 21	2.5	W	8 12	2.9	8 47	3.0	S	9 38	3.3	9 52	3.2	19
20	M	7 25	2.4	8 16	2.7	T	9 06	3.0	9 34	3.1	S	10 24	3.3	10 35	3.1	20
21	T	8 24	2.6	9 08	2.8	F	9 57	3.1	10 20	3.1	M	11 09	3.2	11 18	3.0	21
22	W	9 21	2.7	9 58	2.9	S	10 47	3.1	11 05	3.0	T	11 55	3.0	...	...	22
23	T	10 15	2.8	10 46	2.9	S	11 36	3.0	11 50	2.9	W	12 01	2.8	12 41	2.9	23
24	F	11 08	2.8	11 33	2.9	M	...	...	12 24	2.9	T	12 46	2.7	1 30	2.7	24
25	S	11 59	2.8	...	...	T	12 35	2.7	1 14	2.7	F	1 34	2.5	2 23	2.5	25
26	S	12 21	2.8	12 53	2.7	W	1 22	2.5	2 06	2.5	S	2 29	2.4	3 21	2.4	26
27	M	1 08	2.6	1 46	2.5	T	2 11	2.4	3 03	2.4	S	3 30	2.4	4 21	2.4	27
28	T	1 57	2.4	2 42	2.4	F	3 06	2.3	4 04	2.3	M	4 34	2.4	5 18	2.5	28
29	W	2 49	2.3	3 40	2.3	S	4 07	2.2	5 06	2.3	T	5 34	2.5	6 08	2.6	29
30	T	3 44	2.1	4 41	2.2	S	5 09	2.2	6 02	2.3	W	6 26	2.6	6 52	2.7	30
31	F	4 42	2.1	5 42	2.2	M	6 07	2.3	6 51	2.4						31

Dates when Ht. of **Low** Water is below Mean Low with Ht. of lowest given for each period and Date of lowest in ( ):

18th - 27th: -0.6' (21st - 22nd)   18th - 23rd: -0.3' (19th - 20th, 22nd)

Average Rise and Fall 2.5 ft.

When a high tide exceeds av. ht., the *following* low tide will be lower than av.

2009 HIGH WATER
MIAMI HARBOR ENTRANCE, FL
25°45.8'N, 80°07.8'W

| | Daylight Saving Time | | Standard Time starts Nov. 1 at 2 a.m. | | | Standard Time | |

DAY OF MONTH	DAY OF WEEK	OCTOBER				DAY OF WEEK	NOVEMBER				DAY OF WEEK	DECEMBER				DAY OF MONTH
		a.m.	Ht.	p.m.	Ht.		a.m.	Ht.	p.m.	Ht.		a.m.	Ht.	p.m.	Ht.	
1	T	7 13	2.7	7 33	2.8	S	*7 09	3.0	*7 15	2.8	T	7 31	2.9	7 32	2.7	1
2	F	7 57	2.9	8 12	2.9	M	7 53	3.1	7 57	2.9	W	8 19	3.0	8 22	2.7	2
3	S	8 38	3.0	8 50	2.9	T	8 38	3.2	8 41	2.9	T	9 08	3.0	9 13	2.8	3
4	S	9 19	3.1	9 28	2.9	W	9 24	3.2	9 27	2.9	F	9 58	3.0	10 06	2.8	4
5	M	10 00	3.1	10 06	2.9	T	10 12	3.1	10 17	2.9	S	10 48	2.9	11 02	2.7	5
6	T	10 43	3.1	10 47	2.9	F	11 03	3.0	11 11	2.8	S	11 40	2.8	...	...	6
7	W	11 28	3.1	11 30	2.8	S	11 58	3.0	...	...	M	12 01	2.7	12 33	2.7	7
8	T	...	...	12 17	3.0	S	12 11	2.7	12 55	2.9	T	1 02	2.6	1 29	2.6	8
9	F	12 21	2.8	1 12	2.9	M	1 17	2.7	1 56	2.8	W	2 08	2.6	2 28	2.5	9
10	S	1 17	2.7	2 11	2.8	T	2 25	2.7	2 56	2.8	T	3 12	2.5	3 26	2.5	10
11	S	2 23	2.7	3 16	2.8	W	3 33	2.8	3 56	2.8	F	4 15	2.6	4 24	2.4	11
12	M	3 35	2.7	4 21	2.8	T	4 36	2.9	4 52	2.8	S	5 14	2.6	5 19	2.4	12
13	T	4 46	2.8	5 22	2.9	F	5 34	3.0	5 44	2.9	S	6 08	2.6	6 10	2.4	13
14	W	5 52	3.0	6 19	3.0	S	6 26	3.1	6 33	2.9	M	6 56	2.6	6 57	2.4	14
15	T	6 50	3.1	7 10	3.1	S	7 14	3.1	7 18	2.9	T	7 41	2.6	7 41	2.4	15
16	F	7 43	3.3	7 57	3.2	M	7 58	3.1	8 01	2.8	W	8 22	2.6	8 22	2.4	16
17	S	8 31	3.4	8 42	3.2	T	8 40	3.0	8 43	2.8	T	9 01	2.6	9 02	2.3	17
18	S	9 17	3.4	9 25	3.1	W	9 21	2.9	9 23	2.7	F	9 38	2.5	9 42	2.3	18
19	M	10 01	3.3	10 07	3.0	T	10 01	2.8	10 04	2.6	S	10 15	2.4	10 22	2.2	19
20	T	10 44	3.2	10 48	2.9	F	10 42	2.7	10 45	2.5	S	10 52	2.4	11 03	2.2	20
21	W	11 26	3.0	11 30	2.8	S	11 23	2.6	11 30	2.4	M	11 30	2.3	11 46	2.1	21
22	T	...	...	12 10	2.9	S	...	...	12 05	2.5	T	...	...	12 08	2.2	22
23	F	12 13	2.7	12 55	2.7	M	12 18	2.3	12 50	2.4	W	12 33	2.0	12 49	2.1	23
24	S	1 00	2.5	1 43	2.6	T	1 11	2.2	1 38	2.3	T	1 24	2.0	1 34	2.0	24
25	S	1 52	2.4	2 35	2.5	W	2 08	2.2	2 28	2.3	F	2 19	2.0	2 24	2.0	25
26	M	2 51	2.4	3 30	2.5	T	3 07	2.3	3 20	2.3	S	3 20	2.1	3 20	2.0	26
27	T	3 53	2.4	4 25	2.5	F	4 05	2.4	4 13	2.3	S	4 21	2.2	4 19	2.1	27
28	W	4 53	2.5	5 17	2.5	S	5 00	2.5	5 04	2.4	M	5 21	2.3	5 20	2.2	28
29	T	5 48	2.6	6 05	2.6	S	5 52	2.6	5 54	2.5	T	6 18	2.4	6 18	2.3	29
30	F	6 38	2.7	6 49	2.7	M	6 42	2.8	6 43	2.6	W	7 11	2.6	7 14	2.4	30
31	S	7 25	2.9	7 33	2.8						T	8 02	2.7	8 08	2.5	31

*Standard Time starts

Dates when Ht. of **Low** Water is below Mean Low with Ht. of lowest given for each period and Date of lowest in ( ):

2nd - 5th: -0.3' (3rd - 4th)
28th: - 0.2'
30th - 31st: -0.5' (31st)

Average Rise and Fall 2.5 ft.

When a high tide exceeds av. ht., the *following* low tide will be lower than av.

163

The Shore Angler and Alongshore Currents
By Dave Anderson

Many shore anglers or surfcasters neglect to take into account the importance of current when making a trip into the littoral zone. It plays a major role in shore fishing success.

Currents Alongshore

If one day you happen upon a pod of baitfish dimpling the surface with gamefish on their tails, it's a great angling experience to be sure. But when that pod suddenly disappears or disperses, do you know which way to go to set yourself up for a possible round two? Knowing the direction of the alongshore current will give you that direction, and after that you simply have to walk, cast, watch and listen. Baitfish will fall in downstream of any obstruction or rocky structure, and gamefish on the hunt will travel upstream toward it. To make a natural presentation here, use a lure that mimics a crippled baitfish and make it struggle against the tide. A hunting striped bass will not pass up the easy meal.

Inlets

For currents at an inlet or river mouth, shore and boat anglers alike have to know the concept of tide "lag." This can occur anywhere water from a larger body is forced through a smaller opening. Inlets demonstrate the most extreme lags and offer an exaggerated example to describe how to fish current.

It's a rule of thumb that inlets fish best on a dropping tide. If you set yourself up at the mouth of any river or inlet to catch the high tide, you had better pack a lunch. As the tide in the ocean floods and fills in the area surrounding your inlet, the narrow mouth cannot quickly accommodate the massive amount of water that the tide is bringing. Therefore, the water backs up on the outside and then rushes into the pond or river at an increased speed for a period of hours until the tide inside "catches up" and equalizes. (See p. 155) It is this period of equalization or "slack tide" that an angler should note. This marks a time when bait that has been holding out of the current will make a move, and also when gamefish in the inlet will move out and feed on the dispersing schools of bait.

After the short slack period, the current will turn and begin to dump back into the ocean. Just like the incoming, the ebbing current may not stop until several hours after low tide outside. Some research will allow you to make your own tide chart for your inlet, by simply noting the hours and minutes difference from the posted tide. This same scenario takes place in harbors, bays and sounds all over the coast.

Using Current Diagrams

The current charts in this book will pay big dividends for all shore saltwater anglers.

Look at pp. 66-77 of this book, the Tidal Current Chart for Buzzard's Bay, Vineyard and Nantucket Sounds. Notice how the water speed picks up considerably when it rushes out of Nantucket Sound and into Vineyard Sound. This is due to the massive amount of water trying to escape Nantucket Sound through the markedly narrower Vineyard Sound. As the water backs up between Falmouth and Oak Bluffs, it rushes through at great speed until the water level equalizes. These ferocious currents are what draw in the monster striped bass that have made the Elizabeth Islands famous.

So where does the ocean current make the biggest difference for a shore angler? It's no secret or coincidence that many of the most famous striper outposts are points and islands. Montauk Point, Block Island, Cuttyhunk Island, and Race Point are possibly the four most famous striper destinations in the world. What makes them so attractive to anglers and gamefish is the current and structure.

Points and islands represent obstacles for the ocean currents, so as the currents run into them, the water backs up and the current speed increases, forcing large amounts of water around and through a reduced opening. This gives all inshore gamefish a major advantage. It's nothing for them to power through a screaming tiderip, but the smaller baitfish are in major trouble if they get sucked through. Their little bodies are powerless when matched up against a raging current, leaving them at the mercy of the sea until they make it out the other side. That is, if they make it. Striped bass and other inshore species set up in these areas, counting on that very scenario to take place over and over on each tide. They will lie in wait, shaded from the current by a piece of structure (a boulder, bar, or drop-off) and wait for helpless prey to struggle by.

Fatal Attraction

Angling presentations under these circumstances should imitate a baitfish that has been swept into the rips. Plastic swimmers, allowed to swing with the tide and worked with short bursts of speed followed by a struggling loss of ground, are often irresistible to cruising bass. The forward jerks trigger a kill instinct when it looks like their meal may escape. In deeper water, where the current is often reduced and often holds larger fish, bucktails or weighted rubber shads will be effective. Live baits such as eels or menhaden, drifted through the rips, are often too good to be ignored. By far the most exciting way to snatch a bass from these roiled tiderips is to toss out a topwater offering, such as a pencil popper, when the current begins to slow. This may replicate bait that was wounded during the height of the tide and is now making a last-ditch effort to find the safety of the shallow stones. No gamefish with any sense of pride will allow that to happen.

Current Logic

Current strength has a direct correlation to how bait and gamefish will concentrate, react and feed. It is an accepted fact that the largest specimens will wait out the strongest currents in favor of feeding under less strenuous conditions. With that in mind, we can now draw conclusions from Eldridge by simply selecting an area and watching the tide strength by the hour. If your location happens to be a structure-laden stretch of coast that is battered by powerful current, then a concentrated effort around the structure will often produce the best results during peak current periods. Making note of the periods of slack water and fishing within an hour or so on either side of it will often represent your best chance at connecting with the largest specimens, especially during spring tides when the current is the strongest.

Beyond that the book will help you determine the best areas to fish by using logic in conjunction with the facts within the book. If you see a powerful current pushing toward a major obstacle, such as a point or island, it would pay to investigate. Work the shorelines, look for deep holes or large obstructions that will provide relief from the currents, and fish them during all tide stages until you can decode their ins and outs. No matter how you slice it, knowing more about the current and tides in your home waters is going to make you a better angler. We can't ask for much better than that!

Dave Anderson is the Managing Editor for the New England Fisherman Magazine. He has been published in several outdoor publications spanning nearly the entire Atlantic coast and has given seminars for some of the Northeast's most prestigious fishing clubs. He spends most of his free time surfcasting from the shores of Buzzards and Narragansett Bays in southern New England, writing, or building his line of custom wooden plugs.

CHARACTERISTICS OF LIGHT SIGNALS
(see footnote on next page for abbreviations used.)

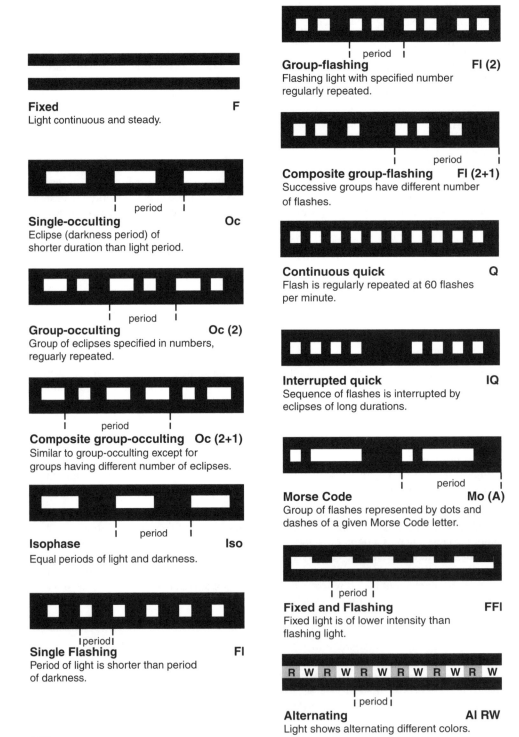

Fixed   **F**
Light continuous and steady.

Single-occulting   **Oc**
Eclipse (darkness period) of
shorter duration than light period.

Group-occulting   **Oc (2)**
Group of eclipses specified in numbers,
regularly repeated.

Composite group-occulting   **Oc (2+1)**
Similar to group-occulting except for
groups having different number of eclipses.

Isophase   **Iso**
Equal periods of light and darkness.

Single Flashing   **Fl**
Period of light is shorter than period
of darkness.

Group-flashing   **Fl (2)**
Flashing light with specified number
regularly repeated.

Composite group-flashing   **Fl (2+1)**
Successive groups have different number
of flashes.

Continuous quick   **Q**
Flash is regularly repeated at 60 flashes
per minute.

Interrupted quick   **IQ**
Sequence of flashes is interrupted by
eclipses of long durations.

Morse Code   **Mo (A)**
Group of flashes represented by dots and
dashes of a given Morse Code letter.

Fixed and Flashing   **FFl**
Fixed light is of lower intensity than
flashing light.

Alternating   **Al RW**
Light shows alternating different colors.

LIGHTS, FOG SIGNALS and OFFSHORE BUOYS

NOVA SCOTIA, EAST COAST

North Canso Lt., W. side of N. entr. to Strait of Canso – Fl. W. ev. 3 s., Obscured S. of 120°, Ht. 36.7 m. (120'), Rge. 13 mi., (45-41-29.8N/61-29-18.1W)

Cranberry Is. Lt., off Cape Canso, S. part of Is. – Fl. W. ev. 15 s., 2 Horns 2 bl. ev. 60 s., Horns point 066° and 141°, Ht. 16.9 m. (56'), Rge. 23 mi., Racon (B), (45-19-29.6N/60-55-38.2W)

White Head Is. Lt., SW side of Is. – Fl. W. ev. 5 s., Horn 1 bl. ev. 30 s., Horn points 190°, Ht. 18.2 m. (60'), Rge. 12 mi., (45-11-49.1N/61-08-10.8W)

Country Is. Lt., S. side of Is. – Fl. W. ev. 20 s., Ht. 16.5 m. (54'), Rge. 10 mi., (45-05-59.8N/61-32-31.9W)

Liscomb Is. Lt., near Cranberry Pt. – Fl. W. ev. 10 s., Horn 1 bl. ev. 30 s., Ht. 21.9 m. (72'), Rge. 14 mi., (44-59-15.8N/61-57-58.4W)

Beaver Is. Lt., E. end of Is. – Fl. W. ev. 7 s., Horn 1 bl. ev. 60 s., Horn points 144°, Ht. 19.9 m. (66'), Rge. 14 mi., (44-49-29.2N/62-20-16W)

Ship Harbour Lt., on Wolfes Pt. – Fl. G. ev. 12 s., Ht. 22.9 m. (76'), Rge. 5 mi., (44-44-55.6N/62-45-24.2W)

Owls Head Lt., at end of head – Fl. W. ev. 4 s., Ht. 25.8 m. (84'), Rge. 6 mi., (44-43-14.7N/62-47-59.4W)

Egg Is. Lt., center of Is. – Fl. W. ev. 15 s., Ht. 25.6 m. (84'), Rge. 9 mi., (44-39-53.1N/62-51-48.3W)

Jeddore Rock Lt., summit of rock – Fl. W. ev. 12 s., Ht. 29.3 m. (97'), Rge. 8 mi., (44-39-46.8N/63-00-39.6)

Bear Cove Lt. & Bell By. "H6," NE of cove, Q. R., Racon (N), Red, (44-32-36.3N/63-31-19.6W)

Sambro Harbor Lt. & Wh. By. "HS," S. of SW breaker, Halifax Hbr. app. – Mo(A)W ev. 6 s., RWS, (44-24-30N/63-33-36.5W)

Chebucto Head Lt., on summit, Halifax Hbr. app. – Fl. W. ev. 20 s., Horn 2 bl. ev. 60 s., Horn points 113°, Ht. 47.8 m. (157'), Rge. 13 mi., Racon (Z), (44-30-26.6N/63-31-21.8W)

Halifax Alpha Lt. & Wh. By. "HA," Halifax app. – Mo(A)W 6 s., RWS, (44-21-45N/63-24-15W)

Sambro Is. Lt., center of Is. – Fl. W. ev. 5 s., 2 Horns 3 bl. ev. 60 s. in unison, Horns point 068° and 178°, Ht. 42.7 m. (145'), Rge. 22 mi., (44-26-12N/63-33-48W)

Ketch Harbour Lt. By. "HE 19," Ketch Harbour entr. – Fl. G. ev 4 s., Green (44-28-19.6N/63-32-16W)

Betty Is. Lt., on Brig Pt. – Fl. W. ev. 15 s., Horn 1 bl. ev. 60 s., Ht. 19.2 m. (63'), Rge. 13 mi., (44-26-19.7N/63-46-00.4W)

Pearl Is. Lt., off St. Margaret's & Mahone Bays – Fl. W. ev. 10 s., Ht. 19.0 m. (63'), Rge. 8 mi., (44-22-57.2N/64-02-54W)

East Ironbound Is. Lt., center of Is. – Iso. W. ev. 6 s., Ht. 44.5 m. (147'), Rge. 13 mi., (44-26-22.4N/64-04-59.7W)

East Point Island Lt., Mahone Bay – F.G., Ht. 9.6 m. (31'), Rge. 7 mi., (44-20-59.2N/64-12-15W)

Abbreviations: **Alt.**, Alternating; **App.**, Approach; **By.**, Buoy; **Ch.**, Channel; **Entr.**, Entrance; **ev.**, every; **F.**, Fixed; **fl.**, flash; **Fl.**, Flashing; **Fl(2)**, Group Flashing; **G.**, Green; **Hbr.**, Harbor or Harbour, **Ht.**, height; **Is.**, Island; **Iso.**, Isophase (Equal interval); **Iso. W.**, Isophase White (Red sector(s) of Lights warn of dangerous angle of approach. Bearings and ranges are <u>from</u> the observer <u>to</u> the aid.); **Jct.**, Junction; **Lt.**, Light; **Ltd.**, Lighted.; **mi.**, miles; **Mo(A)**, Morse Code "A," **Mo(U)**, Morse Code "U"; **Oc.**, Occulting; **Pt.**, Point; **Q.**, Quick (Flashing); **RaRef.**, Radar Reflector; **R.**, Red; **rge.**, range; **RWS**, R.&W. Stripes; **RWSRST**, RWS with R. Spherical Topmarks; **s.**, seconds; **Wh.**, Whistle; **W.**, White; **Y.**, Yellow

Notices To Mariners: Keep informed of important changes. Visit **www.navcen.uscg.gov/lnm/** to receive Local Notices to Mariners via email. When reporting discrepancies in navigational aids, contact nearest C.G. unit and give official name of the aid.

Table for **Converting Seconds to Decimals of a Minute**, p. 262, for standard **GPS** input of Lat/Lon.

Striped Bass and Tides

by Jay MacLaughlin

The Striped Bass is a fish of moving waters. The most popular gamefish on the Atlantic coast, stripers have evolved to hunt, breed, and thrive where tide and current shift and stir.

In the turning of the natural year, Native Americans caught and ate stripers in warm seasons long before the arrival of European settlers. The first white men who came to the northeast coast of America wrote back to England with reports of these game and tasty fish in stupefying abundance. "At the turning of the tyde, I have seen so many go out of a river that I thought I could cross over them dry shod," said Thomas Morton of the Massachusetts Bay Colony in 1637. In 1669, funds from the sale of striped bass were used in Massachusetts to build the nation's first public schools. Striped bass have been a part of the fabric of coastal life in North America for time out of memory.

Today's anglers can't count on keeping their feet dry standing on the backs of the very fish they're fishing for, but its abundance and size still makes the striper fishery an extraordinary engine of economic activity and a source of recreation for millions.

Why the fascination with stripers? Bluefish fight harder. False albacore and bonito provide more excitement. Flounder and fluke have a more delicate taste. But the striper has it all. The striper is the largest gamefish available to anglers fishing from shore or from boats close to shore. Fifty-pound fish are caught every year. They are determined fighters. Those that aren't released to fight again provide delicious table fare. And stripers swim in every marine habitat from the St. John River in Maine to the northern coast of Florida, in open water, piled up in rips and breachways, at the mouths of rivers and streams, in brackish marshes upstream from the sea, and on sandy flats and in the churning surf close to shore.

The striped bass is a picture of evolutionary success. Bronze, purple, and green above, and pearly white below its signature seven lateral pinstripes, the bass is slab-sided, with powerful muscles and large fins to turn and sprint, and a broad tail to overtake and slash through prey. Big eyes help the bass see and hunt in the dark, and its bucket mouth is designed to engulf and crush food.

In early spring, the stripers spawn in the streams that flow into the Chesapeake Bay, and in the Delaware Bay, and in the Hudson river above the influence of the tide. The adult fish, slim and hungry, swarm out of the mouths of the bays and around Cape Charles and Cape May and point north along the beaches of Maryland and Delaware and New Jersey and on around Long Island and Cape Cod. With warming water, schools of smaller fish arrive first, just behind the herring and shad. Then come what John Cole calls the "middleweights," and finally the matriarchs and patriarchs, the thirty-, forty-, and fifty-pound fish that stir the winter dreams of fishermen all along the coast.

To grow and repeat the cycle of migration, the migrating bass feed on an unusually varied diet: crabs and lobster; squid; all the herring family, the alewives, bluebacks, and bunkers; and then when they settle in their summer haunts, eel and scup; sandeels and and silversides.

To know the striper's menu is to help an angler know *what* to fish. Happily, baitshops and luremakers supply the real thing and all manner of artificials in deadly imitation of the natural prey. As important as is the question of what to fish, the question of *when* to fish presents the greater challenge.

Striped bass feed in moving water. Knowledge of tides and currents and wave patterns is a powerful tool in any striper angler's mental tackle box. As David Ross of the Woods Hole Oceanographic Institution says in his essential book, *The Fisherman's Ocean,* "Most saltwater fish are wanderers, following the currents or the food that is carried with currents. Therefore, an area with moving water is generally a good place to fish... Simply put, tides give us a timetable with which to deduce when the water is moving, which is when the fish are most likely to be moving."

Continued p. 170

Cross Is. Lt., E. Pt. of Is. – Fl. W. ev. 10 s., Ht. 24.9 m. (82′), Rge. 10 mi., (44-18-43.7N/64-10-06.4W)

West Ironbound Is. Lt., Entr. to La Have R. – Fl. W. ev. 12 s., Ht. 24.3 m. (80′), Rge. 8 mi., (44-13-43.7N/64-16-28W)

Mosher Is. Lt., W. side Entr. to La Have R. – F.W., Horn 1 bl. ev. 20 s., Ht. 23.3 m. (77′), Rge. 13 mi., (44-14-14.6N/64-18-59.1W)

Cherry Cove Lt., betw. Little Hbr. & Back Cove – F.G., Horn 1 bl. ev. 30 s., Horn points 055°46′, Ht. 6.4 m. (21′), Rge. 7 mi., (44-09-29.8N/64-28-53.3W)

Medway Head Lt., W. side entr. to Pt. Medway – Fl. W. ev. 12 s., Ht. 24.2 m. (80′), Rge. 11 mi., (44-06-10.6N/64-32-23.3W)

Western Head Lt., W. side entr. to Liverpool Bay – Fl. W. ev. 15 s., Horn 1 bl. ev. 60 s., Horn points 104°, Ht. 16.8 m. (55′), Rge. 14 mi., (43-59-20.8N/64-39-44.5W)

Lockeport Lt., on Gull Rock, entr. to hbr. – Fl. W. ev. 15 s., Horn 1 bl. ev. 30 s., Ht. 16.7 m. (56′), Rge. 12 mi., (43-39-18.3N/65-05-55.9W)

Cape Roseway Lt., near SE Pt. of McNutt Is. – Fl. W. ev. 10 s., Ht. 33.1 m. (109′), Rge. 10 mi., (43-37-21.4N/65-15-50W)

Cape Negro Is. Lt., on SE end of Is. – Fl(2) W. ev. 15 s., Horn 1 bl. ev. 60 s., Ht. 28.3 m. (92′), Rge. 10 mi., (43-30-26.2N/65-20-44.2W)

The Salvages Lt., SE end of Is. – Fl. W. ev. 12 s., Horn 3 bl. ev. 60 s., Ht. 15.6 m. (51′), Rge. 10 mi., (43-28-08.1N/65-22-44W)

Baccaro Point Lt., E. side entr. to Barrington Bay – Mo(D)W 10 s., Horn 1 bl. ev. 20 s., Horn points 200°, Ht. 15.0 m. (49′), Rge. 15 mi., (43-26-59N/65-28-15W)

Cape Sable Lt., on cape – Fl. W. ev. 5 s., Horn 1 bl. ev. 60 s., Horn points 150°, Ht. 29.7 m. (97′), Rge. 18 mi., Racon (C), (43-23-24N/65-37-16.9W)

West Head Lt., Cape Sable Is. – F.R., Horn 2 bl. ev. 60 s., Horn points 254°, Ht. 15.6 m. (51′), Rge. 7 mi., (43-27-23.8N/65-39-16.9W)

Outer Island Lt., on S. Pt. of Outer Is. – Fl. W. ev. 10 s., Ht. 13.7 m. (46′), Rge. 10 mi., (43-27-23.2N/65-44-36.2W)

Seal Is. Lt., S. Pt. of Is. – Fl. W. ev. 10 s., Horn 3 bl. ev. 60 s., Horn points 183°, Ht. 33.4 m. (110′), Rge. 19 mi., Radiobeacon, (43-23-40N/66-00-51W)

NOVA SCOTIA, WEST COAST

Peases Is. Lt., S. Pt. of one of the Tusket Is. – Fl. W. ev. 6 s., Horn 2 bl. ev. 60 s., Ht. 16 m. (53′), Rge. 9 mi., (43-37-42.6N/66-01-34.9W)

Cape Forchu Lt., E. Cape S. Pt. Yarmouth Sd. – Fl. W. ev. 12 s., Ht. 34.6 m. (113′), Rge. 12 mi., Racon (B), (43-47-38.8N/66-09-19.3W)

Lurcher Shoal Bifurcation Light By. "NM," W. of SW shoal – Fl.(2+1) R. ev. 6 s., Racon (K), R.G.R. marked "NM," (43-49-00.3N/66-29-58W)

Cape St. Marys Lt., E. side of Bay – Fl. W. ev. 5 s., Horn 1 bl. ev. 60 s., Horn points 251° 30′, Ht. 31.8 m (105′), Rge. 13 mi., (44-05-09.2N/66-12-39.6W)

Brier Is. Lt., on W. side of Is. R. & W. Tower – Fl(3) W. ev. 18 s., 2 Horns 2 bl. ev. 60 s., Horns point 270° and 315°, Ht. 22.2 m. (72′), Rge. 14 mi., (44-14-55N/66-23-32W)

Boars Head Lt., W. side of N. entr. to Petit Passage – Fl. W. ev. 5 s., Horn 3 bl. ev. 60 s., Horn points 315°, Ht. 28.0 m. (91′), Rge. 16 mi., (44-24-14.5N/66-12-55W)

Prim Pt. Lt., Digby Gut, W. Pt. of entr. to Annapolis Basin – Iso. W. ev. 6 s., Horn 1 bl. ev. 30 s., Horn points 318°, Ht. 24.8 m. (82′), Rge. 12 mi., (44-41-28N/65-47-10.8W)

Ile Haute Lt., on highest Pt. – Fl. W. ev. 4 s., Rge. 6 mi., Ht. 112 m. (367′), (45-15-03.3N/65-00-19.8W)

NEW BRUNSWICK COAST

Cape Enrage Lt., at pitch of cape – Fl. G. ev. 6 s., Horn 3 bl. ev. 60 s., Horn points 220°, Ht. 40.7 m. (134′), Rge. 10 mi., (45-35-38.1N/64-46-47.7W)

Quaco Lt., tower on head – Fl. W. ev. 10 s., Horn 1 bl. ev. 30 s., Horn points 130°, Ht. 26.0 m. (86′), Rge. 21 mi., (45-19-25.3N/65-32-08.8W)

For abbreviations see footnote p. 167

Continued from p. 168

Easy to say; hard to put into practice. As master striper fisherman Lou Tabory reminds us, there is no substitute for time on the water. And an aid to memory in the form of a fishing log is a superb way to capture and organize what you learn in order to capitalize on your time on the water. So, which tide, which current will produce fish for you?

If moving water is better, then spring tides should be the best to fish. Simply put, spring tides are the stronger tides that occur around the full moon and the new moon. *See p. 234* for **Phases of the Moon**. More movement of water means more bait on the move, and in violent movement from rips and currents, the bait become confused or unable to hold position in the water—optimum feeding time for opportunistic stripers.

There's more. Within each tidal cycle, the movement of water isn't constant from ebb to flood and back again. In some places, tidal movement begins slowly, then accelerates, sometimes twice in one tidal cycle. The landforms that surround harbors, inlets, and bays, all prime fishing grounds for stripers, also affect current speed. If that's not enough, there is the effect of wind and barometric pressure, as any boater in Nantucket or Vineyard Sound in a late-summer souwester can testify.

At Great Point on Nantucket, the east and west movement of the Pollock Rip Channel through Nantucket Sound changes direction independently of the time of tides. *See pp. 66-77.* The long underwater ridge stretching out from the point creates a powerful rip. There are particular combinations of tide and current that spill baitfish over that rip at the mercy of the moving water. Using their powerful tails to hold in the rip and conserve energy, stripers and other predatory gamefish just wait on the down-current side of the rip for the food to come to them. Drifting a tin lure or bucktail deceiver through the rip is a sure way of using local tidal and current knowledge to catch more fish.

There is some argument to fishing "up the tide," that is, to fish in the days before the full or new moon—on the theory that bait is likely to be moving in higher tides. Within tides, angler's opinions vary. Do I fish a whole tide? Do I fish the last three hours of a rising tide? Do I fish just after the turn? There are as many answers as there are fishermen and places to fish. One safe generalization is that at the times of slack tide, with water moving little or not at all, fish, both prey and predator, are likely to be dispersed and fishing is typically least.

All of this argues for being observant, for capturing your fishing experience and learning from each outing. Mark structure at low tide, then see how it fishes as water fills in. Record your trips, transfer waypoints to a fishing log and add wind and weather and tide and current. Over time, you'll accumulate a fund of knowledge about the intersection of moving water and moving fish that will make each trip better and better. After all, as John Buchan, a diplomat and author and inveterate angler said, "The charm of fishing is that it is the pursuit of what is elusive but attainable, a perpetual series of occasions for hope." Tight lines.

Jay MacLaughlin is a surfcaster and flyfisherman. He fishes from boats and from shore in the waters of Cape Cod and Nantucket. He is a writer and editor and the principal of his own marketing communications firm west of Boston.

The following are particularly useful in knowing how to read water:

The Fisherman's Ocean, David A. Ross; Stackpole Books, 2000
Inshore Fly Fishing, Lou Tabory; Lyons & Burford, 1992

The Internet communities that focus on striper fishing in the Northeast include:
Stripers Online http://www.stripersonline.com/surftalk/
Flyfish Saltwaters http://www.flyfishsaltwaters.com/ubb/ultimatebb.php
ReelTime http://reel-time.com/forum/index.php
StriperSurf http://www.striped-bass.com/Stripertalk/index.php
FlyFishing Forum http:www.flyfishingforum.com/flytalk/index.htm

Cape Spencer Lt., pitch of cape – Fl. W. ev. 11 s., Horn 3 bl. ev. 60 s., Horn points 165°, Ht. 61.6 m. (203'), Rge. 14 mi., (45-11-42.5N/65-54-35.5W)

Partridge Is. Lt., highest pt. of Is., Saint John Harbour – Fl. W. ev. 7.5 s., Ht. 35.3 m. (116'), Rge. 19 mi., (45-14-21N/66-03-13.8W)

Musquash Head Lt., E. side entr. to Musquash Hbr. – Fl. W. ev. 3 s., Horn 1 bl. ev. 60 s., Horn points 180°, Ht. 35.1 m. (116'), Rge. 20 mi., (45-08-37.1N/66-14-14.2W)

Pt. Lepreau Lt., on point – Fl. W. ev. 5 s., Horn 3 bl. ev. 60 s., Horn points 190°, Ht. 25.5 m. (84'), Rge. 14 mi., (45-03-31.7N/66-27-31.3W)

Pea Pt. Lt., E. side entr. to Letang Hbr. – F.W. visible 251° thru N & E to 161°, Horn 2 bl. ev. 60 s., Horn points 180°, Ht. 17.2 m. (56'), Rge. 12 mi., (45-02-20.4N/66-48-28.2W)

Head Harbour Lt., outer rock of E. Quoddy Head – F.R., Horn 1 bl. ev. 60 s., Horn points 116°, Ht. 17.6 m. (58'), Rge. 13 mi., (44-57-28.6N/66-54-00.2W)

Swallowtail Lt., NE Pt. of Grand Manan – Oc. W. ev. 6 s., Horn 1 bl. ev. 20 s., Horn points 100°, Ht. 37.1 m. (122'), Rge. 12 mi., (44-45-51.1N/66-43-57.5W)

Great Duck Is. Lt., S. end of Is. – Fl. W. ev. 10 s., Horn 1 bl. ev. 60 s., Horn points 120°, Ht. 15.3 m. (50'), Rge. 18 mi., (44-41-03.5N/66-41-34.3W)

Southwest Head Lt., S. end of Grand Manan – Fl. W. ev. 10 s., Horn 1 bl. ev. 60 s., Horn points 240°, Ht. 47.5 m. (156'), Rge. 16 mi., (44-36-02.9N/66-54-19.8W)

Gannet Rock Lt., S. of Grand Manan – Fl. W. ev. 5 s., Horn 3 bl. ev. 60 s., Horn omni-directional, Ht. 28.2 m. (93'), Rge. 19 mi., Racon (G), (44-30-37.1N/66-46-52.9W)

MAINE

West Quoddy Head Lt., Entr. Quoddy Roads – Fl(2) W. ev. 15 s., Horn 2 bl. ev. 30 s., Ht. 83', Rge. 18 mi., (44-48-54N/66-57-02W)

Libby Island Lt., Entr. Machias Bay – Fl(2) W. ev. 20 s., Obscured from 208°-220°, Horn 1 bl. ev. 15 s., Ht. 91', Rge. 18 mi., (44-34-06N/67-22-03W)

Moose Peak Lt., E. end Mistake Is. – Fl. W. ev. 30 s., Horn 2 bl. ev. 30 s., Ht. 72', Rge. 20 mi., (44-28-30N/67-31-54W)

Petit Manan Lt., E. Pt. of Is. – Fl. W. ev. 10 s., Horn 1 bl. ev. 30 s., Ht. 123', Rge. 19 mi., (44-22-03N/67-51-52W)

Prospect Harbor Point Lt. – Fl. R. ev. 6 s., (2 W. sect.), Ht. 42', Rge. R. 7 mi., W. 9 mi., ltd. 24 hrs., (44-24-12N/68-00-48W)

Mount Desert Lt., 20 mi. S. of island – Fl. W. ev. 15 s., Horn 2 bl. ev. 30 s., Ht. 75', Rge. 20 mi., (43-58-07N/68-07-42W)

Great Duck Island Lt., S. end of island – Fl. R. ev. 5 s., Horn 1 bl. ev. 15 s., Ht. 67', Rge. 19 mi., (44-08-31N/68-14-45W)

Southwest Head Lt., S. end of Grand Manan – Fl. W. ev. 10 s., Horn 1 bl. ev. 60 s., Horn points 240°, Ht. 47.5 m. (156'), Rge. 16 mi., (44-36-02.9N/66-54-19.8W)

Frenchman Bay Ltd. By. "FB," Fl. (2+1) R. ev. 6 s., Rge. 4 mi., R&G Bands, Racon (B), (44-19-24N/68-07-24W)

Egg Rock Lt., Frenchman Bay – Fl. R. ev. 5 s., Horn 2 bl. ev. 30 s., Ht. 64', Rge. 18 mi., (44-21-14N/68-08-18W)

Baker Island Lt., SW Entr. Somes Sound – Fl. W. ev. 10 s., Ht. 105', Rge. 10 mi., (44-14-30N/68-11-54W)

Bass Harbor Head Lt., SW Pt. Mt. Desert Is. – Oc. R. ev. 4 s., Ht. 56', Rge. 13 mi., ltd. 24 hrs., (44-13-19N/68-20-14W)

Blue Hill Bay Lt. #3, on Green Is. – Fl. G. ev. 4 s., Ht. 25', Rge. 5 mi., SG on tower, (44-14-54N/68-29-54W)

Burnt Coat Harbor Lt. – Oc. W. ev. 4 s., Ht. 75', Rge. 9 mi., (44-08-03N/68-26-50W)

Halibut Rocks Lt., Jericho Bay – Fl. W. ev. 6 s., Horn 1 bl. ev. 10 s., Ht. 25', Rge. 6 mi., NR on tower, (44-08-00N/68-31-30W)

Eggemoggin Ltd. Bell By. "EG" – Mo(A)W, Rge. 5 mi., RWSRST, (44-19-13N/68-44-34W)

For abbreviations see footnote p. 167

Eggemoggin Reach Bell By. "ER" – RWSRST, (44-18-00N/68-46-29W)

Crotch Island Lt. #21, Deer Is. Thorofare – Fl. G. ev. 4 s., Ht. 20', Rge. 5 mi., SG on tower, (44-08-48N/68-40-36W)

Saddleback Ledge Lt., Isle au Haut Bay – Fl. W. ev. 6 s., Horn 1 bl. ev. 10 s., Ht. 54', Rge. 9 mi., (44-00-54N/68-43-36W)

Isle Au Haut Lt., Isle au Haut Bay – Fl. R. ev. 4 s., W. Sect. 034°-060°, Ht. 48', Rge. R. Fl. 6 mi., W. 8 mi., (44-03-53N/68-39-05W)

Deer Island Thorofare Lt., W. end of thorofare – Fl. W. ev. 6 s., Horn 1 bl. ev. 15 s., Ht. 52', Rge. 8 mi., (44-08-04N/68-42-12W)

Goose Rocks Lt., E. Entr. Fox Is. Thorofare – Fl. R. ev. 6 s., W. Sect. 301°-304°, Horn 1 bl. ev. 10 s., Ht. 51', Rge. R. 11 mi., W. 12 mi., (44-08-08N/68-49-50W)

Eagle Island Lt., E. Penobscot Bay – Fl. W. ev. 4 s., Ht. 106', Rge. 9 mi., (44-13-04N/68-46-04W)

Green Ledge Lt. #4, E. Penobscot Bay – Fl. R. ev. 6 s., Ht. 31', Rge. 5 mi., TR on tower, (44-17-25N/68-49-42W)

Heron Neck Lt., E. Entr. Hurricane Sound – F.R., W. Sect. 030°-063°, Horn 1 bl. ev. 30 s., Ht. 92', Rge. R. 6 mi., W. 9 mi., (44-01-30N/68-51-44W)

Matinicus Rock Lt., Penobscot Bay App. – Fl. W. ev. 10 s., Horn 1 bl. ev. 15 s., Ht. 90', Rge. 20 mi., (43-47-00N/68-51-18W)

Grindel Pt. Lt., West Penobscot Bay – Fl. W. ev. 4 s., Ht. 39', Rge. 5 mi., (44-16-53N/68-56-35W)

Two-Bush Island Lt., Two-Bush Ch. – Fl. W. ev. 5 s., R. Sect. 061°-247°, Horn 1 bl. ev. 15 s., Ht. 65', Rge. W. 15 mi., R. 21 mi., (43-57-51N/69-04-26W)

Two Bush Island Ltd. Wh. By. "TBI" – Mo(A)W, Rge. 6 mi., RWS, (43-58-18N/69-00-18W)

Whitehead Lt., W. side of S. entr. Muscle Ridge Ch. – Oc.G. ev. 4 s., Horn 2 bl. ev. 30 s., Ht. 75', Rge. 6 mi., (43-58-43N/69-07-27W)

Owl's Head Lt., S. side Rockland Entr. – F.W., Horn 2 bl. ev. 20 s., Ht. 100', Rge. 16 mi., Obscured from 324°-354° by Monroe Island, ltd. 24 hrs., (44-05-32N/69-02-38W)

Rockland Harbor Breakwater Lt., S. end of breakwater – Fl. W. ev. 5 s., Horn 1 bl. ev. 15 s., Ht. 39', Rge. 17 mi., (44-06-15N/69-04-39W)

Lowell Rock Lt. #2, Rockport Entr. – Fl. R. ev. 6 s., Ht. 25', Rge. 5 mi., TR on spindle, (44-09-48N/69-03-36W)

Browns Head Lt., W. Entr. Fox Is. Thorofare – F. W., 2 R. Sect. 001°-050° and 061°-091°, Horn 1 bl. ev. 10 s., Ht. 39', Rge. R. 11 mi., F.W. 14 mi., ltd. 24 hrs., (44-06-42N/68-54-34W)

Curtis Island Lt., S. side Camden Entr. – Oc.G. ev. 4 s., Ht. 52', Rge. 6 mi., (44-12-05N/69-02-56W)

Northeast Point Lt. #2, Camden Entr. – Fl. R. ev. 4 s., Ht. 20', Rge. 5 mi., TR on white tower, (44-12-31N/69-02-47W)

Dice Head Lt., N. side Entr. to Castine – Fl. W. ev. 6 s., Ht. 134', Rge. 11 mi., White tower, (44-22-58N/68-49-08W)

Fort Point Lt., W. side Entr. to Penobscot R. – F.W., Horn 1 bl. ev. 10 s., Ht. 88', Rge. 15 mi., ltd. 24 hrs., (44-28-02N/68-48-42W)

Marshall Point Lt., E. side of Pt. Clyde Hbr. S. Entr. – F.W., Horn 1 bl. ev. 10 s., Ht. 30', Rge. 13 mi., ltd. 24 hrs., (43-55-03N/69-15-41W)

Marshall Point Ltd. By. "MP" – Mo(A)W, Rge. 6 mi., RWSRST, (43-55-18N/69-10-54W)

Monhegan Island Lt., Penobscot Bay – Fl. W. ev. 15 s., Ht. 178', Rge. 20 mi., (43-45-53N/69-18-57W)

Manana Island Fog Signal Station, Penobscot Bay – Horn 2 bl. ev. 60 s., (43-45-48N/69-19-36W)

For abbreviations see footnote p. 167

Franklin Is. Lt., Muscongus Bay – Fl. W. ev. 6 s., Ht. 57′, Rge. 8 mi., (43-53-32N/69-22-29W)

Pemaquid Pt. Lt., W. side Muscongus Bay Entr. – Fl. W. ev. 6 s., Ht. 79′, Rge. 14 mi., (43-50-12N/69-30-21W)

Ram Is. Lt., Fisherman Is. Passage S. side – Iso. R. ev. 6 s., 2 W. Sect. 258°-261° and 030°-046°, Covers fairways, Horn 1 bl. ev. 30 s., Ht. 36′, Rge. W. 11 mi., R. 9 mi., (43-48-14N/69-35-57W)

Burnt Is. Lt., Boothbay Hbr. W. side Entr. – Fl. R. ev. 6 s., 2 W. Sect. 307°-316° and 355°-008°, Covers fairways. Horn 1 bl. ev. 10 s., Ht. 61′, Rge. R. 12 mi., W. 15 mi., ltd. 24 hrs., (43-49-31N/69-38-25W)

The Cuckolds Lt., Boothbay – Fl(2) W. ev. 6 s., Horn 1 bl. ev. 15 s., Ht. 59′, Rge. 12 mi., (43-46-46N/69-39-00W)

Seguin Lt., 2 mi. S. of Kennebec R. mouth – F.W., Horn 2 bl. ev. 20 s., Ht. 180′, Rge. 18 mi., (43-42-27N/69-45-29W)

Hendricks Head Lt., Sheepscot R. mouth E. side – F.W., R. Sect. 180°-000°, Ht. 43′, Rge. R. 7 mi., F.W. 9 mi., (43-49-21N/69-41-23W)

Pond Is. Lt., Kennebec R. mouth W. side – Iso. W. ev. 6 s., Horn 2 bl. ev. 30 s., Ht. 52′, Rge. 9 mi., (43-44-24N/69-46-12W)

Perkins Is. Lt., Kennebec R. – Fl. R. ev. 2.5 s., 2 W. Sect. 018° – 038°, 172° – 188°, Covers fairways, Ht. 41′, Rge. R. 5 mi., W. 6 mi., (43-47-12N/69-47-07W)

Squirrel Pt. Lt., Kennebec R. – Iso. R. ev. 6 s., W. Sect. 321° - 324°, Covers fairway, Ht. 25′, Rge. R. 7 mi., W. 9 mi., (43-48-59N/69-48-09W)

Fuller Rock Lt., off Cape Small – Fl. W. ev. 4 s., Ht. 39′, Rge. 6 mi., NR on tower, (43-41-45N/69-50-01W)

White Bull Ltd. Gong By. "WB" – Mo(A)W, Rge. 6 mi., RWS, (43-42-48N/69-55-12W)

Whaleboat Island Lt., Broad Sd., Casco Bay – Fl. W. ev. 6 s., Ht. 47′, Rge. 6 mi., NR on tower, (43-44-31N/70-03-40W)

Cow Island Ledge Lt., Portland to Merepoint – Fl. W. ev. 6 s., Ht. 23′, Rge. 8 mi., RaRef., NR on spindle, (43-42-12N/70-11-18W)

Halfway Rock Lt., midway betw. Cape Small Pt. and Cape Eliz. – Fl. R. ev. 5 s., Horn 2 bl. ev. 30 s., Ht. 77′, Rge. 19 mi., (43-39-21N/70-02-12W)

Portland Ltd. Wh. By. "P", Portland Hbr. App. – Mo(A)W, Rge. 6 mi., Racon (M), RWSRST, (43-31-36N/70-05-28W)

Ram Island Ledge Lt., N. side of Portland Hbr. Entr. – Fl. (2) W. 2 ev. 6 s., Horn 1 bl. ev. 10 s., Ht. 77′, Rge. 8 mi., (43-37-54N/70-11-12W)

Cape Elizabeth Lt., S. of Portland Hbr. Entr. – Fl(4) W. ev. 15 s., Horn 2 bl. ev. 60 s., Ht. 129′, Rge. 15 mi., (43-34-00N/70-12-00W)

Portland Head Lt., SW side Portland Hbr. Entr. – Fl. W. ev. 4 s., Horn 1 bl. ev. 15 s., Ht. 101′, Rge. 24 mi., ltd. 24 hrs., (43-37-23N/70-12-28W)

Spring Pt. Ledge Lt., Portland main ch. W. side – Fl. W. ev. 6 s., 2 R. Sect., 2 W. Sectors 331°-337° and 074°-288°, Horn 1 bl. ev. 10 s., Ht. 54′, Rge. R. 10 mi., W. 12 mi., ltd. 24 hrs., (43-39-08N/70-13-26W)

Wood Island Lt., S. Entr. Wood Is. Hbr. N. side – Alt. W. and G. ev. 10 s., Horn 2 bl. ev. 30 s., Ht. 71′, Rge. W. 18 mi., G. 16 mi., (43-27-24N/70-19-42W)

Goat Is. Lt., Cape Porpoise Hbr. Entr. – Fl. W. ev. 6 s., Horn 1 bl. ev. 15 s., Ht. 38′, Rge. 12 mi., (43-21-28N/70-25-30W)

Cape Neddick Lt., On N. side of Nubble – Iso. R. ev. 6 s., Horn 1 bl. ev. 10 s., Ht. 88′, Rge. 13 mi., (43-09-55N/70-35-28W)

Jaffrey Point Lt. #4 – Fl. R. ev. 4 s., Ht. 22′, rge. 5 mi., TR on tower, (43-03-18N/70-42-48W)

Boon Is. Lt., 6.5 mi. off coast – Fl. W. ev. 5 s., Horn 1 bl. ev. 10 s., Ht. 137′, Rge. 19 mi., (43-07-18N/70-28-36W)

York Harbor Ltd. Bell By. "YH" – Mo(A)W, Rge. 5 mi., RWSRST, (43-07-45N/70-37-01W)

For abbreviations see footnote p. 167

Looking Back ... from the 1980 Eldridge

FOR BETTER RESULTS WITH YOUR R.D.F.

The Radio Direction Finder (R.D.F.), or at least its loop, should be above the mass of boat (as on top of cabin), as rigging, etc. deflect Bearings from true direction. It should also be mounted on the center line, so rigging will be symmetrically disposed, relative to the loop.

The sharpest bearings are taken dead ahead, or astern. They are usually more accurate if earphones are used, to reduce extraneous noise.

The "null" (bearing at which sound signal is tuned out by rotation of the loop) is ideally only about one degree wide. In most R.D.F.s, however, this null may be from 5 to 10 or more degrees in width. For best results, oscillate loop back and forth through the whole area of the null; the point at which you bisect the null will be your best Bearing.

The angle of the boat's heel creates another type of error. Zero when bearing is dead ahead or astern and at maximum when 45° from abeam. Auxiliaries should be pointed directly at Radio Beacon Station for best results.

Long distance bearings at night are frequently unreliable due to deflection from the ionosphere. Radio signals are also generally deflected by intervening land masses.

Have a spare R.D.F. battery aboard if cruising.

NEW HAMPSHIRE

Whaleback Lt., Portsmouth Entr. NE side –Fl(2) W. ev. 10 s., Horn 2 bl. ev. 30 s., Ht. 59', Rge. 17 mi., (43-03-32N/70-41-47W)

Portsmouth Harbor Lt. (New Castle), on Fort Point – F. G., Horn 1 bl. ev. 10 s., Ht. 52', Rge. 12 mi., (43-04-18N/70-42-30W)

Rye Harbor Entr. Ltd. Wh. By. "RH" – Mo(A)W, Rge. 6 mi., RWSRST, (42-59-38N/70-43-45W)

Isles Of Shoals Lt., 5.5 mi. off coast – Fl. W. ev. 15 s., Horn 1 bl. ev. 30 s., Ht. 82', Rge. 13 mi., (42-58-02N/70-37-24W)

MASSACHUSETTS

Newburyport Harbor Lt., N. end of Plum Is. – Oc.(2) G. ev. 15 s., Obscured from 165°-192° and 313°-344°, Ht. 50', Rge. 10 mi., (42-48-54N/70-49-06W)

Merrimack River Entr. Ltd. Wh. By. "MR"– Mo(A)W, Rge. 6 mi., RWSRST, (42-48-34N/70-47-03W)

Ipswich Lt., Ipswich Entr. S. side – Oc.W. ev. 4 s., Ht. 30', Rge. 5 mi., NR on tower, (42-41-07N/70-45-58W)

Rockport Breakwater Lt. #6, W. side Entr. Rockport inner hbr. – Fl. R. ev. 4 s., Ht. 32', Rge. 5 mi., TR on tower, (42-39-36N/70-36-42W)

Annisquam Harbor Lt., E. side Entr. – Fl. W. ev. 7.5 s., R. Sector 180°-217°, Horn 2 bl. ev. 60 s., Ht. 45', Rge. R. 11 mi., W. 14 mi., (42-39-42N/70-40-54W)

Straitsmouth Lt., Rockport Entr. S. side – Fl. G. ev. 6 s., Horn 1 bl. ev. 15 s., Ht. 46', Rge. 6 mi., (42-39-42N/70-35-18W)

Cape Ann Lt., E. side Thacher Is. – Fl. R. ev. 5 s., Horn 2 bl. ev. 60 s., Ht. 166', Rge. 17 mi., (42-38-12N/70-34-30W)

Eastern Point Ltd. Wh. By. #2 – Fl. R. ev. 4 s., Rge. 3 mi., Red, (42-34-12N/70-39-48W)

Eastern Point Lt., Gloucester Entr. E. side – Fl. W. ev. 5 s., Ht. 57', Rge. 22 mi., ltd. 24 hrs., (42-34-49N/70-39-52W)

Gloucester Breakwater Lt., W. end – Oc.R. ev. 4 s., Horn 1 bl. ev. 10 s., Ht. 45', Rge. 6 mi., (42-34-57N/70-40-20W)

Bakers Island Lt., Salem Ch. – Alt. Fl. W. and R. ev. 20 s., Horn 1 bl. ev. 30 s., Ht. 111', Rge. R. 14 mi., W. 16 mi., (42-32-11N/70-47-09W)

Hospital Point Range Front Lt., Beverly Cove W. side – F.W., Ht. 70', (42-32-48N/70-51-24W)

The Graves Ltd. Wh. By. #5 – Fl. G. ev. 4 s., Rge. 4 mi., Green, (42-22-33N/70-51-28W)

Marblehead Lt., N. point Marblehead Neck – F.G., Ht. 130', Rge. 7 mi., (42-30-18N/70-50-00W)

The Graves Lt., Boston Hbr. S. Ch. Entr. – Fl(2) W. ev. 12 s., Horn 2 bl. ev. 20 s., Ht. 98', Rge. 15 mi., (42-21-54N/70-52-09W)

Boston App. Ltd. By. "BG"– Mo(A)W, Rge. 6 mi., RWS, (42-23-24N/70-51-30W)

Deer Island Lt., President Roads, Boston Hbr. – Alt. W. and R. ev. 10 s., R. Sect. 198°-222°, Obscured 112°-186°, Horn 1 bl. ev. 10 s., Ht. 53', Rge. R. 11 mi., (42-20-23N/70-57-16W)

Deer Island Danger Lt., On south end of spit – F.R., R. Sect. 198°-222°, Ht. 15', Rge. 6 mi., (42-20-24N/70-57-18W)

Long Island Head Lt., President Roads, Boston Hbr. – Fl. W. ev. 2.5 s., Ht. 120', Rge. 6 mi., (42-19-48N/70-57-30W)

Boston Ltd. Wh. By. "B", Boston Hbr. Entr. – Mo(A)W, Rge. 6 mi., Racon (B), RWSRST, (42-22-42N/70-46-58W)

Boston App. Ltd. By. "BF" (NOAA) –Fl(4) Y. ev. 20 sec, Rge. 7 mi., Yellow, (42-20-44N/70-39-04W)

Boston Entr. Ltd. Wh. By. "NC", Boston N. Ch. – Mo(A)W, Rge. 6 mi., RWSRST, Racon (N), (42-22-30N/70-54-18W)

For abbreviations see footnote p. 167

ELDRIDGE TIDE AND PILOT BOOK

A beautiful island. A world class marina. A special rate.

What *moor* could one ask?

On Nantucket Island you'll find **beauty and relaxation** at every turn come spring and fall. **The shops and boutiques** are offering the season's best sales. Summertime crowds are **noticeably absent**, which means no long lines at the **finest restaurants**. And the boating weather? **Perfect.**

Nantucket Boat Basin, just steps away from Nantucket Town, a 240-slip, **full-service marina** that caters to some of the world's most discerning travelers.

SPRING & FALL OFFER
Moor 2 nights and get the 3rd night FREE

Offer is valid May 3-June 18 and Sept 7-Oct 29, 2009. Sun-Thurs only. Subject to availability. Not valid for groups, previous reservations or for Memorial Day Weekend.

Mention Code: ETP

800.NAN-BOAT   nantucketboatbasin.com   Open year round
Nantucket Boat Basin is owned and operated by Nantucket Island Resorts.

178

Minots Ledge Lt., Boston Hbr. Entr. S. side – Fl(1+4+3) W. ev. 45 s., Horn 1 bl. ev. 10 s., Ht. 85′, Rge. 10 mi., (42-16-12N/70-45-30W)

Boston Lt., SE side Little Brewster Is. – Fl. W. ev. 10 s., Horn 1 bl. ev. 30 s., Ht. 102′, Rge. 27 mi., (42-19-42N/70-53-24W)

Scituate App. Ltd. Gong By. "SA" – Mo(A)W, Rge. 6 mi., RWSRST, (42-12-06N/70-41-48W)

Plymouth Lt. (Gurnet), N. side Entr. to hbr. – Fl(3) W. ev. 30 s., R. Sect. 323°-352°, Horn 2 bl. ev. 15 s., Ht. 102′, Rge. R. 15 mi., W. 17 mi., (42-00-12N/70-36-00W)

Race Point Lt., NW Point of Cape Cod – Fl. W. ev. 10 s., Obscured 220°-292°, Ht. 41′, Rge. 16 mi., (42-03-44N/70-14-35W)

Wood End Lt., Entr. to Provincetown – Fl. R. ev. 10 s., Horn 1 bl. ev. 30 s., Ht. 45′, Rge. 13 mi., (42-01-17N/70-11-37W)

Long Point Lt., Provincetown Entr. SW side – Oc.G. ev. 4 s., Horn 1 bl. ev. 15 s., Ht. 36′, Rge. 8 mi., (42-01-59N/70-10-07W)

Mary Ann Rocks Ltd. Wh. By. #12 – Fl. R. ev. 2.5 s., Rge. 4 mi., Red, (41-55-06N/70-30-24W)

Mary Ann Rocks Ltd. Bell By. "CC" – Mo(A)W, Rge. 6 mi., RWSRST, (41-48-54N/70-27-36W)

Cape Cod Canal Breakwater Lt. #6, E. Entr. – Fl. R. ev. 5 s., Horn 1 bl. ev. 15 s., Ht. 43′, Rge. 15 mi., (41-46-47N/70-29-23W)

Highland Lt., NE side of Cape Cod – Fl. W. ev. 5 s., Ht. 170′, Rge. 18 mi., ltd. 24 hrs., (42-02-22N/70-03-39W)

Nauset Beach Lt., E. side of Cape Cod – Alt(2) W. R. ev. 10 s., (Tides divide and run in opposite directions abreast of light), Ht. 120′, Rge. W. 24 mi., R. 20 mi., (41-51-36N/69-57-12W)

Chatham Beach Ltd. Wh. By. "C" – Mo(A)W, Rge. 5 mi., RWSRST, (41-39-12N/69-55-30W)

Chatham Lt., W. side of hbr. –Fl(2)W. ev. 10 s., Ht. 80′, Rge. 24 mi., ltd. 24 hrs., (41-40-17N/69-57-01W)

Chatham Inlet Bar Guide Lt., Fl. Y. ev. 2.5 s., Ht. 62′, Rge. 11 mi., (41-40-17N/69-57-00W)

Hyannis Harbor App. Ltd. Bell By. "HH" – Mo(A)W, Rge. 6 mi., RWSRST, (41-36-00N/70-17-24W)

Pollock Rip Ch. Ltd. By. #8 – Fl. R. ev. 6 s., Rge. 4 mi., (41-32-43N/69-58-56W)

Cape Wind Meteorlogical Tower Lts. "MT(2)", Mo(U)Y, Ht. 35′, Rge. 4 mi., Horn Mo(U), (41-28-19N/70-18-53W)

Nantucket Traffic Lane Ltd. Wh. By. "NA" – Fl.Y. ev. 6 s., Rge. 7 mi., Racon (N), Yellow, (40-25-42N/73-11-28W)

Nantucket Traffic Lane Ltd. Wh. By. "NB" – Fl. Y. ev. 4 s., Rge 6 mi., Yellow, (40-26-26N/73-38-47W)

Nantucket Shoals Ltd. Wh. By. "N" – Fl. Y. ev. 6 s., Rge. 6 mi. Racon (N), Yellow, (40-30-09N/69-14-48W)

Nantucket Lt., (Great Point), Nantucket, N. end of Is., – Fl. W. ev. 5 s., R. sect. 084°-106° (Covers Cross Rip & Tuckernuck Shoals), Ht. 71′, Rge. W. 14 mi., R. 12 mi., (41-23-25N/70-02-54W)

Sankaty Head Lt., E. end of Is. – Fl. W. ev. 7.5 s., Ht. 158′, Rge. 24 mi., (41-17-04N/69-57-58W)

Nantucket Sound Ltd. Wh. By. "NS" – Mo(A)W, Rge. 6 mi., RWSRST, (41-27-42N/70-23-42W)

Nantucket East Breakwater Lt. #3, Outer Entr. to hbr. – Fl. G. ev. 4 s., Horn 1 bl. ev. 10 s., Ht. 30′, Rge. 12 mi., (41-18-37N/70-06-00W)

Brant Point Lt., Hbr. Entr. W. side – Oc.R. ev. 4 s., Horn 1 bl. ev. 10 s., Ht. 26′, Rge. 10 mi., (41-17-24N/70-05-25W)

For abbreviations see footnote p. 167

Cape Poge Lt., NE point of Chappaquiddick Is. – Fl. W. ev. 6 s., Ht. 65', Rge. 9 mi., (41-25-10N/70-27-08W)

Muskeget Ch. Ltd. Wh. By. "MC" – Mo(A)W, Rge. 6 mi., RWSRST, (41-15-00N/70-26-10W)

Edgartown Harbor Lt., Inner end of hbr. W. side – Fl. R. ev. 6 s., Ht. 45', Rge. 5 mi., (41-23-27N/70-30-11W)

East Chop Lt., E. side Vineyard Haven Hbr. Entr. – Iso. G. ev. 6 s., Ht. 79', Rge. 9 mi., (41-28-13N/70-34-03W)

West Chop Lt., W. side Vineyard Haven Hbr. Entr. – Oc.W. ev. 4 s., R. Sect. 281°-331°, Horn 1 bl. ev. 30 s., Ht. 84', Rge. R. 10 mi., W. 14 mi., (41-28-51N/70-35-59W)

Nobska Point Lt., Woods Hole E. Entr. – Fl. W. ev. 6 s., R. Sect. 263°-289°, Horn 2 bl. ev. 30 s., Ht. 87', Rge. R. 11 mi., W. 13 mi., ltd 24 hrs., (41-30-57N/70-39-18W)

Tarpaulin Cove Lt., SE side Naushon Is. – Fl. W. ev. 6 s., Ht. 78', Rge. 9 mi., (41-28-08N/70-45-27W)

Menemsha Creek Entr. Jetty Lt. #3 – Fl. G. ev. 4 s., Ht. 25', Rge. 5 mi., (41-21-16N/70-46-07W)

Cuttyhunk East Entr. Ltd. Bell By. "CH" – Mo(A)W, Rge. 5 mi., RWSRST, (41-26-36N/70-53-24W)

Gay Head Lt., W. point of Martha's Vineyard – Alt. W. and R. ev. 15 s., Ht. 170', Rge. W. 24 mi., R. 20 mi. Obscured 342°-359° by Nomans Land, ltd. 24 hrs., (41-20-54N/70-50-06W)

BUZZARDS BAY

Canapitsit Ch. Entr. Bell By. "CC", – RWSRST, (41-25-01N/70-54-23W)

Narragansett - Buzzards Bay App. Ltd. Wh. By. "A" – Mo(A)W, Rge. 6 mi., Racon (N), RWSRST, (41-06-00N/71-23-24W)

Buzzards Bay Entr. Lt., W. Entr. – Fl. W. ev. 2.5 s., Horn 2 bl. ev. 30 s., Ht. 67', Rge. 17 mi., Racon (B), (41-23-48N/71-02-01W)

Dumpling Rocks Lt. #7, off Round Hill Pt. – Fl. G. ev. 6 s., Ht. 52', Rge. 8 mi., (41-32-18N/70-55-17W)

Buzzards Bay Midch. Ltd. Bell By. "BB" (east of Wilkes Ledge) – Mo(A)W, Rge. 6 mi., RWSRST, (41-30-33N/70-49-54W)

New Bedford West Barrier Lt. – Q.G., Horn 1 bl. ev. 10 s., Ht. 48', Rge. 8 mi., (41-37-27N/70-54-22W)

New Bedford East Barrier Lt. – Q. R., Ht. 48', Rge. 5 mi., (41-37-29N/70-54-19W)

Padanaram Breakwater Lt. #8 – Fl. R. ev. 4 s., Ht. 25', Rge. 5 mi., (41-34-27N/70-56-21W)

Butler Flats Lt. – Fl. W. ev. 4 s., Ht. 53', (41-36-12N/70-53-42W)

Cleveland Ledge Lt., Cape Cod Canal App. E. side of S. Entr. – Fl. W. ev. 10 s., Horn 1 bl. ev. 15 s., Ht. 74', Rge. 15 mi., Racon (C), (41-37-51N/70-41-39W)

Ned Point Lt. – Iso. W. ev. 6 s., Ht. 41', Rge. 12 mi., (41-39-03N/70-47-44W)

Westport Harbor Entr. Lt. #7, W. side – Fl. G. ev. 6 s., Ht. 35', Rge. 9 mi., (41-30-24N/71-05-18W)

RHODE ISLAND

Sakonnet River Entr. Ltd. Wh. By. "SR" – Mo(A)W, Rge. 6 mi., RWSRST, (41-25-45N/71-13-23W)

Sakonnet Lt. – Fl. W. ev. 6 s., R. sect. 195°-350°, Ht. 58', Rge. W. 7 mi., R. 5 mi., (41-27-11N/71-12-09W)

Sakonnet Breakwater Lt. #2, Entr. to hbr. – Fl. R. ev. 4 s., Ht. 34', Rge. 8 mi., (41-28-00N/71-11-42W)

Narragansett Bay Entr. Ltd. Wh. By. "NB" – Mo(A)W, Rge. 6 mi., Racon (B), RWS, (41-23-00N/71-23-24W)

For abbreviations see footnote p. 167

Beavertail Lt. – Narrag. Bay E. passage – Fl. W. ev. 9 s., Obscured 175°-215°, Horn 1 bl. ev. 30 s., Ht. 64', Rge. 15 mi., ltd. 24 hrs., (41-26-58N/71-23-59W)

Castle Hill Lt. – Iso R. 6 s., Horn 1 bl. ev. 10 s., Ht. 40', Rge. 12 mi., (41-27-44N/71-21-47W)

Fort Adams Lt. #2, Narrag. Bay E. passage – Fl. R. ev. 6 s., Horn 1 bl. ev. 15 s., Ht. 32', Rge. 7 mi., (41-28-54N/71-20-12W)

Newport Harbor Lt., N. end of breakwater – F.G., Ht. 33', Rge. 11 mi., (41-29-36N/71-19-38W)

Rose Is. Lt., Fl W. ev. 6 s., Ht. 48', (41-29-42N/71-20-36W)

Prudence Is. Lt. (Sandy Pt.), Narrag. Bay E. passage – Fl. G. ev. 6 s., Ht. 28', Rge. 6 mi., (41-36-21N/71-18-13W)

Hog Island Shoal Lt., N. side Entr. to Mt. Hope Bay – Iso. W. ev. 6 s., Horn 2 bl. ev. 30 s., Ht. 54', Rge. 12 mi., (41-37-56N/71-16-24W)

Musselbed Shoals Lt.#6A, Mt. Hope Bay Ch. – Fl. R. ev. 6 s., Ht. 26', Rge. 6 mi., (41-38-12N/71-15-36W)

Castle Is. Lt. #2, N. of Hog Is. – Fl. R. ev. 6 s., Ht. 26', Rge. 3 mi., (41-39-14N/71-17-10W)

Bristol Harbor Lt. #4 – F.R., Ht. 25', Rge. 11 mi., (41-39-58N/71-16-42W)

Conimicut Lt., Providence R. App. – Fl. W. ev. 2.5 s., R. Sect. 322°-349°, Horn 2 bl. ev. 30 s., Ht. 58', Rge. W. 8 mi., R. 5 mi., (41-43-01N/71-20-42W)

Bullock Point Lt. "BP", Prov. R. – Oc.W. ev. 4 s., Ht. 29', Rge. 6 mi., (41-44-12N/71-21-54W)

Pomham Rocks Lt., Prov. R. – F.R., Ht. 67', Rge. 6 mi., (41-46-39N/71-22-10W)

Providence River Ch. Lt. #42, off rock – Iso. R. ev. 6 s., Ht. 31', Rge. 4 mi., (41-47-39N/71-22-47W)

Mt. Hope Bay Jct. Ltd. Gong By. "MH" – Fl(2+1) R. 6 s., Rge., 3 mi., R. & G. Bands, (41-39-32N/71-14-03W)

Borden Flats Lt., Mt. Hope Bay – Fl. W. ev. 2.5 s., Horn 1 bl. ev. 10 s., Ht. 47', Rge. 11 mi., (41-42-18N/71-10-30W)

Wickford Harbor Lt. #1, Narrag. Bay W. passage – Fl. G. ev. 6 s., Ht. 40', Rge. 6 mi., (41-34-24N/71-26-10W)

Warwick Lt., Greenwich Bay App. – Oc.G. ev. 4 s., Horn 1 bl. ev. 15 s., Ht. 66', Rge. 12 mi., ltd. 24 hrs., (41-40-00N/71-22-42W)

Point Judith Lt., Block Is. Sd. Entr. – Oc(3)W. ev. 15 s., Horn 1 bl. ev. 15 s., Ht. 65', Rge. 16 mi., (41-21-42N/71-28-54W)

Block Island North Lt., N. end of Is. – Fl. W. ev. 5 s., Ht. 58', Rge. 13 mi., (41-13-39N/71-34-33W)

Block Island Southeast Lt., SE end of Is. – Fl. G. ev. 5 s., Horn 1 bl. ev. 30 s., Ht. 261', Rge. 20 mi., ltd. 24 hrs., (41-09-10N/71-33-04W)

Pt. Judith Harbor of Refuge W. Entr. Lt. #3 – Fl. G. ev. 6 s., Horn 1 bl. ev. 30 s., Ht. 35', Rge. 5 mi., (41-21-56N/71-30-53W)

Watch Hill Lt., Fishers Is. Sd. E. Entr. – Alt. W. and R. ev. 5 s., Horn 1 bl. ev. 30 s., Ht. 61', Rge. 14 mi., ltd. 24 hrs., (41-18-14N/71-51-30W)

FISHERS ISLAND SOUND

Latimer Reef Lt., Fishers Is. Sd. main ch. – Fl. W. ev. 6 s., Bell 2 strokes ev. 15 s., Ht. 55', Rge. 9 mi., (41-18-18N/71-56-00W)

N. Dumpling Lt., Fishers Is. Sd. main ch. – F.W., Horn 1 bl. ev. 30 s., R. Sect. 257°-023°, Ht. 94', Rge. R. 7 mi., F.W. 9 mi., (41-17-17N/72-01-10W)

Stonington Outer Breakwater Lt. #4 – Fl. R. ev. 4 s., Horn 1 bl. ev. 10 s., Ht. 46', Rge. 5 mi., (41-19-00N/71-54-28W)

LONG ISLAND SOUND, NORTH SIDE

Race Rock Lt., SW end of Fishers Is. – Fl. R. ev. 10 s., Horn 2 bl. ev. 30 s., Ht. 67', Rge. 16 mi., (41-14-37N/72-02-49W)

For abbreviations see footnote p. 167

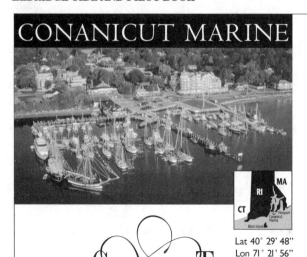

Valiant Rock Ltd. Wh. By. #11 (northerly of rock) Q.G., Rge. 4 mi. Racon (B), (41-13-46N/72-04-00W)

Bartlett Reef Lt., S. end of reef – Fl. W. ev. 6 s., Horn 2 bl. ev. 60 s., Ht. 35', Rge. 8 mi., (41-16-28N/72-08-14W)

New London Ledge Lt., W. side of Southwest ledge –Fl(3) W. R. ev. 30 s., Horn 2 bl. ev. 20 s., Ht. 58', Rge. W. 17 mi., R. 14 mi., (41-18-18N/72-04-42W)

New London Harbor Lt., W. side Entr. – Iso. W. ev. 6 s., R. Sect. 000°-041°, Ht. 89', Rge. R. 14 mi., W. 17 mi., (41-19-00N/72-05-24W)

Saybrook Breakwater Lt., W. jetty – Fl. G. ev. 6 s., Horn 1 bl. ev. 30 s., Ht. 58', Rge. 14 mi., (41-15-48N/72-20-34W)

Lynde Pt. Lt., Conn. R. mouth W. side – F.W., Ht. 71', Rge. 14 mi., (41-16-17N/72-20-35W)

Twenty-Eight Foot Shoal Ltd. Wh. By. "TE" – Fl.(2+1) R. ev. 6 s., , Rge. 4 mi., R&G Bands, (41-09-18N/72-30-24W)

Falkner Is. Lt., off Guilford Hbr. – Fl. W. ev. 10 s., Ht. 94', Rge. 13 mi., (41-12-43N/72-39-13W)

Branford Reef Lt., SE Entr. New Haven – Fl. W. ev. 6 s., Ht. 30', Rge. 7 mi., (41-13-18N/72-48-18W)

Southwest Ledge Lt., E. side Entr. New Haven – Fl. R. ev. 5 s., Horn 1 bl. ev. 15 s., Ht. 57', Rge. 14 mi., (41-14-04N/72-54-44W)

New Haven Lt. – Fl. W. ev. 4 s., Ht. 35', Rge. 7 mi., (41-13-16N/72-56-32W)

Stratford Pt. Lt., W. side Entr. Housatonic R. –Fl(2)W. ev. 20 s., Ht. 52', Rge. 16 mi., (41-09-07N/73-06-12W)

Stratford Shoal Lt., Middle Ground – Fl. W. ev. 5 s., Horn 1 bl. ev. 15 s., Ht. 60', Rge. 13 mi., (41-03-35N/73-06-05W)

Tongue Pt. Lt., at Bridgeport Breakwater – Fl. G. ev. 4 s., Ht. 31', Rge. 5 mi., (41-10-00N/73-10-39W)

Penfield Reef Lt., S. side Entr. to Black Rock – Fl. R. ev. 6 s., Horn 1 bl. ev. 15 s., Ht. 51', Rge. 15 mi., (41-07-00N/73-13-18W)

Peck Ledge Lt., E. App. to Norwalk – Fl. G. ev. 2.5 s., Ht. 61', Rge. 5 mi., (41-04-39N/73-22-11W)

Greens Ledge Lt., W. end of ledge – Alt. Fl. W. and R. ev. 24 s., Horn 2 bl. ev. 20 s., Ht. 62', Rge. W. 18 mi., R. 15 mi., (41-02-30N/73-26-38W)

Stamford Harbor Ledge Obstruction Lt., on SW end of Harbor Ledge – Fl. W. ev. 4 s., Ht. 80', (41-00-06N/73-32-06W)

Great Captain Is. Lt., SE Pt. of Is. – Alt. Fl. W. and R. ev. 12 s., Horn 1 bl. ev. 15 s., Ht. 62', Rge. W. 17 mi., R. 15 mi., (40-58-54N/73-37-24W)

Larchmont Harbor Lt. #2, East Entr. – Fl. R. ev. 4 s., Ht. 26', Rge. 5 mi., (40-55-05N/73-43-52W)

LONG ISLAND SOUND, SOUTH SIDE

Little Gull Is. Lt., E. Entr. L.I. Sd. – Fl(2) W. ev. 15 s., Horn 1 bl. ev. 15 s., Ht. 91', Rge. 18 mi., (41-12-23N/72-06-25W)

Plum Gut Lt. – Fl. W. ev. 2.5 s., Ht. 50' Rge. 5 mi., (41-10-26N/72-12-42W)

Plum Island Ltd. Wh. By. "PI" – Mo(A)W, Rge. 6 mi., RWSRST, (41-13-17N/72-10-48W)

Plum Is. Hbr. West Dolphin Lt., W. end of Is. – Fl. G., Horn 1 bl. ev. 10 s., Ht. 12', Rge. 6 mi., (Maintained by U.S. Agr. Dept.), (41-10-18N/72-12-24W)

Orient Pt. Lt., outer end of Oyster Pond Reef – Fl. W. ev. 5 s., Horn 2 bl. ev. 30 s., Ht. 64', Rge. 17 mi., (41-09-48N/72-13-24W)

Horton Pt. Lt., NW point of Horton Neck – Fl. G. ev. 10 s., Ht. 103', Rge. 14 mi., (41-05-06N/72-26-44W)

Mattituck Breakwater Lt. "MI" – Fl. W. ev. 4 s., Ht. 25', Rge. 6 mi., (41-00-55N/72-33-40W)

For abbreviations see footnote p. 167

Old Field Pt. Lt. – Alt. Fl. R. and Fl. G. ev. 24 s., Ht. 74', Rge. 14 mi., (40-58-37N/73-07-07W)

Eatons Neck Lt., E. side Entr. Huntington Bay – F. W., Horn 1 bl. ev. 30 s., Ht. 144', Rge. 18 mi., (40-57-14N/73-23-43W)

Cold Spring Hbr. Lt., on Pt. of shoal – F.W., R. Sect. 039°-125°, Ht. 37', Rge. W. Sect. 9 mi., R. Sect. 7 mi., (40-54-48N/73-29-36W)

Glen Cove Breakwater Lt. #5, E. side Entr. to hbr. – Fl. G. ev. 4 s., Ht. 24', Rge. 5 mi., (40-51-42N/73-39-36W)

Port Jefferson App. Ltd. Wh. By. "PJ" – Mo(A)W, Rge. 6 mi., RWSRST, (40-59-16N/73-06-27W)

Huntington Harbor Lt. – Iso. W. ev. 6 s., Horn 1 bl. ev. 15 s., Ht. 42', Rge. 9 mi., (40-54-39N/73-25-52W)

LONG ISLAND, OUTSIDE

Montauk Point Ltd. Wh. By. "MP" – Mo(A)W, Rge. 6 mi., Racon (M), RWSRST, (41-01-48N/71-45-40W)

Montauk Pt. Lt., E. end of L.I. – Fl. W. ev. 5 s., Horn 1 bl. ev. 15 s., Ht. 168', Rge. 18 mi., (41-04-15N/71-51-26W)

Montauk Hbr. Entr. Ltd. Bell By. "M" – Mo(A)W, Rge. 6 mi., RWSRST, (41-05-07N/71-56-23W)

Shinnecock Inlet App. Ltd. Wh. By. "SH" – Mo(A)W, Rge. 6 mi., RWSRST, (40-49-00N/72-28-36W)

Moriches Inlet App. Ltd. Wh. By. "M" – Mo(A)W, Rge. 6 mi., RWS, (40-44-08N/72-45-12W)

Shinnecock Lt., W. side of Inlet – Fl(2) W. ev. 15 s., Horn 1 bl. ev. 15 s., Ht. 75', Rge. 11 mi., (40-50-31N/72-28-42W)

Jones Inlet Lt., end of breakwater – Fl. W. ev. 2.5 s., Ht. 33', Rge. 6 mi., (40-34-24N/73-34-32W)

Jones Inlet Ltd. Wh. By. "JI" – Mo(A)W, Rge. 6 mi., RWSRST, (40-33-30N/73-35-00W)

E. Rockaway Inlet Ltd. Bell By. ER" – Mo(A), Rge. 6 mi., RWSRST, (40-34-17N/73-45-49W)

Fire Is. Lt., 5.5 mi. E. of inlet – Fl. W. ev. 7.5 s., Ht. 167', Rge. 24 mi., ltd. 24 hrs., (40-37-57N/73-13-07W)

Rockaway Point Breakwater Lt. #4, end of breakwater – Fl. R. ev. 4 s., Ht. 34', Rge. 5 mi., (40-32-25N/73-56-27W)

NEW YORK HARBOR & APPROACHES

Execution Rocks Lt. – Fl. W. ev. 10 s., Ht. 62', Rge. 15 mi., Racon (X), (40-52-41N/73-44-16W)

Hart Is. Lt. #46, off S. end of Is. – Fl. R. ev. 4 s., Ht. 23', Rge. 6 mi., (40-50-42N/73-46-00W)

Stepping Stones Lt., outer end of reef – Oc.G. ev. 4 s., Ht. 46', Rge. 8 mi., (40-49-28N/73-46-29W)

Throgs Neck Lt., Fort Schuyler – F. R., Ht. 60', Rge. 11 mi., (40-48-16N/73-47-26W)

Whitestone Pt. Lt. #1, East R. main ch. – Q.G., Ht. 56', Rge. 3 mi., (40-48-06N/73-49-10W)

Kings Pt. Lt. – (Private Aid), Iso. W. ev. 2 s., Ht. 102', (40-48-42N/73-45-48W)

Hell Gate Lt. #15, East R. Hallets Pt. – Fl. G. ev. 2.5 s., Ht. 33', Rge. 4 mi., (40-46-42N/73-56-06W)

Mill Rock South Lt. #16, East R., main ch. – Fl. R. ev. 4 s., Ht. 37', Rge. 4 mi., (none given)

Governors Is. Extension Lt., SW Pt. of Is. – F. R., Horn 1 bl. ev. 15 s., Ht. 47', Rge. 9 mi., (40-41-09N/74-01-35W)

For abbreviations see footnote p. 167

Governors Is. Lt. #2 – NW pt of Is. – 2 F.R. arranged vertically, Lower Lt. Obscured from 240°-243°, 254°-256°, 264°-360°, Horn 2 bl. ev. 20 s., Ht. 75′ and 60′, Rge. 7 mi., (40-41-35N/74-01-11W)

Verrazano-Narrows Bridge Fog Signal – (Private Aid), 2 Horns on bridge 1 bl. ev. 15 s., (none given)

Coney Is. Lt., N.Y. Hbr. main ch. – Fl. R. ev. 5 s., Ht. 75′, Rge. 16 mi., ltd. 24 hrs., (40-34-36N/74-00-42W)

Romer Shoal Lt., N.Y. Hbr. S. App. – Fl(2) W. ev. 15 s., Horn 2 bl. ev. 30 s., Ht. 54′, Rge. 16 mi., (40-30-47N/74-00-49W)

West Bank (Range Front) Lt., Ambrose Ch. outer sect. – Iso. W. ev. 6 s., R. Sect. 004°-181° and W from 181° - 004°, Horn 2 bl. ev. 20 s., Ht. 69′, Rge. W. 10 mi., R. 8 mi., ltd. 24 hrs., (40-32-17N/74-02-34W)

Staten Island (Range Rear) Lt., Ambrose Ch. outer sect. – F. W. , Visible on range line only, Ht. 231′, Rge. 18 mi., ltd. 24 hrs., (40-34-30N/74-08-36W)

Old Orchard Shoal Lt., N.Y. Hbr. – Fl. W. ev. 6 s., R. Sect. 087°-203°, Ht. 51′, Rge. W. 5 mi., R. 5 mi., (40-30-44N/74-05-55W)

Sandy Hook Lt. – F. W., Ht. 88′, Rge. 19 mi., ltd. 24 hrs., (40-27-42N/74-00-07W)

Sandy Hook Ch. (Range Front) Lt. – Q. W., G., and R. sectors, Red from 063°-073° and Green from 300.5°-315.5°, Ht. 44′, Racon (C), (40-29-15N/73-59-35W)

Southwest Spit Jct. Ltd. Gong By. "SP" – Fl(2+1) R. ev. 6 s., Rge. 3 mi., R. & G. Bands, (40-28-42N/74-03-00W)

Sandy Hook Pt. Lt. – Iso W. ev. 6 s., Ht. 38′, Rge. 15 mi., Bell 1 stroke ev. 10 s., "NB" on Skeleton Tower, (40-28-15N/74-01-07W)

Scotland Ltd. Wh. By. "S", Sandy Hook Ch. App. – Mo(A)W, Rge. 7 mi., Racon (M), RWSRST, (40-26-33N/73-55-01W)

Ambrose Lt., Fl. W. ev. 5 s., Horn 1 bl. ev. 15 s., Ht. 76′, Rge. 18 mi., Racon (N), (40-27-00N/73-48-00W)

Ambrose Ch. Ltd. Wh. By. "A" – Mo(A)W, Rge. 6 mi., RWSRST, (40-28-49N/73-53-37W)

NEW JERSEY

Highlands Lt. – Oc. W. ev. 10 s., Obscured 334°-140°, Ht. 246′, (40-23-48N/73-59-12W)

Atlantic Highlands Breakwater Lt. – Fl. W. ev. 4 s., Ht. 33′, Rge. 7 mi., (40-25-07N/74-01-10W)

Kill Van Kull Ch. Jct. Ltd. Wh. By. "KV"– Fl (2+1) R. ev. 6 s., Rge. 4 mi., R. & G. Bands, Racon (K), (40-39-01N/74-03-52W)

Kill Van Kull Ch. Jct. Ltd. By. "A"– Fl (2+1) G. ev. 6 s., Rge. 3 mi., G. & R. Bands (40-38-45N/74-10-07W)

Kill Van Kull Ch. East Jct. Ltd. By. "E"– Fl (2+1) G. ev. 6 s., Rge. 3 mi. (40-38-31N/74-09-15W)

Manasquan Inlet S. Breakwater Lt. #3 - Fl. G. ev. 6 s., Horn 1 bl. ev. 30 s., Ht. 35′ Rge. 8 mi., (40-06-01N/74-01-54W)

Shark River Inlet Ltd. Wh. By. "SI" – Mo(A)W, Rge. 5 mi., RWSRST, (40-11-09N/74-00-03W)

Barnegat Inlet S. Breakwater Lt. #7 – Q. G., Ht. 37′, Rge. 6 mi., Horn 1 bl. ev. 30 s., (39-45-26N/74-05-36W)

Barnegat Ltd. By. "B" – Fl. Y. ev. 6 s., Rge. 7 mi., Racon (B), Yellow, (39-45-48N/73-46-04W)

Barnegat Inlet Outer Ltd. Wh. By. "BI" – Mo(A)W, Rge. 5 mi., RWSRST, (39-44-28N/74-03-51W)

Little Egg Inlet Outer Ltd. Wh. By. "LE" – Mo(A)W, Rge. 5 mi., RWSRST, (39-26-47N/74-17-18W)

Brigantine Inlet Wreck Ltd. By. "WR2" (100 yards, 090° from wreck) – Q. R., Rge. 4 mi., Red, (39-24-48N/74-13-47W)

For abbreviations see footnote p. 167

Great Egg Harbor Inlet Outer Ltd. Wh. By. "GE" – Mo(A)W, Rge. 5 mi., RWSRST, (39-16-14N/74-31-56W)

Hereford Inlet Lt., S. side – Fl. W. ev. 10 s., Ht. 57', Rge. 24 mi., (39-00-24N/74-47-28W)

Hereford Inlet Ltd. Wh. By. "H" – Mo(A)W, Rge. 5 mi., RWSRST, (38-58-59N/74-46-11W)

Five Fathom Bank Ltd. By. "F", Delaware Bay Entr. – Fl. Y. ev. 2.5 s., Rge. 6 mi., Racon (M), Yellow, (38-46-49N/74-34-32W)

Cape MayLt. – Fl. W. ev. 15 s., Ht. 165', Rge. 24 mi., (38-55-59N/74-57-37W)

NEW JERSEY, DELAWARE AND MARYLAND

Delaware Ltd. By. "D", Del. Bay – Fl. Y. ev. 6 s., Rge. 6 mi., Racon (K), Yellow, (38-27-18N/74-41-47W)

Delaware Traffic Lane Ltd. By. "DA" – Fl. Y. ev. 2.5 s., Rge. 6 mi., Yellow, (38-32-45N/74-46-56W)

Delaware Traffic Lane Ltd. By. "DB" – Fl. Y. ev. 4 s., Rge. 6 mi., Yellow, (38-38-12N/74-52-11W)

Delaware Traffic Lane Ltd. By. "DC" – Fl. Y. ev. 2.5 s., Rge. 6 mi., Yellow, (38-43-47N/74-57-33W)

Brown Shoal Lt. – Fl. W. ev. 2.5 s., Ht. 23', Rge. 7 mi., Racon (B), (38-55-20N/75-06-03W)

Brandywine Shoal Lt., Del. Bay main ch. 9 mi. from S. end of shoal – Fl. W. ev. 10 s., R. Sect. 151°-338°, Horn 1 bl. ev. 15 s., Ht. 60', Rge. W. 19 mi., R. 13 mi., (38-59-10N/75-06-47W)

Harbor of Refuge Lt., Del. Bay main ch., S. end of breakwater – Fl. W. ev. 10 s., 2 R. Sect. 325°-351° and 127°-175°, Horn 2 bl. ev. 30 s., Ht. 72', Rge. W. 19 mi., R. 16 mi., (38-48-52N/75-05-33W)

Fourteen Foot Bank Lt., Del. Bay main ch. W. side – Fl. W. ev. 9 s., R. Sect. 332.5°-151°, Horn 1 bl. ev. 30 s., Ht. 59', Rge. W. 13 mi., R. 10 mi., (39-02-54N/75-10-56W)

Miah Maull Shoal Lt., Del Bay main ch. – Oc. W. ev. 4 s., R. Sect. 137.5°-333°, Horn 1 bl. ev. 10 s., (Mar. 15-Dec. 15). Ht. 59', Rge. W. 15 mi., R. 12 mi., Racon (M), (39-07-36N/75-12-31W)

Elbow of Cross Ledge Lt., Del. Bay main ch. – Iso. W. ev. 6 s., R. Sect. 155°-325°, Horn 2 bl. ev. 20 s., (Mar. 15-Dec. 15), Ht. 61', Rge. W. 15 mi., R. 11 mi., (39-10-56N/75-16-06W)

Ship John Shoal Lt., Del. Bay main ch. – Fl. W. ev. 5 s., R. Sect. 138°-321.5°, Horn 1 bl. ev. 15 s. (Mar. 15 - Dec. 15) , Ht. 50', Rge. W. 16 mi., R. 12 mi., Racon (O), (39-18-19N/75-22-36W)

Egg Island Point Lt. – Fl. W. ev. 4 s., Ht. 27', Rge. 7 mi., (39-10-21N/75-07-55W)

Ben Davis Pt. Lt. "BD" – Fl. W. ev. 6 s., Ht. 30', Rge. 6 mi., (39-17-27N/75-17-18W)

Old Reedy Is. Lt. – Iso. W. ev. 6 s., R. Sect. 353°-014°, Ht. 20', Rge. W. 8 mi., R. 6 mi., (39-30-03N/75-34-08W)

Fenwick Is. Lt. – Oc.W. ev. 13 s., Ht. 83', Rge. 8 mi., (38-27-06N/75-03-18W)

Ocean City Inlet Jetty Lt., on end of jetty – Iso. W. ev. 6 s., Horn 1 bl. ev. 15 s., Ht. 28', Rge. 8 mi., (38-19-27N/75-05-06)

VIRGINIA

Assateague Lt., S. side of Is. – Fl(2) W. ev. 5 s., Ht. 154', Rge. 22 mi., (37-54-40N/75-21-22W)

Wachapreague Inlet Ltd. Wh. By. "W" – Mo(A)W, Rge. 5 mi., RWSRST, (37-34-54N/75-33-37W)

Quinby Inlet Ltd. Wh. By. "Q" – Mo(A)W, Rge. 5 mi., RWSRST, (37-28-06N/75-36-05W)

For abbreviations see footnote p. 167

Distance Table in Nautical Miles

*Approximate

Bar Harbor to
Halifax, N.S. 259
Yarmouth, N.S. 101
Saint John, N.B. 122
Machiasport 52
Rockland 62
Boothbay Harbor 86
Portland 115
Marblehead 169

Rockland to
Boothbay Harbor 42
Belfast 22
Bucksport 33

Boothbay Harbor to
Kennebec River 11
Monhegan 15
Portland 36

Portland Ltd. Buoy "P" to
Biddeford 17
Portsmouth 54
Cape Cod Light 99
Cape Cod Canal (E. Entr.) 118
Pollock Rip Slue 141

Portsmouth (Whaleback) to
York River 7
Biddeford Pool 30
Newburyport Entr. 15
Gloucester – via Annisquam 28

Gloucester to
Boston 26
Scituate 26
Plymouth 43
Cape Cod Canal (E. Entr.) 52
Provincetown 45

Marblehead to
Portsmouth 43
Biddeford Pool 68
Portland 87
Boothbay Harbor 104
Rockland 133
Plymouth 38
Cape Cod Canal (E. Entr.) 47

Boston (Commonwealth Pier)
Marblehead 17
Isles of Shoals 52
Portsmouth 58
Portland 95
Kennebec River 107
Boothbay Harbor 116
Rockland 149
North Haven 148
Bangor 194
St. John, N.B. 286
Halifax, N.S. 380
Cohasset 14
Cape Cod Canal, E. Entr. 50
Provincetown 50
Vineyard Haven 77
New Bedford 81
Fall River 107
Newport 122
New London 140
New York 234

**Western Entr., Cape Cod Canal to
East Entrance 8
Woods Hole 15
Quicks Hole 20
New Bedford 24
Newport 50
New London 83

Woods Hole to
Hyannis 19
Chatham 32
Cuttyhunk 14
Marion 11

Vineyard Haven to
Edgartown 9
Marblehead – around Cape 114
Canal – via Woods Hole 20
Newport 45
New London 77
New Haven 114
South Norwalk 140
City Island 153

*Each distance is by the shortest route that safe navigation permits between the two ports concerned.

**Western entr., The beginning of the "land cut" at Bourne Neck, 7.3 nautical miles up the channel from Cleveland Ledge Lt.

Continued p. 194

Great Machipongo Inlet Ltd. Wh. By. "GM" – Mo(A)W, Rge. 5 mi., RWSRST, (37-23-36N/75-39-06W)

Great Machipongo Entr. Lt. #5, S. side – Fl. G. ev. 4 s., Ht. 15', Rge. 4 mi., (none given)

Cape Charles Lt., N. side of Entr. to Ches. Bay – Fl. W. ev. 5 s., Ht. 180', Rge. 18 mi., (37-07-23N/75-54-23W)

Chesapeake Light, off Entr. to Ches. Bay – Fl(2) W. ev. 15 s., Horn 1 bl. ev. 30 s., Ht. 117', Rge. 19 mi., Blue Tower, Racon (N), (36-54-17N/75-42-46W)

Chesapeake Bay Entr. Ltd. Wh. By. "CH" – Mo(A)W, Rge. 7 mi., Racon (C), (36-56-08N/75-57-27W)

Cape Henry Lt., S. side of Entr. to Ches. Bay – Fl. W. Mo (U) W ev. 20 s., R. Sect. 154°-233°, Ht. 164', Rge. W.17 mi., R. 15 mi., (36-55-35N/76-00-26W)

CHESAPEAKE BAY

Thimble Shoal Lt., Thimble Shoal Ch. – Fl. W. ev. 10 s., Horn 1 bl. ev. 30 s., Ht. 55', Rge. 20 mi., (37-00-55N/76-14-23W)

Worton Pt. Lt., Fl. W. ev. 6 s., Ht. 93' Rge. 6 mi., (39-19-06N/76-11-11W)

Old Point Comfort Lt., N. side Entr. to Hampton Roads – Fl(2) W. ev. 12 s., W. Sect. 265°-038°, Ht. 54', Rge. W. 16 mi., R. 14 mi., (37-00-06N/76-18-23W)

York Spit Lt., N. side Entr. to York R. – On pile, Fl. W. ev. 6 s., Ht. 30', Rge. 8 mi., (37-12-35N/76-15-15W)

Wolf Trap Lt., Ches. Ch. – Fl. W. ev. 15 s., Ht. 52', Rge. 14 mi., (37-23-24N/76-11-24W)

Stingray Pt. Lt., Ches. Ch. – Fl. W. ev. 4 s., Ht. 34', Rge. 9 mi., (37-33-41N/76-16-23W)

Windmill Pt. Lt., Ches. Ch. – On pile. Fl. W. ev. 6 s., 2 R. Sectors 293°-082° and 091.5°-113°, Ht. 34', Rge. W. 9 mi., R. 7 mi., (37-35-49N/76-14-10W)

Tangier Sound Lt., Ches. Ch. – Fl. W. ev. 6 s., R. Sect. 115°-193°, Ht. 45', Rge. W. 12 mi., R. 9 mi., (37-47-15N/75-58-25W)

Smith Pt. Lt., Ches. Ch. – Fl. W. ev. 10 s., R. Sect. 003°-156°, Horn 1 bl. ev. 15 s. (operates continuously) , Ht. 52', Rge. W. 22 mi., R. 18 mi., (37-52-47N/76-11-01W)

Point Lookout Lt., Ches. Ch. – Fl(2) W. ev. 5 s., Ht. 39', Rge. 8 mi., (38-01-30N/76-19-24W)

Holland Is. Bar Lt., Ches. Ch. – Fl. W. ev. 2.5 s., Horn 1 bl. ev. 30 s. (operates continuously Sept. 15 - June 1), Ht. 37', Rge. 7 mi., (38-04-07N/76-05-45W)

Point No Point Lt., Ches. Ch. – Fl. W. ev. 6 s., Ht. 52', Rge. 9 mi., (38-07-42N/76-17-24W)

Hooper Is. Lt., Ches. Ch. – Fl. W. ev. 6 s., Horn 1 bl. ev. 30 s. (operates continuously Sept. 15 - June 1), Ht. 63', Rge. 9 mi., (38-15-24N/76-15-00W)

Drum Pt. Lt.#4, Ches. Ch. – Fl. R. ev. 2.5 s., Ht. 22', Rge. 5 mi., (38-19-08N/76-25-15W)

Cove Pt. Lt., Ches. Ch. – Fl. W. ev. 10 s., Obscured from 040°-110°, Ht. 45', Rge. 12 mi., (38-23-11N/76-22-54W)

Sharps Is. Lt., Ches. Ch. – Fl. W. ev. 6 s., R. Sect. 159°-262°, Ht. 54', Rge. W. 9 mi., R. 7 mi., (38-38-20N/76-22-39W)

Bloody Point Bar Lt., Ches. Ch. – Fl. W. ev. 6 s., 2 R. Sectors 003°-022° and 183°-202°, Ht. 54', Rge. W. 9 mi., R. 7 mi., (38-50-02N/76-23-30W)

Thomas Pt. Shoal Lt., Ches. Ch. – Fl. W. ev. 5 s., 2 R. Sectors 011°-051.5° and 096.5°-202°, Horn 1 bl. ev. 15 s., Ht. 43', Rge. W. 16 mi., R. 11 mi., (38-53-56N/76-26-09W)

Sandy Pt. Shoal Lt., Ches. Ch. – Fl. W. ev. 6 s., Ht. 51', Rge. 9 mi., (39-00-57N/76-23-04W)

Baltimore Lt. – Fl. W. ev. 2.5 s., R. Sector 082° - 160°, Ht. 52', Rge. W. 7 mi., R. 5 mi., (39-03-30N/76-23-56W)

Chesapeake Bay Bridge West Ch. Fog Signal, on main ch. span – Horn (2) 1 bl. ev. 15 s., 5 s. bl., Horn Points 017°-197°, (38-59-36N/76-22-54W)

Chesapeake Bay Bridge East Ch. Fog Signal, on main ch. span – Horn 1 bl. ev. 20 s., 2 s. bl., (38-59-18N/76-21-30W)

For abbreviations see footnote p. 167

Distance Table in Nautical Miles

Continued from p. 192   *Approximate

Nantucket Entr. Bell NB to
Boston – around Cape 105
Boston – via Canal 94
Chatham 23
Edgartown................................... 23
Hyannis....................................... 21
Woods Hole.................................. 30
Cape Cod Canal (W. Entr.) 45
Newport....................................... 71

New Bedford (State Pier) to
Woods Hole.................................. 14
Newport....................................... 38
New London 74
New York (Gov. Is.)................. 166

Newport to
Providence................................... 21
Stonington................................... 34
New London 48
New Haven................................... 84
City Island 122

Block Island (FR Horn) to
Nantucket 79
Vineyard Haven......................... 52
Cleveland Ledge Lt.................. 50
New Bedford 44
Newport....................................... 22
Race Point Lt.............................. 21
New London 29

New London to
Greenport.................................... 25
New Haven................................... 49
Bridgeport 60
City Island 86

Port Jefferson to
Larchmont................................... 30
So. Norwalk................................. 15
Milford... 14
Old Saybrook............................... 43
New London 53

City Island to
Governors Island 17
Execution Rocks.......................... 3

Execution Rocks to
Port Chester 8
Stamford...................................... 12
Oyster Bay Harbor 14
So. Norwalk................................. 19
Bridgeport................................... 29

Port Jefferson............................. 30
Milford... 37
New Haven................................... 49
Conn. River 69
Mystic 84
Montauk Point 87

New York (Battery) to
Jones Inlet................................... 34
Fire Island Inlet......................... 47
Moriches Inlet............................ 74
Shinnecock Inlet 88
Montauk Point 117
Keyport....................................... 22
Asbury Park 35
Manasquan 40
Little Egg Inlet 81
Atlantic City............................... 97
Philadelphia............................... 235
Chesapeake Lt. Stn................ 247
Cape Henry Lt. 262
Norfolk 288
Baltimore 418

Brielle-Manasquan to
E. Rockaway Inlet..................... 32
Jones Inlet................................... 35
Fire Island Inlet........................ 45
Montauk Point 117
Barnegat Inlet............................ 21
Atlantic City............................... 51

Delaware Breakwater to
Reedy Pt. Entr. (C&D Canal) .. 51
Annapolis – via Canal............... 97
Norfolk 167
New York 150
New London 242
Providence................................. 275
New Bedford 278
Boston (outside) 399
Portland (outside) 443

Old Point Comfort to
Baltimore 163
Philadelphia............................... 240
New York 276
New London 363
Providence................................. 392
New Bedford 397
Boston (outside) 512
Portland...................................... 553

NORTH CAROLINA

Currituck Beach Lt. – Fl. W. ev. 20 s., Ht. 158', Rge. 18 mi., (36-22-37N/75-49-47W)

Bodie Is. Lt. – Fl(2) W. ev. 30 s., Ht. 156', Rge. 18 mi., (35-49-07N/75-33-48W)

Oregon Inlet Jetty Lt. – Iso. W. ev. 6 s., Ht. 28', Rge. 7 mi., (35-46-26N/75-31-30W)

Cape Hatteras Lt., – Fl. W. ev. 7.5 s., Ht. 192', Rge. 24 mi., (35-15-02N/75-31-44W)

Hatteras Inlet Lt. – Iso. W. ev. 6 s., Ht. 48', Rge. 10 mi., (35-11-52N/75-43-56W)

Ocracoke Lt., on W. part of island – F.W., Ht. 75', Rge. 15 mi., (35-06-32N/75-59-10W)

Cape Lookout Lt., on N. pt. of cape – Fl. W. ev. 15 s., Ht. 156', Rge. 25 mi., (34-37-22N/76-31-28W)

Beaufort Inlet Ch. Ltd. Wh. By. "BM" – Mo(A)W, Rge. 6 mi., Racon (M), RWSRST, (34-34-49N/76-41-33W)

New River Inlet Ltd. Wh. By. "NR" – Mo(A)W, Rge. 6 mi., RWSRST, (34-31-02N/77-19-33W)

Oak Is. Lt., on SE pt. of island – Fl(4) W. ev. 10 s., Ht. 169', Rge. 24 mi., (33-53-34N/78-02-06W)

Cape Fear River Entr. Ltd. Wh. By. "CF" – Mo(A)W, Rge. 6 mi., Racon (C), RWSRST, (33-46-17N/78-03-02W)

SOUTH CAROLINA

Little River Inlet Entr. Ltd. Wh. By. "LR" – Mo(A)W, Rge. 5 mi., RWSRST, (33-49-48N/78-32-30W)

Little River Inlet North Jetty Lt. #2 – Fl. R. ev. 4 s., Ht. 24', Rge. 4 mi., (33-50-30N/78-32-42W)

Winyah Bay Ltd. Wh. By. "WB" – Mo(A)W, Rge. 6 mi., RWSRST, (33-11-36N/79-05-12W)

Georgetown Lt., E. side Entr. to Winyah Bay – Fl(2) W. ev. 15 s., Ht. 85', Rge. 15 mi., (33-13-24N/79-11-06W)

Charleston Entr. Ltd. By. "C" – Mo(A), Rge. 6 mi., Racon (K), RWSRST, (32-37-05N/79-35-30W)

Charleston Lt., S. side of Sullivans Is. – Fl(2) W. ev. 30 s., Ht. 163', Rge. 26 mi., (32-45-28N/79-50-35W)

GEORGIA

Tybee Lt., NE end of Is. – F. W., Ht. 144', Rge. 19 mi., ltd. 24 hrs., (32-01-20N/80-50-44W)

Tybee Lighted Buoy "T" – Mo(A)W, Rge. 6 mi., Racon (G), RWSRST, (31-57-52N/80-43-09W)

St. Simons Ltd. By. "STS" – Mo(A)W, Rge. 6 mi., Racon (B), RWSRST, (31-02-49N/81-14-25W)

St. Simons Lt., N. side Entr. to St. Simons Sd. – F. Fl. W. ev. 60 s., Ht. 104', Rge. F. W. 18 mi., Fl. W. 23 mi., (31-08-03N/81-23-37W)

FLORIDA

Amelia Is. Lt., 2 mi. from N. end of Is. – Fl. W. ev. 10 s., R. Sect. 344°-360°, Ht. 107', Rge. W. 23 mi., R. 19 mi., (30-40-24N/81-26-30W)

St. Johns Lt., on shore – Fl(4) W. ev. 20 s., Obscured 179°-354°, Ht. 83', Rge. 19 mi., (30-23-06N/81-23-54W)

St. Johns Ltd. By. "STJ" – Mo(A)W, Rge. 6 mi., Racon (M), RWSRST, (30-23-35N/81-19-08W)

St. Augustine Lt., N. end of Anastasia Is. – F. Fl. W. ev. 30 s., Ht. 161', Rge. F. W. 19 mi., Fl. W. 24 mi., (29-53-08N/81-17-19W)

Ponce De Leon Inlet Lt., S. side on inlet – Fl(6) W. ev. 30 s., Ht. 159', Rge. 22 mi., (29-04-48N/80-55-42W)

Cape Canaveral Lt., on Cape – Fl(2) W. ev. 20 s., Ht. 137', Rge. 24 mi., (28-27-37N/80-32-36W)

For abbreviations see footnote p. 167

The Ship's Bell Code

Telling time by ship's bell has a romantic background that goes back hundreds of years. It is based in the workday routine of the ship's crew. A ship at sea requires a constant watch throughout the whole twenty-four hours of the day. To divide the duty, the day is broken up into six watches of four hours each and the crew into three divisions, or watches.

Each division of the crew stands two four-hour watches a day. In order to rotate the duty, so that a division does not have to stand the same watch day in and day out, the 4 to 8 watch in the afternoon is divided into two watches known as the dog watches.

The Mid-Watch - Midnight to 4 A.M.
The Morning Watch - 4 A.M. to 8 A.M.
The Forenoon Watch - 8 A.M. to 12 Noon
The Afternoon Watch - 12 Noon to 4 P.M.

The 1st Dog Watch - 4 P.M. to 6 P.M.
The 2nd Dog Watch - 6 P.M. to 8 P.M.
The First Watch - 8 P.M. to Midnight

To apprise the crew of the time, the ship's bell was struck by the watch officer at half hour intervals, the first half hour being one bell, the first hour two bells, hour and a half three bells, and so on up to eight bells, denoting time to relieve the watch. By this method of timekeeping eight bells marks 4, 8, or 12 o'clock.

8 Bells	4:00	8:00	12:00
1 Bell	4:30	8:30	12:30
2 Bells	5:00	9:00	1:00
3 Bells	5:30	9:30	1:30
4 Bells	6:00	10:00	2:00
5 Bells	6:30	10:30	2:30
6 Bells	7:00	11:00	3:00
7 Bells	7:30	11:30	3:30

Courtesy of Chelsea Clock Co., Chelsea, MA

Sebastian Inlet N. Jetty Lt. – Fl. W. ev. 4 s., R. Sect. 104°-154°, Ht. 27', Rge. W. 5 mi., R. 4 mi., (27-51-41N/80-26-51W)

Jupiter Inlet Lt., N. side of inlet – Fl(2) W. ev. 30 s., Obscured 231°-234°, Ht. 146', Rge. 25 mi., (26-56-55N/80-04-55W)

Hillsboro Inlet Entr. Lt., N. side of inlet – Fl. (2) W. ev. 20 s., Obscured 015°-186°, Ht. 136', Rge. 28 mi., (26-15-33N/80-04-51W)

Port Everglades Ltd. By. "PE" - Mo(A)W, Rge. 7 mi., Racon (T), RWSRST, (26-05-30N/80-04-48W)

Miami Ltd. By. "M" – E. end of Miami Beach, Mo(A)W, Rge. 7 mi., Racon (M), RWSRST, (25-46-06N/80-05-00W)

Fowey Rocks Lt., Hawk Ch. – Fl. W. ev. 10 s., 2 R. Sect., W. Sectors 188°-359° and 022°-180°, R. in intervening sectors, Ht. 110', Rge. W. 15 mi., R. 10 mi. Racon (O), (25-35-24N/80-05-48W)

Carysfort Reef Lt., outer line of reefs – Fl(3) W. ev. 60 s., 3 W. Sectors 211°-018° and 049°-087° and 145°-184°, R. in intervening sectors, Ht. 100', Rge. W. 15 mi., R. 13 mi., Racon (C), (25-13-18N/80-12-42W)

Alligator Reef Lt., outer line of reefs – Fl(4) W. ev. 60 s., 2 R. sectors 223°-249° and 047°-068°, Ht. 136', Rge. W. 16 mi., R. 13 mi., Racon (G), (24-51-06N/80-37-06W)

Sombrero Key Lt., outer line of reefs – Fl(5) W. ev. 60 s., 3 W. Sectors 222°-238° and 264°-066° and 094°-163°, R. in intervening sectors, Ht. 142', Rge. W. 15 mi., R. 12 mi., Racon (M), (24-37-36N/81-06-36W)

American Shoal Lt., outer line of reefs – Fl(3) W. ev. 15 s., W. sectors 270°-067° and 125°-242°, Obscured 90°-125°, R. in intervening sectors, Ht. 109', Rge. W. 10 mi., R. 13 mi., Racon (Y), (24-31-30N/81-31-12W)

Sand Key Lt., Fl(2) W. ev. 15 s., 2 Red Sectors 072°-086° and 248°-270°, Ht. 109', Rge. W. 14 mi., R. 14 mi., Racon (N), (24-27-14N/81-52-39W)

Dry Tortugas Lt., on Loggerhead Key – Fl. W. ev. 20 s., Ht. 151', Rge. 20 mi., Racon (K), (24-38-00N/82-55-12W)

BERMUDA – APPROACH LIGHTS FROM SEAWARD

North Rock Beacon – Fl(4)W. ev. 20 s., Ht. 70', Rge. 12 mi., RaRef, (32-28.5N/64-46.1W)

North East Breaker Beacon – Fl. W. ev. 2.5 s., Ht. 45', Rge. 12 mi., Racon (N), RaRef, (Red tower on red tripod base reading "Northeast," (32-28.7N/64-41.0W)

Kitchen Shoal Beacon – Fl(3)W. ev. 15 s., Ht. 45', Rge. 12 mi., RaRef, RWS, Red "Kitchen" on White background, (32-26.1N/64-37.6W)

Eastern Blue Cut Beacon – Fl.W. (U) ev. 10 s., Ht. 60', Rge. 12 mi., RaRef, B&W Tower "Eastern Blue Cut" on white band, (32-23.9N/64-52.6W)

Chub Heads – Q. Fl(9) W. ev. 15 s., Ht. 60', Rge. 12 mi., RaRef, Yellow and Black Horizontal Stripe Tower with "Chub Heads" in White on Black Central band, Racon (C), (32-17.2N/64-58.9W)

Mills Breaker By. – Q. Fl(3)W. ev. 5 s., Black "Mills" on yellow background, (32-23.9N/64-36.9W)

Spit By. – Q. Fl(3) W. ev. 10 s., Black "Spit" on yellow, (32-22.7N/64-38.5W)

Sea By. – Mo(A)W ev. 6 s., RWS, Red "SB" in white on side, (32-22.9N/64-37.1W)

St. David's Is. Lt. – F. R. and G. Sectors below Fl(2) W. ev. 20 s., Ht. 212', Rge. W. 15 mi., R. and G. 20 mi., (32-21.8N/64-39.1W) Your bearing from seaward of G. Sector is 221°-276° True; remaining Sector is R. and partially obscured by land 044°-135° True.

Kindley Field Aero Beacon – Alt. W and G.; 1 White, 1 Green (rotating Aero Beacon), Ht. 140', Rge. 15 mi., (32-21.95N/64-40.55W)

Gibb's Hill Lighthouse – Fl. W. ev. 10 s., Ht. 354', Rge. 26 mi., (32-15.2N/64-50.1W)

☆ Note: The information in this volume has been compiled from U. S. Government sources and others, and carefully checked. The Publishers cannot assume any liability for errors, omissions, or changes.

Foregoing information checked to date, September, 2008. See page 4 for free Supplement in May 2009.

LOOKING BACK AT LEADLINES

While it is not known when leadlines were first used, it seems apparent that it was long before Columbus's time. They are certainly one of the earliest navigation instruments. As recently as 50 years ago, the sounding lead was regarded as invaluable for coastwise piloting. In his book The Coastwise Navigator, Captain Warwick Tompkins says of the sounding lead: "Today it is second only to the magnetic compass in the list of the ship's essential equipment." That was 1938.

The sounding lead of many years ago was most often attached to a very long line, as there were no electronic sounders to probe great depths. Instead of a knot every fathom, more elaborate markings were used to indicate unambiguously a variety of depths:

2 fathoms - a 2-ended scrap of leather	13 fathoms - a piece of thick blue serge
3 fathoms - a 3-ended scrap of leather	15 fathoms - a piece of white calico
5 fathoms - a scrap of white calico	17 fathoms - a piece of red wool bunting
7 fathoms - a strip of red wool bunting	20 fathoms - a cord with 2 knots
10 fathoms - leather with a round hole in it	30 fathoms - a cord with 3 knots

Why did sailors of old need to know about such great depths? For one, fishing schooners needed to know when they were over the relatively shallow banks offshore where great schools of fish congregated. (Sailors would sometimes "arm the lead" by putting tallow into the "cavity" to bring up a bottom sample.) Also, charts were often not so reliable in places, and it was prudent to check depths. Consider the recent groundings of ocean-going ships, incidents which underline the prudence every skipper should exercise when proceeding in "thin" waters.

TIDAL HEIGHTS AND DEPTHS

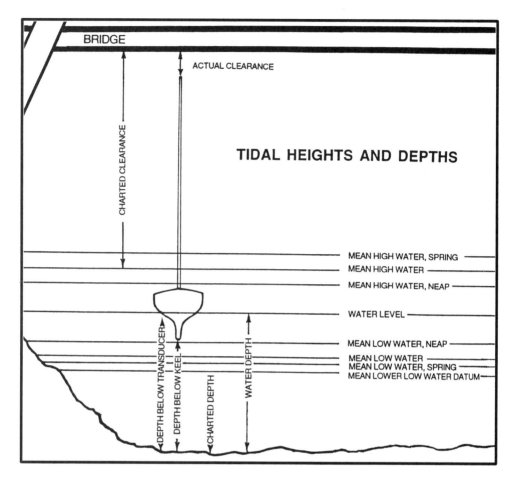

Mean High Water, Spring - the mean of high waters heights of spring tides
Mean High Water - the mean of all high water heights; the charted clearance of bridges is measured from this height
Mean High Water, Neap - the mean of high water heights of neap tides
Mean Low Water, Neap - the mean of low water heights of neap tides
Mean Low Water - the mean of all low water heights
Mean Low Water, Spring - the mean of low water heights of spring tides
Mean Lower Low Water Datum - the mean of lower low water heights; charted depths originate from this reference height or datum

---

Spring Tides - tides of increased range, occurring twice a month, around the times of the new and full moons
Neap Tides - tides of decreased range, occurring twice a month, around the times of the half moons
Diurnal Inequality - the difference in height of the two daily low waters or the two daily high waters, a result of the moon's (and to a lesser extent the sun's) changing declination above and below the Equator

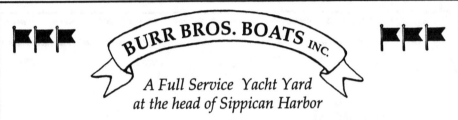
Duration of Slack Water

The tables below give, for various maximum currents, the approximate period of time during which weak currents not exceeding 0.5 knot will be encountered. This duration includes the last of the Flood or Ebb and the beginning of the following Ebb or Flood; i.e. half of the duration will be before and half after the time of Slack Water.

Duration of Weak Current for Cape Cod Canal, Hell Gate, Ches. & Del. Canal:

If the Max. Curr. will be:	Total Period of Time with Vel. not more than 0.5 knot:
1.0 kts.	89 min.
1.5 kts.	52 min.
2.0 kts.	36 min.
3.0 kts.	22 min.
4.0 kts.	17 min.
5.0 kts.	13 min.

Duration of Weak Current for all other places:

If the Max. Curr. will be:	Total Period of Time with Vel. not more than 0.5 knot:
1.0 kts.	120 min.
1.5 kts.	78 min.
2.0 kts.	58 min.
3.0 kts.	38 min.
4.0 kts.	29 min.
5.0 kts.	23 min.

THE RULE OF TWELFTHS
Figuring Changes in Tidal Height by the Hour

Since the average interval between high and low (or low and high) is just over six hours, we can safely divide the cycle into six segments of one hour each.

On average the tide drops according to the following fractions:

1st hour - 1/12; **2nd hour** - 2/12; **3rd hour** - 3/12; **4th hour** - 3/12; **5th hour** - 2/12; **last hour** - 1/12.

Both fractions and approximate percentages appear below.

Hour after HIGH	Drop Fraction	Drop %	% of HIGH
HIGH	------	------	100%
1 hour	1/12	8%	92%
2 hours	2/12	17%	75%
3 hours	3/12	25%	50%
4 hours	3/12	25%	25%
5 hours	2/12	17%	8%
6 hours	1/12	8%	LOW

Applying the formula to get heights by the hour at four random ports:

TIDE	FRACTION	SW. HBR. ME	STMFD. CT	NRFK. VA	CHLSTN. SC
MHW		10.2 ft.	7.2 ft.	2.8 ft.	5.2 ft.
1 hour	1/12	9.4	6.6	2.6	4.8
2 hours	2/12	7.7	5.4	2.1	3.9
3 hours	3/12	5.1	3.6	1.4	2.6
4 hours	3/12	2.6	1.8	0.7	1.3
5 hours	2/12	0.8	0.6	0.5	0.4
MLW	1/12	0.0	0.0	0.0	0.0

Using GPS to Create a Deviation Table

Once you have a GPS, it's easy to pay less attention to your compass. After all, GPS is very accurate, and compasses - unless they have been properly adjusted - have error (deviation). But have you tried to steer by a GPS? It's not easy.

You can make your compass more useful by using your GPS to create a deviation table. Choose a day when the wind is light and sea as calm as possible. Find a large open area with little or no current and a minimum of boat traffic. Bring aboard an assistant. In a notebook create two columns: in pencil, label the left column GPS and the right column COMPASS. Down the right column, number each successive line using intervals of either 10° or 20° up to 360°.

Choose a speed which provides responsive steering and which will make any current or leeway a negligible factor. Proceed on any of the numbered courses for at least 15 seconds, giving the GPS time to report a consistent direction. (This number will be "magnetic," with variation accounted for automatically by your GPS.) Once you have held a steady course long enough to get a repeated reading, record it in the left column. Proceed to the next heading.

Completing the circle results in a deviation table for your steering compass. Now, erase the penciled column headings and relabel the GPS column TO GO, and the COMPASS column STEER. *Example: TO GO 094°, STEER 090°.* If you used intervals of 20°, you can safely interpolate and fill in the missing numbers for every 10°, if you wish.

A deviation table admittedly falls far short of the ideal of a compensated compass; however, such a table will allow you to use your compass with a measure of confidence before an adjuster comes aboard. And that is much better than trying to steer by your GPS.

See pg. 203 for information on other ways to compensate your compass.

TWO WAYS TO ADJUST YOUR COMPASS

Nothing can equal the expert services of a professional Compass Adjuster, but if these services are not available, the following information can help you adjust a compass yourself.

Built-in Correctors -- Most modern compasses are regularly fitted with a magnetic corrector system attached to or inside the bottom of the compass, or inside the bottom of the cylinder, if binnacle-mounted. Such "B.I.C.'s" (Built-in-Correctors) are easy to use and are capable of removing virtually all the deviations of a well-located compass.

How B.I.C's Work -- B.I.C.'s generally consist of two horizontal shafts, one running athwart-ship and one running fore and aft. On each shaft are magnets. When these magnets are horizontal, they are equidistant from the directive magnets of the compass and have zero effect on the compass. When a shaft is rotated to any angle, the magnetic poles are no longer in neutral. You will get a large effect with only a few degrees of rotation, and as you approach the maximum correction you will have a decreasing effect from continued rotation. The usual B.I.C. can remove up to about 15° of deviation. The shaft which runs athwartship corrects on North and South headings and has zero effect on East and West. The fore and aft shaft corrects on East and West headings and has zero effect on North and South. Turning a shaft 90° from neutral creates maximum effect in one direction; another 90° turn brings it back to neutral; a third turn of 90° creates maximum effect in the opposite direction; a fourth turn of 90° brings it back to neutral. When rotating the shaft from neutral, you will get a large effect with only a few degrees of rotation, and as you approach the maximum correction you will have only a very small effect from continued rotation.

Getting Started -- Pick a quiet day and a swinging area with calm conditions. Have someone else at the helm and the engine just ticking over. In addition to the GPS method for creating a Deviation Table (see p. 202), there are two methods to choose from in correcting your steering compass.

Method #1 -- The first is to use **a handbearing compass** - not your helm compass - to find a landmark ashore that is within a few degrees of any cardinal point, say North. First, be sure to stand away from the engine or other objects which might cause deviation to your handheld compass. While your helper steers the boat toward that mark, slowly and with a non-magnetic screwdriver delicately rotate the athwartship shaft until the steering compass reads the same as the handbearing compass. Now do the same on an East or West heading, adjusting the fore-and-aft corrector until your steering compass agrees with the handbearing compass.

Method #2 -- Select **a charted object** which is plainly visible, and which lies generally East or West, North or South from you. Carefully plot its magnetic (not true) bearing from your boat. As you steer toward it, turn the appropriate B.I.C. until the helm compass agrees with the charted bearing. Again, if you used East or West for this first step, now find a charted object which is North or South from you and repeat the correction on the proper B.I.C.

Check for deviation all around -- You're almost done. Compare the steering compass heading to your handbearing compass heading at least every 45° on all the cardinal (N, E, S, W) and intercardinal (NE, SE, SW, NW) points, including the headings already checked. If the steering compass reads too little (lower number of degrees), the deviation is Easterly on that heading, and the number of degrees must always be subtracted when steering that magnetic course. If the steering compass reads too much (higher number of degrees), the deviation is Westerly on that heading, and the number of degrees must always be added when steering that magnetic course

Checking for Misalignment -- If you find, on the 8 headings you have checked, a preponderance of Easterly deviations over Westerly deviations or vice versa, add up the total of Easterly, subtract the total of Westerly and divide by 8. The result tells you the amount by which your compass is misaligned. Let's suppose it works out to 1° Easterly. This means your lubbers line is off to port 1°. If you now rotate the compass 1° clockwise, all your Easterly deviations will be reduced by 1°, your Westerly deviations increased by 1° and your 0 deviations will become 1° Westerly. You will now find that the total of your Easterly deviations is the very same as your Westerly total. You have eliminated misalignment error.

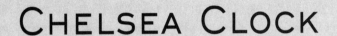

TIME AT SEA

Time is a critical factor in navigation. Navigators of old could always figure their distance north or south of the equator (latitude) by measuring the angular height of the sun at noon with a backstaff, octant, or sextant; however, measuring distance east or west (longitude) requires accurate time. The earliest timekeepers were sandglasses, inverted the moment the sand had all run through, but while they were adequate for approximate timing of intervals, such as the length of a watch, they were very poor for a duration of hours or days.

Around 1759 John Harrison finished his H4, a marine chronometer which successfully coped with the vessel's motion, temperature changes, and varying pressure of the mainspring. A triumph of engineering, it was the first sea-going timepiece which offered accurate, dependable timekeeping, and thus accurate determination of longitude at sea. This, with improvements in instruments measuring celestial altitudes, gave mariners new confidence in position-finding and cartographers much greater accuracy in determining coordinates for charts. Navigation became less risky, and ocean commerce expanded.

There have been refinements to Harrison's chronometers over the years, especially in miniaturization. Mechanical chronometers were used until the development of exceptionally accurate "atomic" time. For U.S. mariners, a transmitter in Fort Collins, CO beams a time signal for a couple of thousand miles, enabling navigators to synchronize their timepieces for highly accurate navigation. See p. 242. Go to http://nist.time.gov/timezone.cgi?Eastern/d/-5/java for accurate East Coast time.

GPS relies on extremely accurate time, depending on differences between the instant a signal is sent and the instant it is received. The most accurate GPS units can pinpoint locations within inches. We can only imagine Harrison's reaction if he were alive today.

What Your GPS Can and Cannot Do

GPS can:

Pinpoint the spot where a person falls overboard. Press the MOB (man overboard) button immediately, then use the "Go To" button to select the new waypoint as your destination. The person on board who is watching the person overboard should continue to do so, but the GPS can help by giving the bearing and the distance to the spot.

For sailors, find optimal sail trim. Sailors can determine ideal sail trim by watching how the SOG (speed over ground) changes after adjusting sheets, travelers, vangs, etc.

For powerboaters, find optimal trim tab setting. With the engine rpms set to a certain speed, watch the GPS readings of SOG as the trim tab angles are gradually changed.

Determine effect of the current. Assuming the boat's knotmeter is correct, then the difference between the GPS reading of SOG and the knotmeter is the positive or negative effect of the current on your boat. Only if you are heading directly into or away from the current will this difference be the current's actual speed.

Determine leeway. If your heading averages 150°, but your COG (course over ground) 155°, then your leeway is 5° to starboard, and your heading should change to 145°.

Find compass deviation. (1) If you can hold a steady compass course for a period of time, and (2) if your GPS is set to read Magnetic, then the difference between what your compass has been reading and the GPS reading is the approximate deviation of your magnetic compass.

Act as an anchor watch. Again, use the MOB button once the hook is down and the rode is out, and let the GPS set off an alarm if your boat wanders too far from the original location.

GPS cannot:

It will not warn you of navigation dangers. *Unless* you have entered waypoints of hazards such as shoals, rocks, etc., your GPS will blindly point you to the next waypoint, possibly over impassable terrain.

It will not warn you when you're off course. *Unless* you set an alarm for a certain cross-track error, you will have no warning other than a silent display that you are off course. The "highway" display on your screen may be most helpful for staying on a determined course.

It will not tell you it's not working properly. If it loses adequate satellite contact, suffers battery failure, or experiences some other malfunction, you won't know *unless* you check it and compare the information displayed to what you should expect it to be.

It will not recall your last position if it loses power. Plot its positions periodically, so when the power fails, you will know a recent position for dead-reckoning plotting.

ATLANTIC COAST GPS/DGPS STATIONS

Revised as of September, 2008

	kHz		Signal microvolts/meter (uV)	Range km	Lat.N.	Lon.W.
CANADA						
*Western Head, NS	312			400	43-59	64-40
*Hartlen Point, NS	298			500	44-36	63-27
*Fox Island, NS	307			325	45-20	61-05
*Pt. Escuminiac, NB	319			475	47-04	64-48
*Partridge Is., NB	295			475	45-14	66-03
UNITED STATES				km.		
*Penobscot, ME	290		100uV	435	44-27.10	68-46.33
*Brunswick, ME	316		75uV	322	43-53.4	69-56.8
*Acushnet, MA	306		100uV	370	41-44.57	70-53.19
*Moriches Pt, NY.	293		75uV	241	40-48.3	72-45.68
*Sandy Hook, NJ	286		100uV	185	40-28.29	74-00.71
*Reedy Point, DE	309		100uV	113	39-33.68	75-34.19
*Hagerstown, MD	307		75uV	250	39-33.19	77-42.79
*Annapolis, MD	301		100uV	290	39-00.37	76-36.33
*Driver, VA	289		75uV	241	36-57.5	76-33.4
*New Bern, NC	294		75uV	259	35-10.50	77-02.92
*Kensington, SC	292		75uV	200	33-28.86	79-20.58
*Savannah, GA	319		75uV	298	32-08.40	81-42.00
*Cape Canaveral, FL	289		75uV	371	28-27.6	80-32.6
*Miami, FL	322		75uV	139	25-43.97	80-09.61
*Key West, FL	286		75uV	204	24-34.94	81-39.18
BERMUDA				mi.		
St. David's Head	323	BSD		150	32-22.0	64-38.

* These stations are land-based receivers and transmitters for Differential GPS (DGPS). The carrier of these Radiobeacons is modulated with a GPS correction (differential) signal, which may be used to greatly improve the accuracy of GPS. Mariners should see no degradation in the usability of the radiobeacon signal for direction finding although a warbling of the identification signal may be noticed. Correction broadcasts are changing from Type 1 format to Type 9-3. Your equipment may need upgrading. For status of GPS broadcasts: call NAVCEN (703) 313-5900.

Please see our May 2009 *Free Supplement* for further updates. To order *see pg. 4.*
*For further information visit U.S.C.G. Navigation Centers website at **http://www.navcen.uscg.gov** or Candian CG at **http://www.ccg-gcc.gc.ca/eng/CCG/DGPS_Beacon_Information***

RACONS

RACONS are Radar Beacons operating in the marine radar frequency bands, 2900-3100 MHz (s-band) and 9300-9500 MHz (x-band). When triggered by a vessel's radar signal they provide a bearing by sending a coded reply (e.g. "T": –). This signal received takes the form of a single line or narrow sector extending radially towards the circumference of the radarscope from a point slightly beyond the spot formed by the echo from the lighthouse, buoy, etc. at the Racon site. Thus *distance off* may be measured to the point at which the Racon coded flash begins. (The figure obtained will be a few hundred feet greater than the actual distance of the ship from the Racon due to the slight response delay in the Racon apparatus.) Hours of transmission are continuous and coverage is all around the horizon unless otherwise stated. Their ranges depend on the effective range of the ship's radar and on the power and elevation of the Racon apparatus. Under conditions of abnormal radio activity, reliance should only be put on a Racon flash that is consistent and when the ship is believed to be within the area of the Racon's quoted range. Mariners are advised to turn off the interference controls of their radar when wishing to receive a Racon signal or else the signal may not come through to the ship.

See p. 209 for list of Racons.

ATLANTIC COAST RACONS

RACON SITE	SIGNAL	RANGE	LAT. N	LONG. W
Cranberry Islands, NS	– • • (B)	under 10 mi.	45-19-29.6	60-55-38.2
Bear Cove Lt. & Bell By "H6"	– • (N)	" "	44-32-36.3	63-31-19.6
Chebucto Head	– – • • (Z)	" "	44-30-26.6	63-31-21.8
Cape Sable	– • – (C)	" "	43-23-24	65-37-16.9
Cape Fourchu	– • • (B)	" "	43-47-38.8	66-09-19.3
Lurcher Shoal Biforcation Ltd. By. "NM"	– • – (K)	" "	43-49-00.3	66-29-58
St. John Harbour. Lt. & Wh. By. "J", NB	– • (N)	" "	45-12-55.3	66-02-36.9
Gannet Rock	– – • (G)	" "	44-30-37.1	66-46-52.9
Portland Ltd. Wh. By. "P", ME	– – (M)	under 16 mi.	43-31-36	70-05-28
Frenchman Bay Ltd. By. "FB"	– • • • (B)	" "	44-19-24	68-07-24
Boston Ltd. Wh. By. "B", MA	– • • • (B)	" "	42-22-42	70-46-58
Boston Entr. Ltd. Wh. By. "NC"	– • (N)	" "	42-22-30	70-54-18
Cleveland Ledge E. Lt.	– • – (C)	" "	41-37-51	70-41-39
Nantucket Shoals Ltd. Wh. By. "N"	– • (N)	" "	40-30-09	69-14-48
Nantucket Traffic Lane Ltd. By. "NA"	– • (N)	" "	40-25-42	73-11-28
Execution Rocks, NY	– • • – (X)	" "	40-52-41	73-44-16
Buzzards Bay Entr. Lt., Horn	– • • • (B)	" "	41-23-48	71-02-01
Narrag. Bay Entr. Ltd. Wh. By. "NB"	– • • • (B)	" "	41-23-00	71-23-24
Narrag.-Buzz. Bay Appr. Ltd. Wh. By. "A", RI	– • (N)	" "	41-06-00	71-23-24
Newport - Pell Bridge Fog Signal	– • (N)	" "	41-30-17	71-20-57
Twenty-Eight Ft. Sh. Ltd. Wh. By. "TE"	– (T)	" "	41-09-18	72-30-24
Ambrose Lt., NY	– • (N)	" "	40-27-00	73-48-00
Kill van Kull Ltd. Wh. Jct. By. "KV"	– • – (K)	" "	40-39-01	74-03-52
Montauk Pt. Ltd. Wh. By. "MP"	– – (M)	" "	41-01-48	71-45-40
Valiant Rock Ltd. Wh. By. #11	– • • • (B)	" "	41-13-46	72-04-00
Southwest Ledge Ltd. Wh. By. #2	– • • • (B)	" "	41-06-23	71-40-14
Hudson Canyon Traffic Lane Ltd. Wh. By. "HA"	– • – • (C)	" "	40-07-36	73-21-24
Scotland Ltd. Wh. By. "S", NJ	– – (M)	" "	40-26-33	73-55-01
Sandy Hook Ch. Rge. Front Lt	– • – • (C)	" "	40-29-15	73-59-35
Barnegat Ltd. By. "B"	– • • • (B)	" "	39-45-48	73-46-04
Del. Bay Appr. Ltd. Wh. By. "CH"	– • – (K)	" "	38-46-14	75-01-20
Del. Ltd. By. "D", DE	– • – (K)	" "	38-27-18	74-41-47
Del. River, Pea Patch Is.	– • • – (X)	" "	39-36-42	75-34-54
Five Fathom Bank Ltd. By. "F"	– – (M)	" "	38-46-49	74-34-32
Brown Shoal Lt.	– • • • (B)	" "	38-55-20	75-06-03
Miah Maull Shoal Lt.	– – (M)	" "	39-07-36	75-12-31
Ship John Shoal Lt.	– – – (O)	" "	39-18-19	75-22-36
Chesapeake Light Tower, VA	– • (N)	" "	36-54-17	75-42-46
Chesapeake Bay Ent. Ltd. Wh. By. "CH"	– • – • (C)	" "	36-56-08	75-57-27
Ches. Ch. Ltd. Bell By. #78	– – • – (Q)	" "	38-33-19	76-25-39
Ches. Ch. Ltd. Bell By. #68	– • • – (X)	" "	37-59-53	76-11-49
Ches. Ch. Ltd. Bell By. #42	– – – (O)	" "	37-25-37	76-05-07
Beaufort Inlet Ch. Ltd. Wh. By. "BM"	– – (M)	" "	34-34-49	76-41-33
Cape Fear River Ent. Ltd. Wh. By. "CF", SC	– • – • (C)	" "	33-46-17	78-03-02
Charleston Entr. Ltd. By. "C"	– • – (K)	" "	32-37-05	79-35-30
Tybee Ltd. By. "T", GA	– – • (G)	" "	31-57-52	80-43-09
St. Simons Ltd. By. "STS"	– • • • (B)	" "	31-02-49	81-14-25
St. Johns Ltd. By. "STJ", FL	– – (M)	" "	30-23-35	81-19-08
Pt. Everglades Appr. Ltd. By. "PE"	– (T)	" "	26-05-30	80-04-48
Miami Ltd. By. "M"	– – (M)	" "	25-46-06	80-05-00
Fowey Rocks Lt.	– – – (O)	" "	25-35-24	80-05-48
Carysfort Rf. Lt.	– • – • (C)	" "	25-13-18	80-12-42
Alligator Rf. Lt.	– – • (G)	" "	24-51-06	80-37-06
Sombrero Key Lt.	– – (M)	" "	24-37-36	81-06-36
American Shoal Lt.	– • – – (Y)	" "	24-31-30	81-31-12
Sand Key Lt.	– • (N)	" "	24-27-14	81-52-39
Dry Tortugas Lt.	– • – (K)	" "	24-38-00	82-55-12
North East Breaker Beacon - Bermuda	– • (N)	" "	32-28.7	64-41.0
Chub Heads Beacon - Bermuda	– • – • (C)	" "	32-17.2	64-58.9

For more on RACONS see p. 207

For more on RACONS see p. 207

DIAL-A-BUOY SERVICE
SEA-STATE & WEATHER CONDITIONS BY TELEPHONE

If you are planning a coastwise voyage, you can rely on a number of sources for weather. A possible source is **Dial-A-Buoy**, offering reports of conditions at numerous coastal and offshore locations along the Atlantic Coast, as well as the coasts of the Gulf of Mexico, the Pacific, and the Great Lakes. In all there over 100 buoy and 60 Coastal-Marine Automated Network (C-MAN) stations. The system is operated by the National Data Buoy Center (NDBC), with headquarters at the Stennis Space Center in Mississippi. The NDBC is part of the National Weather Service (NWS).

The reports from offshore buoys include wind speed, gusts, and direction, wave heights and periods, water temperature, and barometric pressure as recorded within the last hour or so. Reports from land stations cover wind speed and direction, temperature and pressure; some land stations also add water temperature, visibility, and dew point.

The value of this information is apparent. Say someone in your boating party is susceptible to seasickness, and the Dial-A-Buoy report says wave heights are six feet with a period (interval) of eight seconds. Maybe that person would rather stay ashore and experience the gentler motion of a rocking chair. (Wave heights of six feet with a period of twenty seconds, on the other hand, might be tolerable.) Surfers, too, can benefit greatly from wave height reports. Likewise, since actual conditions frequently differ dramatically from forecasts, someone sailing offshore might be interested to know that a Data Buoy ahead is reporting squalls, giving time to shorten sail. And bathers and fishermen might gain from hearing the water temperature reports.

On the next page, we give the station or buoy identifier, location name, and lat/long in degrees and hundredths, as provided by the NWS. To find the station or buoy locations and identifiers using the Internet, you can see maps with station identifiers at **www.ndbc.noaa.gov/**. To find locations by telephone, you can enter a latitude and longitude to receive the locations and identifiers of the closest stations.

To access Dial-A-Buoy using any touch-tone or cell phone, here are the steps:

1. Call 888-701-8992.
2. If you know the identifier of the station or buoy, press 1. Press 2 to get station locations by entering the approximate lat/long of the area you want.
3. Enter the five-digit (or character) station identifier. To enter a Character press the key containing the character.
4. Press 1 to confirm that your entry was correct.
5. If, after hearing the latest report, you wish to hear a forecast for that same location, press 2 then 1. You can jump to the forecast before the end of the station report by pressing 2 then 1 during the reading of the station conditions.
6. If you want to hear the report for another station, press 2 then 2.
7. You do not have to wait for the prompts. For example, you can press "1440271" as soon as you begin to hear the welcome message to hear the report from station 44027.

NOTE: In some cases a buoy may become temporarily unavailable. You should try again later to see if it has come back online. Please be aware that stations that may be adrift and not at the stated location are not reported via the telephone feature. This information is only available on the website by typing in the station identifier at:
http://www.ndbc.noaa.gov/dial.shtml

DIAL-A-BUOY and C-MAN STATION LOCATIONS

Station ID	Location Name	Latitude	Longitude
44027	JONESPORT, ME	44.27N	67.31W
MDRM1	MT DESERT ROCK, ME	43.97N	68.13W
MISM1	MATINICUS ROCK, ME	43.78N	68.86W
44007	PORTLAND, ME	43.53N	70.14W
44005	GULF OF MAINE	43.19N	69.14W
IOSN3	ISLE OF SHOALS, NH	42.97N	70.62W
44013	BOSTON, MA	42.35N	70.69W
BUZM3	BUZZARDS BAY, MA	41.40N	71.03W
44018	SE CAPE COD, MA	41.26N	69.29W
44011	GEORGES BANK, MA	41.11N	66.58W
44017	MONTAUK POINT, NY	40.69N	72.05W
44008	NANTUCKET, MA	40.50N	69.43W
ALSN6	AMBROSE LIGHT, NY	40.45N	73.80W
44025	LONG ISLAND, NY	40.25N	73.17W
TPLM2	THOMAS POINT, MD	38.90N	76.44W
44004	HOTEL, NJ	38.48N	70.43W
44009	DELAWARE BAY, NJ	38.46N	74.70W
CHLV2	CHESAPEAKE LIGHT, VA	36.91N	75.71W
44014	VIRGINIA BEACH, VA	36.61N	74.84W
DUCN7	DUCK PIER, NC	36.18N	75.75W
41025	DIAMOND SHOALS	35.01N	75.40W
41001	E. HATTERAS, NC	34.70N	72.73W
CLKN7	CAPE LOOKOUT, NC	34.62N	76.52W
FPSN7	FRYING PAN SHOAL, NC	33.49N	77.59W
41004	EDISTO, SC	32.50N	79.09W
41002	S. HATTERAS, SC	32.33N	75.44W
41008	GRAYS REEF, GA	31.40N	80.87W
41012	ST. AUGUSTINE, FL	30.04N	80.55W
SAUF1	ST AUGUSTINE, FL	29.86N	81.26W
CDRF1	CEDAR KEY, FL	29.14N	83.03W
41010	CANAVERAL EAST, FL	28.91N	78.47W
41009	CANAVERAL, FL	28.51N	80.17W
LKWF1	LAKE WORTH, FL	26.61N	80.03W
FWYF1	FOWEY ROCKS, FL	25.59N	80.10W
MLRF1	MOLASSES REEF, FL	25.01N	80.38W
LONF1	LONG KEY, FL	24.84N	80.86W
PLSF1	PULASKI SHOAL LIGHT, FL	24.69N	82.77W
SMKF1	SOMBRERO KEY, FL	24.63N	81.11W
SANF1	SAND KEY, FL	24.46N	81.88W

NEW: Most stations have added the ability to access information via RSS feed using your Internet browser. For information regarding how to use this feature please go to: http://www.ndbc.noaa.gov/rss_access.shtml.

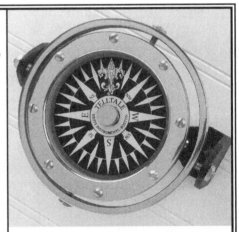

IALA BUOYAGE SYSTEM

Lateral Aids marking the sides of channels seen when entering from Seaward

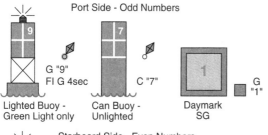

Port Side - Odd Numbers

G "9"
Fl G 4sec
Lighted Buoy -
Green Light only

C "7"
Can Buoy -
Unlighted

Daymark
SG

G
"1"

Port- hand aids are Green, some with Flashing Green Lights.
Daymarks:
1st letter "S" = Square
2nd letter "G" = color Green

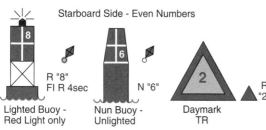

Starboard Side - Even Numbers

R "8"
Fl R 4sec
Lighted Buoy -
Red Light only

N "6"
Nun Buoy -
Unlighted

Daymark
TR

R
"2"

Starboard-hand aids remain Red, some with Flashing Red Lights.
Daymarks:
1st letter "T" = Triangle
2nd letter "R"= color Red

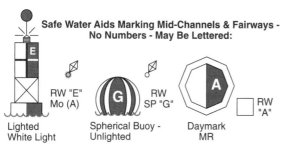

Safe Water Aids Marking Mid-Channels & Fairways - No Numbers - May Be Lettered:

RW "E"
Mo (A)
Lighted
White Light

RW
SP "G"
Spherical Buoy -
Unlighted

Daymark
MR

RW
"A"

Red and White replaces Black and White. Buoys are spherical; or have a Red spherical topmark. Flashing White Light only: Mo (A).
Daymarks:
1st letter "M" = Octagon
2nd letter "R" = color Red

Preferred Channel Aids - Mark Bifurcations - No Numbers -

Preferred Ch. to Starboard (Aid to Port):

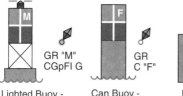

GR "M"
CGpFl G
Lighted Buoy -
Green Light only

GR
C "F"
Can Buoy -
Unlighted

Daymark
JG

GR
"A"

Green replaces Black. Flashing Light (Red or Green) is Composite Gp. Fl. (2 + 1).
Daymarks:
1st letter "J" = Square or Triangle
2nd letter "R" or "G" is color of top band

Preferred CH. to Port (Aid to Starboard):

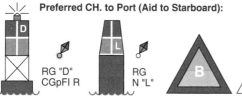

RG "D"
CGpFl R
Lighted Buoy -
Red Light only

RG
N "L"
Nun Buoy -
Unlighted

Daymark
JR

RG
"B"

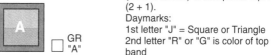

Note: ISOLATED DANGER BUOYS, Black and Red with two Black spherical topmarks - no numbers, may be lettered (if lighted, white light only, Fl (2) 5s). Stay Clear. **SPECIAL AIDS BUOYS** will be all **YELLOW** (if lighted, with yellow light only, Fixed Flashing): Anchorage Areas, Fish Net Areas, Spoil Grounds, Military Exercise Zones, Dredging Buoys (where conventional markers would be confusing), Ocean Data Systems, some Traffic Separations Zone Mid-Channel Buoys.

YACHT FLAGS AND HOW FLOWN

Ensign: 8 AM to sundown only.
   At stern staff of all vessels at anchor, or under way under power or sail.
   At the leech of the aftermost sail, approximately 2/3 of the leech above the clew. When the aftermost sail is gaff-rigged, the Ensign is flown immediately below the peak of the gaff.

Club Burgee:
   At bow staff of launches and power yachts with one mast.
   At main peak of yawls, ketches, catboats, sloops, and cutters.
   At fore peak of schooners and power yachts with two masts.

Private Signal: 8 AM to sundown only, when owner is in command.
   At masthead of catboats, sloops, cutters, and power yachts with one mast.
   At mizzen peak of yawls and ketches.
   At main peak of schooners and power yachts with two masts.

Flag Officers' Flags: Commodore (blue), Vice Commodore (red), and Rear Commodore (white) flags are flown in place of the private signal on all rigs except single-masted sailboats, when it is flown on the masthead replacing the club burgee.

Union Jack: 8 AM to sundown, only at anchor on Sundays, holidays, or occasions for "dressing ship," flown from the bowstaff. Sailing vessels without a bowstaff may lash the Jack to the forestay a few feet above the stem head.

U.S. Power Squadron Ensign: 8 AM to sundown.
   May be flown in place of Ensign, or flown additionally at the starboard yardarm or main spreader, but only when a Squadron member is in command.

INTERNATIONAL SIGNAL FLAGS AND MORSE CODE

CODE FLAG	AND ANSWERING PENNANT

Alpha A · —	**Bravo** B — · · ·	**Charlie** C — · — ·	**Delta** D — · ·	**Echo** E ·	**Foxtrot** F · · — ·
Golf G — — ·	**Hotel** H · · · ·	**India** I · ·	**Juliet** J · — — —	**Kilo** K — · —	
Lima L · — · ·	**Mike** M — —	**November** N — ·	**Oscar** O — — —	**Papa** P · — — ·	
Quebec Q — — · —	**Romeo** R · — ·	**Sierra** S · · ·	**Tango** T —	**Uniform** U · · —	
Victor V · · · —	**Whiskey** W · — —	**XRay** X — · · —	**Yankee** Y — · — —	**Zulu** Z — — · ·	

NUMERAL PENNANTS

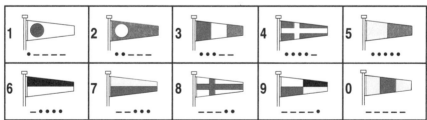

1 · — — — —	2 · · — — —	3 · · · — —	4 · · · · —	5 · · · · ·
6 — · · · ·	7 — — · · ·	8 — — — · ·	9 — — — — ·	0 — — — — —

REPEATERS

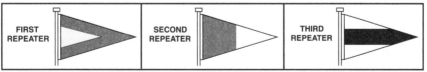

FIRST REPEATER	SECOND REPEATER	THIRD REPEATER

THE INTERNATIONAL CODE OF SIGNALS

The Code comprises 40 flags: 1 Code Flag; 26 letters; 10 numerals; 3 repeaters. With this Code it is possible to converse freely at sea with ships of different countries.

Single Flag Signals

A :: I have a diver down; keep well clear at slow speed.

B :: I am taking in, or discharging, or carrying dangerous goods.

C :: Yes

D :: Keep clear of me; I am maneuvering with difficulty.

E :: I am altering my course to starboard.

F :: I am disabled; communicate with me.

G :: I require a pilot. (When made by fishing vessels when operating in close proximity on the fishing grounds it means; "I am hauling nets.")

H :: I have a pilot on board.

I :: I am altering my course to port.

J :: I am on fire and have dangerous cargo on board; keep well clear of me.

K :: I wish to communicate with you.

L :: You should stop your vessel instantly.

M:: My vessel is stopped and making no way through water.

N :: No

O :: Man overboard.

P :: *In harbor*; All persons should report on board as the vessel is about to proceed to sea.

  At sea; It may be used by fishing vessels to mean "My nets have come fast upon an obstruction."

Q :: My vessel is healthy and I request free pratique.

R :: *nothing currently assigned*

S :: My engines are going astern.

T :: Keep clear of me; I am engaged in pair trawling.

U :: You are running into danger.

V :: I require assistance.

W:: I require medical assistance.

X :: Stop carrying out your intentions and watch for my signals.

Y :: I am dragging my anchor.

Z :: I require a tug. (When made by fishing vessels operating in close proximity on the fishing grounds it means : "I am shooting nets.")

Two Flag Signals – Urgent and Important

NC :: International Distress

AC :: I am abandoning my vessel.

AN :: I need a doctor.

BR :: I require a helicopter.

CD :: I require immediate assistance.

DV :: I am drifting.

EF :: SOS/MAYDAY has been cancelled.

FA :: Will you give me my position?

GW:: Man overboard. Please take action to pick him up.

JL :: You are running the risk of going aground.

LO :: I am not in my correct position. (Used by a light vessel)

PD :: Your navigation lights are not visible.

PP :: Keep well clear of me.

RU :: Keep clear of me, I am maneuvering with difficulty.

QD :: I am going ahead.

QQ :: I require health clearance.

QT :: I am going astern.

QU :: Anchoring is prohibited.

QX :: I request permission to anchor.

SO :: You should stop your vessel instantly.

UM :: The harbor is closed to traffic.

UP :: Permission to enter harbor is urgently requested. I have an emergency.

YU :: I am going to communicate with your station by means of the International Code of Signals.

ZL :: Your signal has been received but not understood.

FLAG ETIQUETTE
or
BE CAREFUL WHERE YOU'RE WAVING THAT THING
by Jan Adkins

THERE ARE ONLY A FEW CRITICAL RULES OF FLAG ETIQUETTE. The rest are less certain, a ragbag of history, tradition, pretension, and practicality.

Flags have always made practical sense. A messy, noisy battlefield was a difficult place to communicate. Flags were bright and easily recognized signals, marking the positions of fighting troops who were usually illiterate. Tanks and armored troop carriers still fly unit guidons.

Flags were even more important at sea as a medium of communication. Crisp, positive use of flag signals was a life-and-death matter, or at least a basic necessity of effective fleet action. Fussiness about flag etiquette afloat is probably due to the importance of naval signals. But dread naught, the basic rules are not as important as Chapman's and yacht club vexillologists (those who study flags) would have you believe. They boil down to priority, placement, and a few time hacks.

As part of any navy, even a dollhouse navy like a yacht club, the first priority is your country, the national ensign, Old Glory. The next in priority is your fleet or the club flag. Next is your own private flag.

An ensign is the naval flag of a country. For some countries the naval flag differs from the national flag, but for the United States it is the Stars and Stripes. You may not have voted for the current president, and you may not agree with the destination of the last cruise missile, but our national ensign symbolizes your right to disagree and remain part of the Union. As such it is given the highest priority and the place of honor.

The next priority is the yacht club burgee. A burgee – a triangular pennant or swallow-tail flag – is not to be confused with burgoo, which in the Royal Navy was an oatmeal porridge and in the southern United States is a thick spicy stew. Both burgoos are difficult to fly.

Your private signal is a further step down in importance, even if you have delusions of grandeur. This is usually a swallowtail but can also be a triangular pennant.

Each flag occupies a position of relative importance, depending on your vessel. The most important flag position of a sailboat is on the aftermost sail. Gaff-rigged vessels fly the national ensign from a block on the gaff peak. Boats with triangular (marconi-rigged) sails can carry the ensign on the leech (trailing edge) of the aftermost sail, whether it is the mainsail or the mizzen. A motor vessel with a gaff extending from a mast flies the national ensign from the gaff peak. A motor vessel or sailboat without a gaff, or a sailboat with its sails furled at anchor or while motoring can fly the ensign from a flagstaff mounted at the stern.

The next most important position on a boat is the truck (tippy-top) of the forward mast. Flags are usually flown above the truck using a "pig stick" (though I have never seen a pig climb to this height). If you belong to a yacht club, the U.S. Power Squadron or U.S. Coast Guard Auxiliary, this is the place for the burgee.

Some power boats have a short flagstaff at the bow called a jack-staff. Underway this is a position of secondary honor and your club burgee flies there. At anchor, on Sundays or holidays, 8 a.m. to sunset, you are permitted to fly the Union Jack (a small blue field with fifty stars) from this otherwise useless, shin-banging piece of nautical tomfoolery.

Next in importance, for the private signal, is the peak of the after mast, whether it is a schooner's main or a yawl's or ketch's mizzen. If, poor thing, you have only a single mast, the next position of honor is a signal halyard block on the starboard spreader.

Continued p. 218

Continued from p. 217

Simple priority then: national ensign at the first place of honor, burgee at the next place, private signal at the next. On a yawl, ketch, or schooner, the ensign flies from the leech of the mizzen, two-thirds up the length, or from a stern flagstaff; the club burgee from the forward masthead; the private signal from the after masthead. On a sloop, the ensign flies from the main leech or a stern flagstaff, the burgee from the main truck, and the private signal from the starboard spreader.

For Old Glory, you have a simple window of opportunity: eight in the morning until sundown. With the yacht club's eight o'clock morning gun you can hear dozens of heads knocking against deck beams above their bunks, curses, then the sudden fluttering of national ensigns at stern flagstaffs. A charming tradition. At sundown or the sunset gun, national ensigns are tucked away, whether underway or at anchor. Burgees, private signals, and most other flags may be flown day and night.

All flags, but especially Old Glory, should be raised briskly and lowered with ceremonious hesitancy.

There are other, more arcane bits of flag folderol to observe. It is considered ill-bred to fly more than one yacht club flag at a time, even if you belong to several. Take your pick and go with it.

Some exhaustive "yachtsmen's guides" offer flag signals for "owner ashore" (a solid blue flag), "guest aboard, owner absent" (solid blue with white diagonal stripe), and "owner's meal flag, go away" (white flag). In the days of crystal fingerbowls and visiting cards, these flags were flown from the starboard spreader. The aristocratic affectation, "crew's meal pennant" (red triangular pennant) was flown at the lowest priority position, the forward port spreader. This is the kind of tradition that occasioned the Bolshevik Revolution.

Another imperative signal, the diver's flag (red with a white diagonal stripe), is flown from any prominent position and indicates "divers below the surface, stay clear of the area."

Letter and number signal flags are more often seen on cocktail napkins that on signal halyards. Nevertheless, they can be useful. An internationally recognized system of one, two, and three-flag messages is part of seagoing language. The most poetic may be the signal flag P, called the Blue Peter, which announces "I am about to sail." Practically, this calls crew members from shore. The O flag signals "man overboard," and vessels sighting this should render assistance. Likewise W, which calls for medical assistance. If a vessel flashes out the U signal and there are no other boats nearby, you are being informed that you are "standing into danger." L is even worse: "stop instantly." K might be flown from a shoreside therapist's office: "desire to communicate." J, "I am on fire, keep clear," seems redundant (smoke, flame, etc.), but T is simply bellicose: "keep clear of me."

Two- and three-flag signals are less emphatic, as LR, "bar is not dangerous." But is the food good? YK, "I am unable to answer your signal." But you just did. ZP, "my last signal was incorrect, I will repeat it correctly." Why trust you now?

There is even etiquette in flying the Jolly Roger. This flag probably got its name from the French *joli rouge*, or "happy red" flag. Original piratical banners were red, signaling the pirates' freedom and promising that quarter (mercy) would be shown to captives. Reluctant prizes were then shown the black flag: "all bets are off." But sailors are a traditional lot, and they probably flew even the dread skull and crossbones from the proper gaff peak. Etiquette is etiquette.

Condensed by the author from an article in Chesapeake Bay Magazine

U.S. STORM SIGNALS

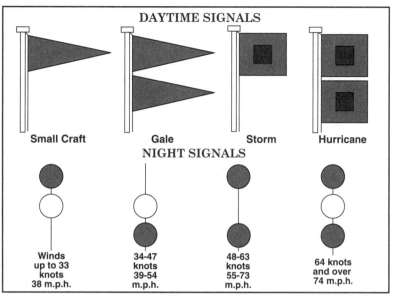

DAYTIME SIGNALS

Small Craft   Gale   Storm   Hurricane

NIGHT SIGNALS

| Winds up to 33 knots 38 m.p.h. | 34-47 knots 39-54 m.p.h. | 48-63 knots 55-73 m.p.h. | 64 knots and over 74 m.p.h. |

The above signals are displayed regularly on Light Vessels, at Coast Guard shore stations, and at many principal lighthouses. Each Coast and Geodetic Survey Chart lists those locations which appear within the area covered by that chart.

When the Rain before the Wind,   Rain before 7
    Topsail Sheets and Halliards Mind,
But when the Wind before the Rain,   Clear before 11
    Then you may set sail again.

Distance of Visibility

Given the curvature of the earth, can you see a 200' high headland from 20 miles away? (Answer below.) How far you can see depends on visibility, which we will assume here to be ideal, and the heights above water of your eye and the object.

To find the theoretical maximum distance of visibility, use the Table below. First, using your height of eye above water (say, 8'), the Table shows that at that height, your horizon is 3.2 n.m. away. Then, from our Lights, Fog Signals and Offshore Buoys (pp. 167-197), your chart, or the Light List, find the height of the object (say, 200'). The Table shows that object can be seen 16.2 n.m. from sea level. Add the two distances: 3.2 + 16.2 = 19.4 n.m. *Answer: not quite!*   (Heights below in feet, distance in nautical miles)

Ht.	Dist.	Ht.	Dist.	Ht.	Dist.	Ht.	Dist.	Ht.	Dist.
4	2.3	30	6.3	80	10.3	340	21.1	860	33.6
6	2.8	32	6.5	90	10.9	380	22.3	900	34.4
8	3.2	34	6.7	100	11.5	420	23.5	1000	36.2
10	3.6	36	6.9	120	12.6	460	24.6	1400	42.9
12	4.0	38	7.1	140	13.6	500	25.7	1800	48.6
14	4.3	40	7.3	160	14.5	540	26.7	2200	53.8
16	4.6	42	7.4	180	15.4	580	27.6	2600	58.5
18	4.9	44	7.6	200	16.2	620	28.6	3000	62.8
20	5.1	46	7.8	220	17.0	660	29.4	3400	66.9
22	5.4	48	8.0	240	17.8	700	30.4	3800	70.7
24	5.6	50	8.1	260	18.5	740	31.1	4200	74.3
26	5.9	60	8.9	280	19.2	780	32.0	4600	77.7
28	6.1	70	9.6	300	19.9	820	32.8	5000	81.0

MEASURED MILE SPEED TABLES

The blank table below should be used for making a permanent record of the speed obtained at any given number of R.P.M. on this yacht.

RPM	SPEED

Secs.	1 Min.	2 Min.	3 Min.	4 Min.	5 Min.	6 Min.	7 Min.	8 Min.	9 Min.	10 Min.	11 Min.
0	60.00	30.00	20.00	15.00	12.00	10.00	8.57	7.50	6.66	6.00	5.45
2	58.06	29.50	19.78	14.87	11.92	9.94	8.53	7.46	6.64	5.98	5.43
4	56.25	29.03	19.56	14.75	11.84	9.89	8.49	7.43	6.61	5.96	5.42
6	54.54	28.57	19.35	14.63	11.76	9.83	8.45	7.40	6.59	5.94	5.40
8	52.94	28.12	19.14	14.51	11.68	9.78	8.41	7.37	6.56	5.92	5.38
10	51.42	27.69	18.94	14.40	11.61	9.73	8.37	7.34	6.54	5.90	5.37
12	50.00	27.27	18.75	14.28	11.53	9.67	8.33	7.31	6.52	5.88	5.35
14	48.64	26.86	18.55	14.17	11.46	9.62	8.29	7.28	6.49	5.86	5.34
16	47.36	26.47	18.36	14.06	11.39	9.57	8.26	7.25	6.47	5.84	5.32
18	46.15	26.08	18.18	13.95	11.32	9.52	8.22	7.22	6.45	5.82	5.30
20	45.00	25.71	18.00	13.84	11.25	9.47	8.18	7.20	6.42	5.80	5.29
22	43.90	25.35	17.82	13.74	11.18	9.42	8.14	7.17	6.40	5.78	5.27
24	42.85	25.00	17.64	13.63	11.11	9.37	8.11	7.14	6.38	5.76	5.26
26	41.86	24.65	17.47	13.53	11.04	9.32	8.07	7.11	6.36	5.75	5.24
28	40.90	24.32	17.30	13.43	10.97	9.27	8.03	7.09	6.33	5.73	5.23
30	40.00	24.00	17.14	13.33	10.90	9.23	8.00	7.05	6.31	5.71	5.21
32	39.13	23.68	16.98	13.23	10.84	9.18	7.96	7.03	6.29	5.69	5.20
34	38.29	23.37	16.82	13.13	10.77	9.13	7.93	7.00	6.27	5.67	5.18
36	37.50	23.07	16.66	13.04	10.71	9.09	7.89	6.97	6.25	5.66	5.17
38	36.73	22.78	16.51	12.95	10.65	9.04	7.86	6.95	6.22	5.64	5.15
40	36.00	22.50	16.36	12.85	10.58	9.00	7.83	6.92	6.20	5.62	5.14
42	35.29	22.22	16.21	12.76	10.52	8.95	7.79	6.89	6.18	5.60	5.12
44	34.61	21.95	16.07	12.67	10.46	8.91	7.76	6.87	6.16	5.59	5.11
46	33.96	21.68	15.92	12.58	10.40	8.86	7.72	6.84	6.14	5.57	5.09
48	33.33	21.42	15.78	12.50	10.34	8.82	7.69	6.81	6.12	5.55	5.08
50	32.72	21.17	15.65	12.41	10.28	8.78	7.66	6.79	6.10	5.53	5.07
52	32.14	20.93	15.51	12.32	10.22	8.73	7.63	6.76	6.08	5.51	5.05
54	31.57	20.69	15.38	12.24	10.16	8.69	7.59	6.74	6.06	5.50	5.04
56	31.03	20.45	15.25	12.16	10.11	8.65	7.56	6.71	6.04	5.48	5.02
58	30.50	20.22	15.12	12.08	10.05	8.61	7.53	6.69	6.02	5.47	5.01
	30.00	20.00	15.00	12.00	10.00	8.57	7.50	6.66	6.00	5.45	5.00
	3 Min.	4 Min.	5 Min.	6 Min.	7 Min.	8 Min.	9 Min.	10 Min.	11 Min.	12 Min.	

Table applies to either nautical or land miles; answer in knots in former case, miles-per-hour in latter. Example: if time elapsed to run measured mile is 4 min. and 2 sec., the 4 min. column opp. 2 sec. gives 14.87, meaning knots or miles-per-hour, depending on type of mile run.

1 nautical mile 6080 ft.
1 land mile 5280 ft.

If a current is involved, run the distance up and downstream, noting the time for each run. Your true boat speed will be found as follows:

$$\frac{60 \times (\text{Distance}) \times (\text{time up} + \text{time downstream})}{2 \times (\text{time up} \times \text{time downstream})}$$

Example: $\dfrac{60 \times 1.0 \ (1 \ \text{mi.}) \times (10 + 8)}{2 \times (10 \times 8)} = \dfrac{1080}{160} = 6.75 = \text{speed}$

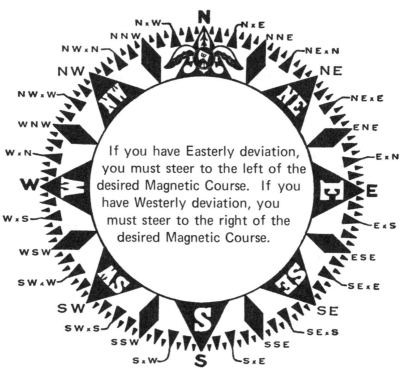

If you have Easterly deviation, you must steer to the left of the desired Magnetic Course. If you have Westerly deviation, you must steer to the right of the desired Magnetic Course.

Table for Turning Compass Points into Degrees, and the Contrary

MERCHANT MARINE PRACTICE

NORTH	**0**	**EAST**	**90**	**SOUTH**	**180**	**WEST**	**270**
N. 1/4E.	2 3/4	E. 1/4S.	92 3/4	S. 1/4W.	182 3/4	W. 1/4N.	272 3/4
N. 1/2E.	5 3/4	E. 1/2S.	95 3/4	S. 1/2W.	185 3/4	W. 1/2N.	275 3/4
N. 3/4E.	8 1/2	E. 3/4S.	98 1/2	S. 3/4W.	188 1/2	W. 3/4N.	278 1/2
N. by E.	11 1/4	E. by S.	101 1/4	S. by W.	191 1/4	W. by N.	281 1/4
N. by E. 1/4E.	14	E. by S. 1/4S.	104	S. by W. 1/4W.	194	W. by N. 1/4N.	284
N. by E. 1/2E.	17	E. by S. 1/2S.	107	S. by W. 1/2W.	197	W. by N. 1/2N.	287
N. by E. 3/4E.	19 3/4	E. by S. 3/4S.	109 3/4	S. by W. 3/4W.	199 3/4	W. by N. 3/4N.	289 3/4
N.N.E.	**22 1/2**	**E.S.E.**	**112 1/2**	**S.S.W.**	**202 1/2**	**W.N.W.**	**292 1/2**
N.E. by N. 3/4N.	25 1/4	S.E. by E. 3/4E.	115 1/4	S.W. by S. 3/4S.	205 1/4	N.W. by W. 3/4W.	295 1/4
N.E. by N. 1/2N.	28 1/4	S.E. by E. 1/2E.	118 1/4	S.W. by S. 1/2S.	208 1/4	N.W. by W. 1/2W.	298 1/4
N.E. by N. 1/4N.	31	S.E. by E. 1/4E.	121	S.W. by S. 1/4S.	211	N.W. by W. 1/4W.	301
N.E. by N.	33 3/4	S.E. by E.	123 3/4	S.W. by S.	213 3/4	N.W. by W.	303 3/4
N.E. 3/4N.	36 1/2	S.E. 3/4E.	126 1/2	S.W. 3/4S.	216 1/2	N.W. 3/4W.	306 1/2
N.E. 1/2N.	39 1/2	S.E. 1/2E.	129 1/2	S.W. 1/2S.	219 1/2	N.W. 1/2W.	309 1/2
N.E. 1/4N.	42 1/4	S.E. 1/4E.	132 1/4	S.W. 1/4S.	222 1/4	N.W. 1/4W.	312 1/4
N.E.	**45**	**S.E.**	**135**	**S.W.**	**225**	**N.W.**	**315**
N.E. 1/4E.	47 3/4	S.E. 1/4S.	137 3/4	S.W. 1/4W.	227 3/4	N.W. 1/4N.	317 3/4
N.E. 1/2E.	50 3/4	S.E. 1/2S.	140 3/4	S.W. 1/2W.	230 3/4	N.W. 1/2N.	320 3/4
N.E. 3/4E.	53 1/2	S.E. 3/4S.	143 1/2	S.W. 3/4W.	233 1/2	N.W. 3/4N.	323 1/2
N.E. by E.	56 1/4	S.E. by S.	146 1/4	S.W. by W.	236 1/4	N.W. by N.	326 1/4
N.E. by E. 1/4E.	59	S.E. by S. 1/4S.	149	S.W. by W. 1/4W.	239	N.W. by N. 1/4N.	329
N.E. by E. 1/2E.	62	S.E. by S. 1/2S.	152	S.W. by W. 1/2W.	242	N.W. by N. 1/2N.	332
N.E. by E. 3/4E.	64 3/4	S.E. by S. 3/4S.	154 3/4	S.W. by W. 3/4W.	244 3/4	N.W. by N. 3/4N.	334 3/4
E.N.E.	**67 1/2**	**S.S.E.**	**157 1/2**	**W.S.W.**	**247 1/2**	**N.N.W.**	**337 1/2**
E. by N. 3/4N.	70 1/4	S. by E. 3/4E.	160 1/4	W. by S. 3/4S.	250 1/4	N. by W. 3/4W.	340 1/4
E. by N. 1/2N.	73 1/4	S. by E. 1/2E.	163 1/4	W. by S. 1/2S.	253 1/4	N. by W. 1/2W.	343 1/4
E. by N. 1/4N.	76	S. by E. 1/4E.	166	W. by S. 1/4S.	256	N. by W. 1/4W.	346
E. by N.	78 3/4	S. by E.	168 3/4	W. by S.	258 3/4	N. by W.	348 3/4
E. 3/4N.	81 1/2	S. 3/4E.	171 1/2	W. 3/4S.	261 1/2	N. 3/4W.	351 1/2
E. 1/2N.	84 1/2	S. 1/2E.	174 1/2	W. 1/2S.	264 1/2	N. 1/2W.	354 1/2
E. 1/4N.	87 1/4	S. 1/4E.	177 1/4	W. 1/4S.	267 1/4	N. 1/4W.	357 1/4
EAST	**90**	**SOUTH**	**180**	**WEST**	**270**	**NORTH**	**0**

2009 SUN'S RISING AND SETTING AT BOSTON
42° 20'N 71°W
Add one hour for Daylight Saving Time, covering summer months

Vernal Equinox: March 20th, 6:44 a.m.   Summer Solstice: June 21st, 12:45 a.m.

Day of Month	Jan. Rises	Jan. Sets	Feb. Rises	Feb. Sets	Mar. Rises	Mar. Sets	April Rises	April Sets	May Rises	May Sets	June Rises	June Sets
	H. M.	H. M.	H. M.	H. M.	H. M.	H. M.	H. M.	H. M.	H. M.	H. M.	H. M.	H. M.
1	7:13	4:22	6:58	4:58	6:19	5:34	5:27	6:09	4:39	6:43	4:11	7:14
2	7:13	4:23	6:57	4:59	6:18	5:35	5:25	6:10	4:38	6:44	4:10	7:15
3	7:13	4:24	6:56	5:01	6:16	5:36	5:24	6:11	4:37	6:45	4:09	7:15
4	7:13	4:25	6:55	5:02	6:15	5:37	5:22	6:13	4:35	6:47	4:08	7:16
5	7:13	4:26	6:54	5:03	6:13	5:39	5:21	6:14	4:34	6:48	4:08	7:17
6	7:13	4:27	6:53	5:04	6:12	5:40	5:19	6:15	4:33	6:49	4:08	7:17
7	7:13	4:28	6:52	5:05	6:10	5:41	5:17	6:16	4:32	6:50	4:08	7:18
8	7:13	4:29	6:50	5:07	6:08	5:42	5:16	6:17	4:31	6:51	4:07	7:19
9	7:13	4:30	6:49	5:08	6:07	5:43	5:14	6:19	4:29	6:52	4:07	7:19
10	7:12	4:31	6:48	5:09	6:05	5:45	5:12	6:20	4:28	6:53	4:07	7:20
11	7:12	4:32	6:46	5:11	6:04	5:46	5:10	6:21	4:27	6:54	4:07	7:20
12	7:12	4:33	6:45	5:12	6:02	5:47	5:09	6:22	4:26	6:55	4:07	7:21
13	7:11	4:34	6:44	5:14	6:00	5:48	5:07	6:23	4:25	6:56	4:07	7:21
14	7:11	4:36	6:43	5:15	5:59	5:49	5:06	6:25	4:24	6:57	4:07	7:22
15	7:10	4:37	6:41	5:16	5:57	5:51	5:04	6:26	4:23	6:58	4:07	7:22
16	7:10	4:38	6:40	5:17	5:55	5:52	5:02	6:27	4:22	6:59	4:07	7:23
17	7:09	4:39	6:38	5:18	5:53	5:53	5:01	6:28	4:21	7:00	4:07	7:23
18	7:09	4:40	6:37	5:20	5:52	5:54	4:59	6:29	4:20	7:01	4:07	7:23
19	7:08	4:42	6:36	5:21	5:50	5:55	4:58	6:31	4:19	7:02	4:07	7:24
20	7:08	4:43	6:34	5:22	5:49	5:56	4:56	6:32	4:18	7:03	4:07	7:24
21	7:07	4:44	6:32	5:24	5:47	5:57	4:54	6:33	4:17	7:04	4:07	7:24
22	7:06	4:45	6:31	5:25	5:45	5:58	4:53	6:34	4:17	7:05	4:07	7:24
23	7:06	4:46	6:29	5:27	5:43	5:59	4:51	6:35	4:16	7:06	4:08	7:25
24	7:05	4:48	6:28	5:28	5:42	6:00	4:50	6:36	4:15	7:07	4:08	7:25
25	7:05	4:49	6:26	5:29	5:40	6:02	4:48	6:37	4:15	7:08	4:08	7:25
26	7:04	4:50	6:24	5:30	5:38	6:03	4:46	6:38	4:14	7:09	4:08	7:25
27	7:03	4:51	6:23	5:31	5:36	6:04	4:45	6:39	4:14	7:10	4:09	7:25
28	7:02	4:52	6:21	5:33	5:34	6:05	4:43	6:40	4:13	7:11	4:09	7:25
29	7:01	4:54			5:33	6:06	4:42	6:41	4:12	7:11	4:10	7:25
30	7:00	4:55			5:31	6:07	4:40	6:42	4:12	7:12	4:10	7:25
31	6:59	4:56			5:29	6:08			4:11	7:13		

For correct SETTING of Sun any day of the year at places specified below, FOR FLAG USE, add or subtract from above table.

	Jan. 15	Feb. 15	Mar. 15	Apr. 15	May 15	June 15
New London	+7	+6	+4	+2	0	−1
Newport	+4	+3	+1	−1	−2	−3
New Bedford	+3	+2	0	−1	−2	−3
Vineyard Haven	+1	−1	−2	−4	−5	−6
Nantucket	−1	−2	−4	−6	−7	−8
Portland	−8	−6	−3	−1	+1	+2
Rockland	−14	−12	−8	−6	−4	−2
Bar Harbor	−18	−15	−11	−8	−5	−3

Note: Times shown in table are first tip of Sun at Sunrise, last tip at Sunset.

2009 SUN'S RISING AND SETTING AT BOSTON

42° 20'N   71°W

Add one hour for Daylight Saving Time, covering summer months

Autumnal Equinox: Sept. 22nd, 4:18 p.m. Winter Solstice: Dec. 21st, 12:47 p.m.

Day of Month	July Rises H. M.	July Sets H. M.	Aug. Rises H. M.	Aug. Sets H. M.	Sept. Rises H. M.	Sept. Sets H. M.	Oct. Rises H. M.	Oct. Sets H. M.	Nov. Rises H. M.	Nov. Sets H. M.	Dec. Rises H. M.	Dec. Sets H. M.
1	4:11	7:25	4:36	7:04	5:09	6:19	5:41	5:27	6:18	4:37	6:53	4:13
2	4:11	7:24	4:37	7:02	5:10	6:17	5:42	5:25	6:19	4:36	6:54	4:12
3	4:12	7:24	4:38	7:01	5:11	6:15	5:43	5:23	6:20	4:35	6:55	4:12
4	4:12	7:24	4:39	7:00	5:12	6:13	5:44	5:21	6:21	4:34	6:56	4:12
5	4:13	7:24	4:40	6:59	5:13	6:11	5:45	5:19	6:23	4:32	6:57	4:12
6	4:13	7:23	4:41	6:58	5:14	6:10	5:46	5:18	6:24	4:31	6:58	4:12
7	4:14	7:23	4:42	6:56	5:15	6:08	5:47	5:16	6:25	4:30	6:59	4:12
8	4:15	7:23	4:43	6:55	5:16	6:06	5:48	5:14	6:26	4:29	7:00	4:12
9	4:16	7:22	4:44	6:54	5:17	6:05	5:49	5:13	6:27	4:28	7:01	4:12
10	4:16	7:22	4:45	6:52	5:18	6:03	5:50	5:11	6:29	4:27	7:02	4:12
11	4:17	7:21	4:46	6:51	5:20	6:02	5:52	5:10	6:30	4:26	7:03	4:12
12	4:18	7:21	4:47	6:49	5:21	6:00	5:53	5:08	6:31	4:25	7:04	4:12
13	4:18	7:20	4:48	6:48	5:22	5:58	5:54	5:06	6:32	4:24	7:04	4:12
14	4:19	7:20	4:49	6:47	5:23	5:56	5:55	5:05	6:33	4:23	7:05	4:12
15	4:20	7:19	4:50	6:45	5:24	5:54	5:56	5:03	6:35	4:23	7:06	4:13
16	4:21	7:19	4:51	6:44	5:25	5:53	5:58	5:01	6:36	4:22	7:06	4:13
17	4:22	7:18	4:53	6:43	5:26	5:51	5:59	4:59	6:37	4:21	7:07	4:13
18	4:22	7:17	4:54	6:41	5:27	5:49	6:00	4:58	6:38	4:20	7:07	4:14
19	4:23	7:17	4:55	6:40	5:28	5:47	6:01	4:56	6:39	4:19	7:08	4:14
20	4:24	7:16	4:56	6:38	5:29	5:45	6:02	4:54	6:41	4:18	7:08	4:14
21	4:25	7:15	4:57	6:36	5:30	5:44	6:04	4:53	6:42	4:18	7:09	4:15
22	4:26	7:14	4:58	6:35	5:31	5:42	6:05	4:51	6:43	4:17	7:09	4:15
23	4:27	7:13	4:59	6:33	5:32	5:40	6:06	4:50	6:44	4:17	7:10	4:16
24	4:28	7:12	5:00	6:32	5:33	5:38	6:07	4:48	6:45	4:16	7:10	4:16
25	4:29	7:11	5:01	6:30	5:34	5:36	6:08	4:47	6:47	4:15	7:11	4:17
26	4:30	7:10	5:02	6:29	5:35	5:35	6:10	4:45	6:48	4:15	7:11	4:17
27	4:31	7:09	5:03	6:27	5:36	5:33	6:11	4:44	6:49	4:14	7:12	4:18
28	4:32	7:08	5:04	6:25	5:37	5:31	6:12	4:43	6:50	4:14	7:12	4:18
29	4:33	7:07	5:05	6:24	5:38	5:30	6:14	4:41	6:51	4:13	7:12	4:19
30	4:34	7:06	5:06	6:22	5:39	5:28	6:15	4:39	6:52	4:13	7:13	4:20
31	4:35	7:05	5:07	6:20			6:17	4:38			7:13	4:21

For correct SETTING of Sun any day of the year at places specified below, FOR FLAG USE, add or subtract from above table.

	July 15	Aug. 15	Sept. 15	Oct. 15	Nov. 15	Dec. 15
New London	0	+1	+2	+5	+6	+7
Newport	−2	−1	0	+2	+4	+5
New Bedford	−2	−1	0	+1	+2	+3
Vineyard Haven	−5	−4	−3	−2	0	+1
Nantucket	−7	−6	−5	−3	−2	−1
Portland	+1	0	−2	−4	−6	−7
Rockland	−4	−5	−7	−10	−12	−14
Bar Harbor	−5	−7	−9	−13	−17	−18

Note: Times shown in table are first tip of Sun at Sunrise, last tip at Sunset.

2009 SUN'S RISING AND SETTING AT NEW YORK (THE BATTERY)

40° 42'N   74°W
Add one hour for Daylight Saving Time, covering summer months

Vernal Equinox: March 20th, 6:44 a.m.   Summer Solstice: June 21st, 12:45 a.m.

Day of Month	Jan. Rises H. M.	Jan. Sets H. M.	Feb. Rises H. M.	Feb. Sets H. M.	Mar. Rises H. M.	Mar. Sets H. M.	April Rises H. M.	April Sets H. M.	May Rises H. M.	May Sets H. M.	June Rises H. M.	June Sets H. M.
1	7:20	4:39	7:06	5:14	6:29	5:48	5:41	6:20	4:55	6:51	4:27	7:21
2	7:20	4:40	7:05	5:15	6:28	5:49	5:39	6:21	4:54	6:52	4:27	7:22
3	7:20	4:41	7:04	5:17	6:26	5:50	5:38	6:22	4:53	6:53	4:27	7:22
4	7:20	4:42	7:03	5:18	6:25	5:51	5:36	6:23	4:51	6:55	4:27	7:23
5	7:20	4:43	7:02	5:19	6:23	5:52	5:34	6:24	4:50	6:56	4:26	7:23
6	7:20	4:44	7:01	5:20	6:22	5:53	5:33	6:25	4:49	6:57	4:26	7:24
7	7:20	4:45	7:00	5:21	6:21	5:54	5:31	6:26	4:48	6:58	4:26	7:24
8	7:20	4:46	6:58	5:23	6:19	5:55	5:29	6:27	4:47	6:59	4:25	7:25
9	7:20	4:47	6:57	5:24	6:18	5:56	5:28	6:28	4:45	7:00	4:25	7:25
10	7:20	4:48	6:56	5:25	6:16	5:58	5:26	6:29	4:44	7:01	4:25	7:26
11	7:20	4:49	6:55	5:26	6:15	5:59	5:25	6:30	4:43	7:02	4:25	7:26
12	7:19	4:50	6:54	5:27	6:13	6:00	5:24	6:31	4:42	7:03	4:24	7:27
13	7:19	4:51	6:52	5:29	6:12	6:01	5:22	6:32	4:41	7:04	4:24	7:27
14	7:19	4:52	6:51	5:30	6:10	6:02	5:20	6:34	4:40	7:05	4:24	7:28
15	7:18	4:53	6:50	5:31	6:08	6:03	5:19	6:35	4:39	7:06	4:24	7:28
16	7:18	4:54	6:49	5:32	6:07	6:04	5:17	6:36	4:38	7:07	4:24	7:29
17	7:17	4:55	6:48	5:33	6:05	6:05	5:16	6:37	4:37	7:08	4:24	7:29
18	7:17	4:56	6:47	5:35	6:04	6:06	5:14	6:38	4:37	7:09	4:24	7:29
19	7:16	4:58	6:46	5:36	6:02	6:07	5:12	6:39	4:36	7:10	4:24	7:30
20	7:16	4:59	6:44	5:37	6:00	6:08	5:10	6:40	4:35	7:11	4:25	7:30
21	7:15	5:00	6:42	5:38	5:59	6:09	5:09	6:41	4:34	7:12	4:25	7:30
22	7:15	5:01	6:41	5:39	5:57	6:10	5:07	6:42	4:33	7:13	4:25	7:30
23	7:14	5:02	6:39	5:41	5:56	6:11	5:06	6:43	4:33	7:14	4:25	7:31
24	7:14	5:04	6:38	5:42	5:54	6:12	5:04	6:44	4:32	7:14	4:26	7:31
25	7:13	5:05	6:36	5:43	5:52	6:13	5:03	6:45	4:31	7:15	4:26	7:31
26	7:12	5:06	6:35	5:44	5:50	6:14	5:02	6:46	4:30	7:16	4:26	7:31
27	7:11	5:07	6:33	5:45	5:49	6:15	5:00	6:47	4:30	7:17	4:27	7:31
28	7:10	5:08	6:31	5:47	5:47	6:16	4:59	6:48	4:29	7:18	4:27	7:31
29	7:09	5:10			5:46	6:17	4:58	6:49	4:29	7:19	4:28	7:31
30	7:08	5:11			5:45	6:18	4:56	6:50	4:28	7:19	4:28	7:31
31	7:07	5:12			5:43	6:19			4:28	7:20		

For correct **SETTING** of Sun any day of the year at places specified below, **FOR FLAG USE**, add or subtract from above table.

	Jan. 15	Feb. 15	Mar. 15	Apr. 15	May 15	June 15
Hampton Rds., Va....	+18	+15	+9	+2	0	−1
Oxford, Md..........	+14	+12	+9	+5	+4	+3
Annapolis, Md.......	+14	+13	+10	+7	+6	+5
Cape May, N. J.......	+8	+7	+4	+1	0	−1
Atlantic City........	+5	+4	+2	0	−1	−2
Manasquan, N. J......	+2	+1	0	−1	−2	−2
Port Jefferson, N. Y...	−5	−4	−4	−3	−3	−3
Bridgeport, Ct.......	−4	−4	−3	−2	−2	−1
New Haven..........	−7	−6	−4	−3	−3	−3

Note: Times shown in table are first tip of Sun at Sunrise, last tip at Sunset.

2009 SUN'S RISING AND SETTING AT NEW YORK (THE BATTERY)

40° 42'N 74°W

Add one hour for Daylight Saving Time, covering summer months

Autumnal Equinox: Sept. 22nd, 4:18 p.m. Winter Solstice: Dec. 21st, 12:47 p.m.

Day of Month	July Rises H.M.	July Sets H.M.	Aug. Rises H.M.	Aug. Sets H.M.	Sept. Rises H.M.	Sept. Sets H.M.	Oct. Rises H.M.	Oct. Sets H.M.	Nov. Rises H.M.	Nov. Sets H.M.	Dec. Rises H.M.	Dec. Sets H.M.
1	4:28	7:31	4:52	7:12	5:22	6:30	5:52	5:40	6:26	4:53	7:00	4:30
2	4:29	7:31	4:53	7:10	5:23	6:28	5:53	5:38	6:27	4:52	7:01	4:29
3	4:30	7:30	4:54	7:09	5:24	6:26	5:54	5:36	6:28	4:51	7:02	4:29
4	4:30	7:30	4:55	7:08	5:25	6:24	5:55	5:34	6:29	4:49	7:03	4:29
5	4:31	7:30	4:56	7:07	5:26	6:22	5:56	5:32	6:31	4:48	7:04	4:29
6	4:31	7:30	4:57	7:06	5:27	6:21	5:57	5:31	6:32	4:47	7:05	4:29
7	4:32	7:29	4:58	7:04	5:28	6:19	5:58	5:29	6:33	4:46	7:06	4:29
8	4:32	7:29	4:59	7:03	5:29	6:17	5:59	5:27	6:34	4:45	7:07	4:29
9	4:33	7:29	5:00	7:02	5:30	6:16	6:00	5:26	6:35	4:44	7:08	4:29
10	4:34	7:28	5:01	7:00	5:31	6:14	6:01	5:24	6:37	4:43	7:08	4:29
11	4:34	7:28	5:02	6:58	5:32	6:13	6:03	5:22	6:38	4:42	7:09	4:29
12	4:35	7:28	5:03	6:57	5:33	6:11	6:04	5:21	6:39	4:41	7:10	4:29
13	4:36	7:27	5:04	6:56	5:34	6:09	6:05	5:19	6:40	4:40	7:11	4:29
14	4:37	7:27	5:05	6:55	5:35	6:07	6:06	5:18	6:41	4:39	7:12	4:30
15	4:37	7:26	5:06	6:54	5:36	6:05	6:07	5:17	6:43	4:39	7:12	4:30
16	4:38	7:26	5:07	6:53	5:37	6:04	6:08	5:15	6:44	4:38	7:13	4:30
17	4:39	7:25	5:08	6:51	5:38	6:02	6:09	5:13	6:45	4:37	7:14	4:30
18	4:39	7:25	5:09	6:50	5:39	6:00	6:10	5:12	6:46	4:36	7:14	4:31
19	4:40	7:24	5:10	6:49	5:40	5:59	6:11	5:10	6:47	4:35	7:15	4:31
20	4:41	7:23	5:11	6:47	5:41	5:57	6:12	5:09	6:49	4:35	7:15	4:31
21	4:42	7:23	5:12	6:46	5:42	5:56	6:14	5:08	6:50	4:34	7:16	4:32
22	4:43	7:22	5:13	6:45	5:43	5:54	6:15	5:06	6:51	4:33	7:16	4:32
23	4:44	7:21	5:14	6:43	5:44	5:52	6:16	5:05	6:52	4:33	7:17	4:33
24	4:45	7:20	5:15	6:42	5:45	5:51	6:17	5:03	6:53	4:32	7:17	4:33
25	4:46	7:19	5:16	6:41	5:46	5:49	6:18	5:02	6:54	4:32	7:18	4:34
26	4:46	7:18	5:16	6:39	5:47	5:48	6:19	5:00	6:55	4:31	7:18	4:34
27	4:47	7:17	5:17	6:38	5:48	5:46	6:20	4:59	6:56	4:31	7:19	4:35
28	4:48	7:16	5:18	6:36	5:49	5:44	6:21	4:58	6:57	4:31	7:19	4:35
29	4:49	7:15	5:19	6:34	5:50	5:43	6:22	4:57	6:58	4:30	7:19	4:36
30	4:50	7:14	5:20	6:33	5:51	5:41	6:23	4:55	6:59	4:30	7:20	4:37
31	4:51	7:13	5:21	6:31			6:25	4:54			7:20	4:38

For correct SETTING of Sun any day of the year at places specified below, FOR FLAG USE, add or subtract from above table.

	July 15	Aug. 15	Sept. 15	Oct. 15	Nov. 15	Dec. 15
Hampton Rds., Va.	−1	+2	+7	+13	+18	+20
Oxford, Md.	+3	+5	+8	+11	+14	+15
Annapolis, Md.	+5	+7	+9	+12	+14	+15
Cape May, N. J.	−1	+1	+3	+6	+8	+9
Atlantic City	−2	0	+2	+4	+6	+6
Mannasquan, N. J.	−2	−1	0	+1	+2	+2
Port Jefferson, N. Y. ..	−3	−3	−4	−4	−5	−5
Bridgeport, Ct.	−1	−2	−3	−4	−4	−5
New Haven	−3	−3	−4	−5	−6	−7

Note: Times shown in table are first tip of Sun at Sunrise, last tip at Sunset.

2009 SUN'S RISING AND SETTING AT JACKSONVILLE, FL

30° 20'N 81° 37'W

Add one hour for Daylight Saving Time, covering summer months

Vernal Equinox: March 20th, 6:44 a.m.   Summer Solstice: June 21st, 12:45 a.m.
Autumnal Equinox: Sept. 22nd, 4:18 p.m. Winter Solstice: Dec. 21st, 12:47 p.m.

	Jan.		**Feb.**		**Mar.**		**April**		**May**		**June**	
	Rises	Sets	Rises	Sets	Rises	Sets	Rises	Sets	Rises	Sets	Rises	Sets
5	7:24	5:39	7:14	6:07	6:48	6:28	6:10	6:47	5:40	7:06	5:24	7:25
10	7:24	5:43	7:10	6:11	6:42	6:31	6:05	6:50	5:35	7:10	5:23	7:27
15	7:24	5:48	7:06	6:15	6:36	6:35	5:59	6:53	5:32	7:13	5:23	7:29
20	7:22	5:52	7:01	6:19	6:30	6:38	5:54	6:56	5:29	7:16	5:24	7:31
25	7:21	5:56	6:56	6:22	6:24	6:41	5:49	6:59	5:27	7:19	5:25	7:32
30	7:17	6:01			6:18	6:44	5:44	7:02	5:25	7:22	5:27	7:32

	July		**Aug.**		**Sept.**		**Oct.**		**Nov.**		**Dec.**	
	Rises	Sets	Rises	Sets	Rises	Sets	Rises	Sets	Rises	Sets	Rises	Sets
5	5:29	7:32	5:46	7:17	6:05	6:44	6:21	6:07	6:43	5:36	7:08	5:25
10	5:32	7:31	5:49	7:12	6:08	6:38	6:25	6:01	6:47	5:32	7:12	5:26
15	5:34	7:30	5:52	7:07	6:10	6:31	6:28	5:55	6:51	5:29	7:16	5:27
20	5:37	7:27	5:55	7:02	6:13	6:25	6:31	5:50	6:55	5:27	7:19	5:29
25	5:40	7:25	5:58	6:57	6:16	6:19	6:35	5:45	6:59	5:25	7:21	5:32
30	5:43	7:21	6:02	6:51	6:19	6:13	6:39	5:40	7:04	5:25	7:23	5:35

 FOR FLAG USE, add or subtract from above table:

		1/15	2/15	3/15	4/15	5/15	6/15	7/15	8/15	9/15	10/15	11/15	12/15
Morehead City —	Rises:	−9	−14	−19	−25	−27	−30	−29	−25	−21	−14	−11	−7
	Sets:	−28	−24	−20	−14	−10	−7	−9	−13	−19	−24	−28	−29
Wilmington —	Rises:	−6	−10	−14	−19	−22	−24	−23	−20	−15	−10	−7	−3
	Sets:	−22	−18	−14	−10	−5	−3	−5	−8	−14	−18	−22	−23
Myrtle Beach —	Rises:	−4	−7	−9	−15	−16	−18	−17	−15	−11	−8	−4	−1
	Sets:	−16	−14	−10	−6	−3	−1	−3	−5	−7	−12	−17	−17
Charleston —	Rises:	−1	−3	−6	−10	−11	−12	−11	−9	−6	−4	−3	+1
	Sets:	−11	−9	−6	−6	+1	0	−1	−3	−7	−9	−11	−12
Savannah —	Rises:	+5	+4	+2	+1	−1	−2	−2	0	+2	+4	+5	+7
	Sets:	−1	0	+2	+4	+6	+6	+5	+4	+3	0	−2	−2
Brunswick —	Rises:	+1	+1	0	−1	−2	−2	−2	−1	0	+1	+1	+2
	Sets:	+1	−1	0	+1	+2	+2	+1	+1	−1	+1	+2	+2
Ponce Inlet —	Rises:	−5	−4	−2	−1	+1	+1	+1	−1	−2	−3	−5	−5
	Sets:	+1	−1	−1	−2	−4	−5	−5	−4	−2	0	+1	+1
Melbourne —	Rises:	−9	−7	−3	−2	+1	+2	+1	−2	−3	−5	−8	−9
	Sets:	+1	−2	−4	−5	−7	−9	−9	−8	−5	−1	0	+1
N.Palm Beach —	Rises:	−14	−10	−10	−6	+1	+3	+2	−2	−4	−8	−12	−13
	Sets:	+2	−2	−6	−8	−11	−14	−14	−11	−7	−4	0	+2
Miami —	Rises:	−15	−10	−4	−1	+3	+6	+5	0	−3	−8	−13	−14
	Sets:	+6	0	−3	−8	−13	−15	−15	−11	−7	−2	+3	+5
Key West —	Rises:	−11	−5	+1	+5	+12	+15	+13	+8	+3	−2	−9	−12
	Sets:	+13	+7	+2	−3	−8	−12	−12	−6	−1	+6	+11	+14

Note: Times shown in table are first tip of Sun at Sunrise, last tip at Sunset.

TIME OF LOCAL APPARENT NOON (L.A.N.) 2009
FOR THE CENTRAL MERIDIAN OF ANY TIME ZONE

LOCAL APPARENT NOON, 2009

	JAN. H:M:S	FEB. H:M:S	MAR. H:M:S	APR. H:M:S	MAY H:M:S	JUN. H:M:S	JUL. H:M:S	AUG. H:M:S	SEP. H:M:S	OCT. H:M:S	NOV. H:M:S	DEC. H:M:S
1	12:03:46	12:13:39	12:12:17	12:03:46	11:57:03	11:57:53	12:03:56	12:06:19	11:59:53	11:49:33	11:43:35	11:49:10
2	12:04:14	12:13:46	12:12:04	12:03:29	11:56:57	11:58:03	12:04:08	12:06:15	11:59:34	11:49:14	11:43:34	11:49:33
3	12:04:42	12:13:52	12:11:52	12:03:11	11:56:50	11:58:12	12:04:18	12:06:10	11:59:14	11:48:55	11:43:34	11:49:56
4	12:05:09	12:13:58	12:11:39	12:02:54	11:56:45	11:58:23	12:04:29	12:06:04	11:58:54	11:48:37	11:43:35	11:50:21
5	12:05:35	12:14:02	12:11:25	12:02:36	11:56:40	11:58:33	12:04:39	12:05:58	11:58:34	11:48:19	11:43:37	11:50:45
6	12:06:02	12:14:06	12:11:11	12:02:19	11:56:35	11:58:44	12:04:49	12:05:51	11:58:14	11:48:01	11:43:39	11:51:11
7	12:06:28	12:14:09	12:10:57	12:02:02	11:56:31	11:58:55	12:04:59	12:05:44	11:57:53	11:47:43	11:43:42	11:51:36
8	12:06:53	12:14:12	12:10:42	12:01:46	11:56:28	11:59:07	12:05:08	12:05:36	11:57:33	11:47:27	11:43:46	11:52:03
9	12:07:17	12:14:13	12:10:27	12:01:29	11:56:25	11:59:18	12:05:17	12:05:27	11:57:12	11:47:10	11:43:51	11:52:29
10	12:07:42	12:14:14	12:10:11	12:01:13	11:56:23	11:59:30	12:05:25	12:05:18	11:56:51	11:46:54	11:43:57	11:52:57
11	12:08:05	12:14:14	12:09:55	12:00:57	11:56:21	11:59:42	12:05:33	12:05:08	11:56:30	11:46:39	11:44:04	11:53:24
12	12:08:28	12:14:13	12:09:39	12:00:41	11:56:20	11:59:55	12:05:41	12:04:58	11:56:09	11:46:24	11:44:12	11:53:52
13	12:08:50	12:14:11	12:09:23	12:00:26	11:56:19	12:00:07	12:05:48	12:04:47	11:55:47	11:46:09	11:44:20	11:54:21
14	12:09:12	12:14:09	12:09:06	12:00:11	11:56:19	12:00:20	12:05:54	12:04:36	11:55:26	11:45:55	11:44:29	11:54:49
15	12:09:33	12:14:06	12:08:49	11:59:57	11:56:20	12:00:33	12:06:00	12:04:24	11:55:05	11:45:42	11:44:40	11:55:18
16	12:09:54	12:14:02	12:08:32	11:59:42	11:56:21	12:00:46	12:06:06	12:04:12	11:54:43	11:45:29	11:44:51	11:55:47
17	12:10:13	12:13:58	12:08:15	11:59:29	11:56:23	12:00:59	12:06:11	12:03:59	11:54:22	11:45:17	11:45:03	11:56:17
18	12:10:32	12:13:53	12:07:57	11:59:15	11:56:26	12:01:12	12:06:16	12:03:46	11:54:01	11:45:06	11:45:15	11:56:46
19	12:10:51	12:13:47	12:07:40	11:59:02	11:56:28	12:01:25	12:06:20	12:03:32	11:53:39	11:44:55	11:45:29	11:57:16
20	12:11:08	12:13:41	12:07:22	11:58:50	11:56:32	12:01:38	12:06:23	12:03:18	11:53:18	11:44:45	11:45:43	11:57:46
21	12:11:25	12:13:34	12:07:04	11:58:38	11:56:36	12:01:51	12:06:26	12:03:03	11:52:57	11:44:35	11:45:58	11:58:16
22	12:11:41	12:13:26	12:06:46	11:58:26	11:56:41	12:02:04	12:06:28	12:02:48	11:52:36	11:44:26	11:46:14	11:58:46
23	12:11:57	12:13:18	12:06:28	11:58:15	11:56:46	12:02:17	12:06:30	12:02:32	11:52:15	11:44:18	11:46:31	11:59:16
24	12:12:11	12:13:09	12:06:10	11:58:04	11:56:51	12:02:30	12:06:32	12:02:16	11:51:54	11:44:10	11:46:48	11:59:46
25	12:12:25	12:13:00	12:05:52	11:57:54	11:56:58	12:02:43	12:06:32	12:02:00	11:51:33	11:44:03	11:47:06	12:00:15
26	12:12:38	12:12:50	12:05:34	11:57:44	11:57:04	12:02:56	12:06:32	12:01:43	11:51:13	11:43:57	11:47:25	12:00:45
27	12:12:50	12:12:39	12:05:16	11:57:35	11:57:11	12:03:08	12:06:31	12:01:26	11:50:52	11:43:51	11:47:45	12:01:14
28	12:13:01	12:12:28	12:04:58	11:57:26	11:57:19	12:03:21	12:06:30	12:01:08	11:50:32	11:43:47	11:48:05	12:01:44
29	12:13:12		12:04:40	11:57:18	11:57:27	12:03:33	12:06:28	12:00:50	11:50:12	11:43:43	11:48:26	12:02:13
30	12:13:22		12:04:22	11:57:10	11:57:35	12:03:45	12:06:26	12:00:31	11:49:53	11:43:39	11:48:48	12:02:41
31	12:13:31		12:04:04		11:57:44		12:06:23	12:00:13		11:43:37		12:03:10

Explanatory Notes: The noon sight and the Sun's Declination (p. 231) results in the vessel's parallel of latitude. It is taken at the time of the sun's meridian passage, when the sun is at maximum altitude.

The moment of meridian passage is called Local Apparent Noon (L.A.N.), and only rarely is it the same time as noon Standard Time or Local Mean Time. Instead, as this Table shows, the sun is either ahead of or behind its theoretical schedule.

Two corrections are involved. 1) To correct for your difference in longitude from the central meridian of your time zone (i.e. 75° for U.S. Atlantic Coast), either a) add 4 minutes of time for each degree West or b) subtract 4 minutes of time for each degree East. 2) If necessary, convert from Daylight Savings Time to Standard Time by subtracting 1 hour from your watch.

Thus for Boston, at 71° West longitude (or 4° East of 75°), L.A.N. occurs 16 minutes before the times listed in the Table.

For New York, at 74° West (1° East of 75°), L.A.N. occurs 4 minutes earlier than times shown.

Converting arc to time:

$$360° = 24 \text{ hours}$$
$$15° = 1 \text{ hour}$$
$$1° = 4 \text{ minutes}$$
$$15' = 1 \text{ minute}$$
$$1' = 4 \text{ seconds}$$

SUN'S TRUE BEARING AT RISING AND SETTING

To find compass deviation using the Sun.
Figures are correct for all Longitudes

Latitudes

Sun's Decl.	38° N Rise	38° N Set	40° N Rise	40° N Set	42° N Rise	42° N Set	44° N Rise	44° N Set	Sun's Decl.
N 23°	60.3°	299.7°	59.3°	300.7°	58.3°	301.7°	57.1°	302.9°	N 23°
22	61.6	298.4	60.7	299.3	59.7	300.3	58.6	301.4	22
21	63.0	297.0	62.1	297.9	61.2	298.8	60.1	299.9	21
20	64.3	295.7	63.5	296.5	62.6	297.4	61.6	298.4	20
19	65.6	294.4	64.9	295.1	64.0	296.0	63.1	296.9	19
18	66.9	293.1	66.2	293.8	65.4	294.6	64.6	295.4	18
17	68.2	291.8	67.6	292.4	66.8	293.2	66.0	294.0	17
16	69.5	290.5	68.9	291.1	68.2	291.8	67.5	292.5	16
15	70.8	289.2	70.3	289.7	69.6	290.4	68.9	291.1	15
14	72.1	287.9	71.6	288.4	71.0	289.0	70.4	289.6	14
13	73.4	286.6	72.9	287.1	72.4	287.6	71.8	288.2	13
12	74.7	285.3	74.3	285.7	73.8	286.2	73.2	286.8	12
11	76.0	284.0	75.6	284.4	75.1	284.9	74.6	285.4	11
10	77.3	282.7	76.9	283.1	76.5	283.5	76.0	284.0	10
9	78.6	281.4	78.2	281.8	77.9	282.1	77.4	282.6	9
8	79.8	280.2	79.5	280.5	79.2	280.8	78.9	281.1	8
7	81.1	278.9	80.9	279.1	80.6	279.4	80.3	279.7	7
6	82.4	277.6	82.2	277.7	81.9	278.1	81.7	278.3	6
5	83.7	276.3	83.5	276.5	83.3	276.7	83.0	277.0	5
4	84.9	275.1	84.8	275.2	84.6	275.4	84.4	275.6	4
3	86.2	273.8	86.1	273.9	86.0	274.0	85.8	274.2	3
2	87.5	272.5	87.4	272.6	87.3	272.7	87.2	272.8	2
N 1°	88.7	271.3	88.7	271.3	88.7	271.3	88.6	271.4	N 1°
0	90.0	270.0	90.0	270.0	90.0	270.0	90.0	270.0	0
S 1°	91.3	268.7	91.3	268.7	91.3	268.7	91.4	268.6	S 1°
2	92.5	267.5	92.6	267.4	92.7	267.3	92.8	267.2	2
3	93.8	266.2	93.9	266.1	94.0	266.0	94.2	265.8	3
4	95.1	264.9	95.2	264.8	95.4	264.6	95.6	264.4	4
5	96.3	263.7	96.5	263.5	96.7	263.3	97.0	263.0	5
6	97.6	262.4	97.8	262.2	98.1	261.9	98.3	261.7	6
7	98.9	261.1	99.1	260.9	99.4	260.6	99.7	260.3	7
8	100.2	259.8	100.5	259.5	100.8	259.2	101.1	258.9	8
9	101.4	258.6	101.8	258.2	102.1	257.9	102.6	257.4	9
10	102.7	257.3	103.1	256.9	103.5	256.5	104.0	256.0	10
11	104.0	256.0	104.4	255.6	104.9	255.1	105.4	254.6	11
12	105.3	254.7	105.7	254.3	106.2	253.8	106.8	253.2	12
13	106.6	253.4	107.1	252.9	107.6	252.4	108.2	251.8	13
14	107.9	252.1	108.4	251.6	109.0	251.0	109.6	250.4	14
15	109.2	250.8	109.7	250.3	110.4	249.6	111.1	248.9	15
16	110.5	249.5	111.1	248.9	111.8	248.2	112.5	247.5	16
17	111.8	248.2	112.4	247.6	113.2	246.8	114.0	246.0	17
18	113.1	246.9	113.8	246.2	114.6	245.4	115.4	244.6	18
19	114.4	245.6	115.1	244.9	116.0	244.0	116.9	243.1	19
20	115.7	244.3	116.5	243.5	117.4	242.6	118.4	241.6	20
21	117.0	243.0	117.9	242.1	118.8	241.2	119.9	240.1	21
22	118.4	241.6	119.3	240.7	120.3	239.7	121.4	238.6	22
S 23°	119.7	240.3	120.7	239.3	121.7	238.3	122.9	237.1	S 23°

Instructions: (1) Knowing the date, find the Sun's Declination from the facing page. Find that Declination down the left column on this page. (2) Find the column with your Latitude, and choose either Rise or Set to determine the True Bearing. (3) Add the local Westerly Variation to the figure. (4) If you are a couple of minutes after sunrise or before sunset, the Sun's bearing changes about 1° each 6 minutes during the first hour after sunrise and before sunset. (5) The deviation found will be correct only for the heading you are on at that time.

THE SUN'S DECLINATION 2009

For celestial navigators, the "noon sight" reading of the Sun's height above the horizon, together with the Sun's Declination from this table, determines latitude.

MEAN NOON – 75° MERIDIAN

THE SUN'S DECLINATION 2009 (1700 G.M.T.)

Day	JAN. South ° '	FEB. South ° '	MAR. South ° '	APR. North ° '	MAY North ° '	JUN. North ° '	JUL. North ° '	AUG. North ° '	SEPT. North ° '	OCT. South ° '	NOV. South ° '	DEC. South ° '
1	-22 57	-16 55	- 7 21	+ 4 47	+15 16	+22 08	+23 04	+17 51	+ 8 03	- 3 26	-14 38	-21 54
2	-22 51	-16 38	- 6 58	+ 5 10	+15 34	+22 16	+22 59	+17 36	+ 7 41	- 3 49	-14 57	-22 03
3	-22 46	-16 20	- 6 35	+ 5 33	+15 52	+22 23	+22 54	+17 20	+ 7 19	- 4 12	-15 15	-22 11
4	-22 39	-16 02	- 6 12	+ 5 56	+16 09	+22 30	+22 49	+17 04	+ 6 57	- 4 35	-15 34	-22 19
5	-22 32	-15 44	- 5 49	+ 6 19	+16 26	+22 37	+22 43	+16 48	+ 6 35	- 4 58	-15 52	-22 27
6	-22 25	-15 25	- 5 25	+ 6 41	+16 43	+22 43	+22 37	+16 31	+ 6 12	- 5 21	-16 10	-22 34
7	-22 17	-15 07	- 5 02	+ 7 04	+16 59	+22 49	+22 31	+16 15	+ 5 50	- 5 44	-16 28	-22 40
8	-22 09	-14 47	- 4 39	+ 7 26	+17 15	+22 54	+22 24	+15 57	+ 5 27	- 6 07	-16 45	-22 47
9	-22 01	-14 28	- 4 15	+ 7 49	+17 31	+22 59	+22 17	+15 40	+ 5 05	- 6 30	-17 02	-22 53
10	-21 52	-14 09	- 3 52	+ 8 11	+17 47	+23 03	+22 09	+15 23	+ 4 42	- 6 53	-17 19	-22 58
11	-21 42	-13 49	- 3 28	+ 8 33	+18 02	+23 08	+22 01	+15 05	+ 4 19	- 7 15	-17 35	-23 03
12	-21 32	-13 29	- 3 04	+ 8 55	+18 18	+23 11	+21 53	+14 47	+ 3 56	- 7 38	-17 52	-23 07
13	-21 22	-13 09	- 2 41	+ 9 16	+18 32	+23 15	+21 44	+14 28	+ 3 33	- 8 00	-18 08	-23 11
14	-21 11	-12 48	- 2 17	+ 9 38	+18 47	+23 18	+21 35	+14 10	+ 3 10	- 8 23	-18 23	-23 15
15	-21 00	-12 28	- 1 53	+ 9 59	+19 01	+23 20	+21 25	+13 51	+ 2 47	- 8 45	-18 38	-23 18
16	-20 49	-12 07	- 1 30	+10 21	+19 15	+23 22	+21 15	+13 32	+ 2 24	- 9 07	-18 53	-23 20
17	-20 37	-11 46	- 1 06	+10 42	+19 28	+23 24	+21 05	+13 13	+ 2 01	- 9 29	-19 08	-23 23
18	-20 25	-11 25	- 0 42	+11 03	+19 41	+23 25	+20 55	+12 54	+ 1 37	- 9 51	-19 22	-23 24
19	-20 12	-11 03	- 0 19	+11 23	+19 54	+23 26	+20 44	+12 34	+ 1 14	-10 12	-19 36	-23 25
20	-19 59	-10 42	+ 0 05	+11 44	+20 07	+23 26	+20 32	+12 14	+ 0 51	-10 34	-19 50	-23 26
21	-19 46	-10 20	+ 0 29	+12 04	+20 19	+23 26	+20 21	+11 54	+ 0 28	-10 55	-20 03	-23 26
22	-19 32	- 9 58	+ 0 53	+12 24	+20 30	+23 26	+20 09	+11 34	+ 0 04	-11 16	-20 16	-23 26
23	-19 18	- 9 36	+ 1 16	+12 44	+20 42	+23 25	+19 56	+11 14	- 0 19	-11 37	-20 28	-23 25
24	-19 03	- 9 14	+ 1 40	+13 04	+20 53	+23 24	+19 44	+10 53	- 0 43	-11 58	-20 40	-23 24
25	-18 48	- 8 51	+ 2 03	+13 24	+21 04	+23 22	+19 31	+10 32	- 1 06	-12 19	-20 52	-23 23
26	-18 33	- 8 29	+ 2 27	+13 43	+21 14	+23 20	+19 18	+10 11	- 1 29	-12 39	-21 03	-23 21
27	-18 18	- 8 06	+ 2 51	+14 02	+21 24	+23 18	+19 04	+ 9 50	- 1 53	-13 00	-21 14	-23 18
28	-18 02	- 7 44	+ 3 14	+14 21	+21 34	+23 15	+18 50	+ 9 29	- 2 16	-13 20	-21 25	-23 15
29	-17 46		+ 3 37	+14 40	+21 43	+23 12	+18 36	+ 9 08	- 2 39	-13 39	-21 35	-23 11
30	-17 29		+ 4 01	+14 58	+21 52	+23 08	+18 21	+ 8 46	- 3 03	-13 59	-21 44	-23 07
31	-17 12		+ 4 24		+22 00		+18 06	+ 8 25		-14 18		-23 03

Vernal Equinox: March 20th, 6:44 a.m. E.S.T.
Summer Solstice: June 21st, 12:45 a.m. E.S.T.

Autumnal Equinox: September 22nd, 4:18 p.m. E.S.T.
Winter Solstice: December 21st, 12:47 p.m. E.S.T.

To find Sun's Declination in the Atlantic Time Zone (1 hour earlier than E.S.T.), take 1/24 of the difference between Day 1 and Day 2. Add or subtract this figure from Day 2 to find the Declination for Day 2.
If Declination is increasing (N. or S.), *subtract*.
If Declination is decreasing (N. or S.), *add*.

2009 MOONRISE AND MOONSET
BOSTON, MA
Add one hour for Daylight Saving Time, March 8 - November 1

Day	JAN Rise h m	JAN Set h m	FEB Rise h m	FEB Set h m	MAR Rise h m	MAR Set h m	APR Rise h m	APR Set h m	MAY Rise h m	MAY Set h m	JUN Rise h m	JUN Set h m	Day
1	0956	2137	0930	....	0803	2304	0910	0025	1042	0042	1320	0032	1
2	1017	2242	0959	0000	0837	....	1020	0120	1157	0114	1428	0055	2
3	1038	2349	1035	0114	0919	0018	1135	0205	1309	0140	1535	0119	3
4	1101	....	1122	0228	1012	0128	1251	0241	1419	0204	1642	0145	4
5	1127	0100	1221	0338	1117	0231	1406	0311	1528	0227	1747	0216	5
6	1200	0214	1332	0440	1230	0323	1519	0336	1636	0250	1848	0252	6
7	1241	0331	1450	0530	1348	0406	1630	0400	1744	0315	1943	0335	7
8	1335	0447	1612	0610	1506	0440	1740	0423	1851	0343	2031	0425	8
9	1442	0557	1731	0642	1622	0509	1849	0447	1956	0415	2112	0521	9
10	1600	0655	1847	0710	1736	0534	1958	0513	2056	0454	2145	0621	10
11	1722	0740	2001	0734	1848	0558	2105	0542	2149	0539	2213	0723	11
12	1843	0817	2111	0757	1959	0621	2209	0617	2235	0632	2237	0826	12
13	2000	0846	2221	0820	2109	0646	2307	0658	2313	0729	2259	0928	13
14	2114	0911	2329	0845	2217	0713	2357	0746	2344	0830	2320	1030	14
15	2223	0934	....	0913	2322	0744	....	0840	....	0933	2341	1133	15
16	2331	0956	0035	0946	....	0821	0040	0940	0011	1036	....	1238	16
17	....	1019	0137	1025	0023	0904	0115	1042	0034	1139	0004	1346	17
18	0038	1044	0235	1111	0118	0954	0144	1145	0056	1242	0029	1458	18
19	0143	1114	0326	1204	0205	1051	0210	1249	0117	1348	0100	1613	19
20	0247	1148	0410	1302	0244	1152	0233	1353	0140	1456	0139	1729	20
21	0348	1229	0447	1405	0317	1255	0255	1459	0204	1607	0228	1841	21
22	0443	1317	0517	1509	0345	1400	0316	1607	0232	1723	0331	1942	22
23	0531	1412	0543	1614	0409	1505	0340	1717	0307	1840	0445	2032	23
24	0612	1513	0607	1719	0431	1610	0405	1832	0351	1955	0605	2112	24
25	0646	1616	0628	1825	0453	1717	0436	1948	0447	2101	0726	2144	25
26	0715	1721	0649	1932	0515	1826	0514	2104	0554	2156	0844	2211	26
27	0740	1825	0711	2040	0539	1938	0602	2214	0710	2240	0959	2235	27
28	0802	1930	0735	2151	0606	2052	0702	2314	0829	2315	1110	2259	28
29	0823	2035	....	....	0638	2207	0811	....	0946	2344	1219	2323	29
30	0843	2141	....	....	0718	2319	0926	0003	1100	....	1327	2348	30
31	0905	2249	....	....	0809	....	....	....	1211	0009	....	....	31

Day	JUL Rise h m	JUL Set h m	AUG Rise h m	AUG Set h m	SEP Rise h m	SEP Set h m	OCT Rise h m	OCT Set h m	NOV Rise h m	NOV Set h m	DEC Rise h m	DEC Set h m	Day
1	1434	....	1626	0017	1650	0200	1602	0257	1538	0502	1526	0616	1
2	1540	0018	1711	0109	1715	0303	1623	0400	1610	0613	1623	0726	2
3	1642	0052	1749	0206	1737	0406	1646	0504	1649	0726	1731	0829	3
4	1739	0133	1820	0307	1758	0509	1710	0611	1738	0837	1846	0922	4
5	1829	0220	1847	0410	1819	0612	1738	0719	1838	0943	2004	1004	5
6	1911	0314	1910	0512	1841	0716	1811	0830	1947	1040	2121	1039	6
7	1947	0413	1931	0615	1906	0822	1853	0941	2101	1127	2236	1108	7
8	2017	0515	1952	0717	1935	0931	1944	1050	2217	1205	2348	1134	8
9	2042	0618	2013	0820	2010	1041	2046	1152	2332	1237	....	1158	9
10	2104	0720	2035	0924	2053	1151	2156	1245	....	1304	0058	1223	10
11	2125	0822	2101	1030	2148	1258	2310	1328	0044	1330	0208	1249	11
12	2146	0924	2131	1139	2253	1357	....	1404	0156	1354	0317	1319	12
13	2207	1027	2209	1250	....	1448	0026	1434	0307	1419	0425	1353	13
14	2231	1133	2257	1401	0006	1530	0142	1501	0417	1447	0531	1434	14
15	2258	1241	2357	1508	0124	1604	0256	1526	0528	1518	0631	1521	15
16	2332	1352	....	1606	0242	1634	0409	1551	0636	1555	0725	1615	16
17	....	1506	0108	1655	0400	1700	0521	1618	0742	1639	0810	1714	17
18	0014	1618	0226	1734	0516	1726	0634	1647	0840	1729	0848	1816	18
19	0109	1724	0348	1807	0630	1751	0745	1721	0931	1825	0919	1918	19
20	0217	1819	0508	1835	0744	1819	0853	1800	1014	1925	0945	2019	20
21	0334	1904	0626	1901	0856	1849	0956	1846	1049	2027	1009	2120	21
22	0456	1940	0741	1926	1006	1925	1052	1939	1118	2129	1030	2221	22
23	0618	2010	0855	1952	1111	2007	1139	2037	1143	2230	1051	2322	23
24	0737	2036	1007	2020	1210	2055	1218	2137	1206	2331	1112	....	24
25	0852	2101	1116	2052	1302	2149	1250	2239	1227	....	1135	0025	25
26	1004	2125	1223	2129	1345	2248	1318	2341	1248	0033	1201	0130	26
27	1115	2151	1325	2212	1421	2350	1342	....	1311	0136	1233	0239	27
28	1224	2220	1421	2303	1451	....	1404	0043	1336	0242	1313	0350	28
29	1331	2252	1508	2358	1517	0052	1426	0146	1405	0351	1403	0502	29
30	1435	2331	1548	....	1540	0154	1447	0249	1441	0503	1506	0609	30
31	1534	....	1622	0058	....	....	1511	0354	....	....	1739	0708	31

Time meridian 75' W.   0000 is midnight.   1200 is noon.   Standard Time.

2009 MOONRISE AND MOONSET

NEW YORK, NY

Add one hour for Daylight Saving Time, March 8 - November 1

Day	JANUARY Rise h m	JANUARY Set h m	FEBRUARY Rise h m	FEBRUARY Set h m	MARCH Rise h m	MARCH Set h m	APRIL Rise h m	APRIL Set h m	MAY Rise h m	MAY Set h m	JUNE Rise h m	JUNE Set h m	Day
1	1007	2150	0945	....	0819	2312	0929	0030	1059	0050	1331	0044	1
2	1028	2254	1016	0009	0854	....	1039	0126	1212	0122	1438	0108	2
3	1051	....	1053	0121	0938	0024	1153	0211	1323	0150	1544	0133	3
4	1115	0000	1141	0234	1031	0133	1308	0249	1431	0216	1649	0201	4
5	1143	0109	1240	0344	1136	0236	1421	0320	1539	0240	1753	0233	5
6	1217	0222	1351	0445	1249	0329	1532	0347	1646	0304	1854	0310	6
7	1300	0338	1508	0536	1405	0413	1642	0412	1752	0330	1949	0354	7
8	1354	0453	1628	0618	1522	0449	1750	0436	1858	0359	2037	0444	8
9	1502	0602	1746	0651	1636	0519	1858	0501	2002	0433	2118	0540	9
10	1619	0701	1901	0720	1749	0545	2006	0528	2102	0512	2152	0639	10
11	1740	0748	2013	0746	1859	0610	2112	0559	2155	0558	2221	0740	11
12	1859	0825	2122	0810	2009	0635	2215	0635	2241	0651	2247	0841	12
13	2015	0856	2230	0835	2117	0701	2312	0717	2319	0748	2310	0943	13
14	2126	0922	2337	0901	2224	0729	....	0805	2352	0848	2332	1044	14
15	2235	0946	....	0930	2329	0801	0003	0859	....	0949	2354	1146	15
16	2341	1010	0041	1004	....	0839	0046	0958	0019	1051	....	1249	16
17	....	1034	0143	1044	0029	0923	0122	1059	0044	1153	0018	1356	17
18	0047	1101	0241	1130	0123	1014	0152	1201	0107	1255	0045	1507	18
19	0151	1131	0332	1223	0210	1110	0219	1304	0129	1359	0117	1621	19
20	0254	1206	0416	1321	0250	1210	0243	1407	0153	1506	0157	1735	20
21	0353	1248	0453	1422	0324	1312	0306	1512	0218	1617	0247	1846	21
22	0448	1336	0525	1526	0353	1415	0329	1618	0248	1731	0350	1948	22
23	0537	1431	0552	1629	0419	1519	0353	1727	0324	1847	0503	2039	23
24	0618	1531	0616	1733	0442	1624	0421	1840	0409	2000	0623	2120	24
25	0653	1633	0639	1838	0505	1729	0453	1955	0506	2107	0742	2153	25
26	0723	1737	0702	1943	0528	1837	0532	2110	0613	2203	0859	2222	26
27	0749	1840	0725	2050	0553	1947	0621	2220	0728	2247	1012	2247	27
28	0812	1943	0750	2200	0621	2100	0721	2320	0846	2323	1122	2312	28
29	0834	2047	....	....	0655	2213	0830	....	1002	2353	1230	2337	29
30	0856	2152	....	....	0737	2325	0944	0010	1114	....	1336	....	30
31	0919	2259	....	....	0828	....	....	....	1224	0020	....	....	31

Day	JULY Rise h m	JULY Set h m	AUGUST Rise h m	AUGUST Set h m	SEPTEMBER Rise h m	SEPTEMBER Set h m	OCTOBER Rise h m	OCTOBER Set h m	NOVEMBER Rise h m	NOVEMBER Set h m	DECEMBER Rise h m	DECEMBER Set h m	Day
1	1442	0004	1632	0036	1658	0217	1613	0311	1553	0512	1545	0623	1
2	1547	0035	1717	0128	1724	0319	1636	0413	1627	0622	1642	0732	2
3	1648	0110	1755	0225	1747	0421	1659	0516	1707	0733	1749	0835	3
4	1745	0151	1827	0325	1810	0522	1725	0621	1757	0844	1904	0928	4
5	1835	0239	1855	0426	1832	0624	1754	0728	1857	0949	2021	1012	5
6	1918	0333	1920	0528	1855	0727	1829	0838	2006	1047	2136	1048	6
7	1954	0432	1942	0629	1921	0832	1911	0948	2119	1134	2249	1118	7
8	2024	0532	2004	0730	1951	0939	2003	1056	2233	1213	....	1145	8
9	2051	0634	2026	0832	2027	1048	2105	1158	2346	1246	0000	1211	9
10	2114	0735	2050	0935	2112	1157	2214	1251	....	1315	0109	1237	10
11	2136	0836	2116	1040	2207	1303	2327	1335	0058	1341	0218	1305	11
12	2158	0937	2148	1148	2311	1403	....	1412	0208	1407	0326	1335	12
13	2221	1039	2227	1257	....	1455	0042	1444	0317	1434	0433	1411	13
14	2245	1143	2316	1407	0024	1537	0156	1512	0427	1502	0537	1452	14
15	2314	1250	....	1513	0141	1613	0309	1539	0536	1535	0637	1540	15
16	2349	1400	0016	1612	0258	1644	0420	1605	0643	1613	0730	1634	16
17	....	1513	0127	1702	0414	1712	0532	1633	0748	1657	0816	1733	17
18	0033	1624	0244	1742	0528	1738	0643	1703	0846	1748	0854	1833	18
19	0128	1730	0404	1816	0641	1805	0753	1738	0937	1844	0927	1934	19
20	0236	1826	0523	1846	0753	1834	0900	1818	1020	1943	0954	2034	20
21	0353	1912	0639	1913	0904	1906	1002	1905	1056	2044	1019	2134	21
22	0513	1949	0753	1940	1013	1943	1057	1958	1126	2145	1041	2233	22
23	0633	2020	0905	2007	1117	2025	1145	2055	1152	2245	1103	2333	23
24	0751	2048	1015	2036	1216	2114	1225	2155	1216	2345	1125	....	24
25	0904	2114	1124	2109	1308	2208	1258	2256	1239	....	1150	0035	25
26	1015	2139	1230	2147	1351	2306	1326	2357	1301	0045	1217	0140	26
27	1124	2206	1331	2231	1428	....	1352	....	1324	0147	1250	0247	27
28	1232	2236	1426	2322	1459	0007	1415	0058	1351	0252	1331	0357	28
29	1338	2310	1514	....	1526	0108	1437	0159	1421	0400	1422	0508	29
30	1441	2350	1555	0017	1550	0210	1500	0301	1458	0510	1525	0615	30
31	1540	....	1629	0116	....	....	1525	0405	....	....	1756	0714	31

Time meridian 75' W. 0000 is midnight. 1200 is noon. Standard Time.

PHASES OF THE MOON 2009 E.S.T.

● New Moon, ◑ 1st Quarter, ○ Full Moon, ◐ Last Quarter, A in Apogee

P in Perigee, N, S Moon farthest North or South of Equator, E on Equator

January	February	March	April	May	June
E 2 2pm	◑ 2 6pm	◑ 4 3am	P 1 9pm	◑ 1 4pm	○ 7 1pm
◑ 4 7am	N 5 10am	N 4 7pm	◑ 2 10am	E 4 2pm	S 7 9pm
N 9 1am	P 7 3pm	P 7 10am	E 7 8am	○ 8 11pm	A 10 11am
P 10 6am	○ 9 10am	○ 10 10pm	○ 9 10am	S 11 3pm	E 15 9am
○ 10 10pm	E 11 1pm	E 11 midn	S 14 8am	A 13 10pm	◐ 15 5pm
E 15 3am	◐ 16 5pm	S 18 midn	A 16 4am	◐ 17 2am	N 21 10pm
◐ 17 10pm	S 18 4pm	◐ 18 1pm	◐ 17 9am	E 19 2am	● 22 3pm
S 22 9am	A 19 noon	A 19 8am	E 21 6pm	● 24 7am	P 23 6am
A 22 7pm	● 24 10pm	E 25 9am	● 24 10pm	N 25 noon	E 28 midn
● 26 3am	E 26 1am	● 26 11am	P 28 1am	P 25 11pm	◑ 29 6am
E 29 7pm		N 31 10pm	N 28 4am	◑ 30 10pm	
				E 31 6pm	

July	August	September	October	November	December
S 5 3am	S 1 9am	○ 4 11am	E 2 11am	○ 2 2pm	○ 2 3am
○ 7 4am	A 3 8pm	E 5 4am	○ 4 1am	N 5 11am	N 2 7pm
A 7 5pm	○ 5 8pm	◐ 11 9am	N 9 5am	P 7 2am	P 4 9am
E 12 4pm	E 8 9pm	N 12 midn	◐ 11 4am	◐ 9 11am	◐ 8 7pm
◐ 15 5am	◐ 13 2pm	P 16 3am	P 13 7am	E 11 7pm	E 9 midn
N 19 8am	N 15 5pm	E 18 5am	E 15 2pm	● 16 2pm	S 15 10pm
P 21 3pm	P 19 midn	● 18 2pm	● 18 1am	S 18 3pm	● 16 7am
● 21 10pm	● 20 5am	S 24 11pm	S 22 7am	A 22 3pm	A 20 10am
E 25 9am	E 21 7pm	◐ 26 midn	A 25 6pm	◑ 24 5pm	E 23 10am
◑ 28 5pm	◑ 27 7am	A 27 11pm	◑ 25 8pm	E 26 3am	◑ 24 1pm
	S 28 4pm		E 29 7pm		N 30 5am
	A 31 4am				○ 31 2pm

Midnight is the *beginning* of the day.

THE TIDES AND THE MOON

Tides are created on the earth by the pull of gravity from the moon, and to a lesser extent from the sun. Gravitational force between two bodies decreases according to the inverse square of the increasing distance between them: doubling a distance results in one-quarter the gravitational force; tripling, one-ninth. Since the moon's pull is stronger on the earth's near side than it is in the center of the earth, a bulge on the side facing the moon is created. However, it is also true that the moon's force at the center of the earth is stronger than it is on the opposite side of the earth. This means that the center is also pulled away from the opposite side, which is the same thing as saying that the opposite side is pulled away from the center. The net effect then is to create a bulge on the opposite side, and so the moon has created two bulges on the earth. High tides are on the sides of the earth where these two bulges are. The rigid body of the earth does not in fact bulge very much, but the water quite easily responds to the changing force and thus builds up at those two places.

Because the earth rotates, different parts of the earth's surface move into and out of these bulges, creating what we call the ocean tides. The moon is also revolving around the earth in the same direction that the earth rotates, but at a slightly slower pace, taking 24 hours and 50 minutes to reappear above the same part of the earth. Accordingly, the usual sequence is for high and low tides to be, on average, about 6 hours 12 1/2 minutes apart. Importantly, the actual time of high tide does not coincide with the time the moon is directly overhead or underneath. The moon's pull is not enough to lift the oceans directly upward (or else everything would weigh measurably less with the moon overhead); instead, the moon drags the fluid oceans somewhat behind it.

The sun also creates tides on the earth, but even the sun's much larger mass (26 million times that of the moon) can't compete with the moon's being 400 times closer than the sun, and so the moon's effect on the earth is 2.17 times that of the sun.

How much the ocean tides rise and fall depends basically on three conditions. (See Phases of the Moon, p. 234.) First, when the sun and moon are in a line with the earth, their gravitational forces work together to produce a greater range of tide than usual. This occurs both at full moon (when the moon is opposite the earth from the sun) and at new moon (when the moon is between the earth and sun). Note in the Boston High and Low Water Tables that just after the dates for Full Moon and New Moon the highest tides of the month occur. These tides are called "spring tides." When the moon is at right angles to the sun (first quarter and last quarter), the forces are working against each other, and the net result is a lower range of tide than usual. These are called "neap tides." There are about 13 "lunar" months in a year, so we have 26 spring tides and 26 neap tides in the year.

Second, the moon's orbit around the earth is not circular, but elliptical, ranging from 252,000 miles at Apogee (A) down to 221,000 miles at Perigee (P), and so the moon's effect on the earth will be greater at "P" than at "A." Note in Boston High and Low Water Tables how much higher the tide is when the Full Moon is at "P" than when the Full Moon is at "A." The position of the moon along its elliptical orbit is thus very important to the height of the tides.

Third, the moon's orbit about the earth is inclined to the plane of the earth's equator. This means that the moon is directly above the equator only twice a month; the rest of the time the moon is either above the northern hemisphere (northern declination) or the southern hemisphere (southern declination). When the moon is directly above the equator, then the day's two high tides will be about the same. But when the moon is north of the equator (N) or south (S), the two high water marks will be different in height.

This effect is known as "semidiurnal inequality." It occurs because the gravitational force of the moon when it has, say, northern declination, is creating the largest bulge in our northern hemisphere while it is passing above us, but on the other side of the world, at the same time, its secondary bulge is directly opposite us, or in the southern hemisphere. For example, when the moon is over China, its greater effect on our side of the earth is in the S. Atlantic, with a lesser effect on the N. Atlantic.

In summary, the moon's phase is most important, with the highest tides at Full and New Moon; second is the moon's position in its elliptical orbit, tides being highest when the moon is at Perigee; and last is the effect of the moon's declination in creating tides of different heights on the same day.

We thank Nelson Caldwell, Smithsonian Astrophysical Observatory, for his contribution to this article.

Daily Moon Phases 2009

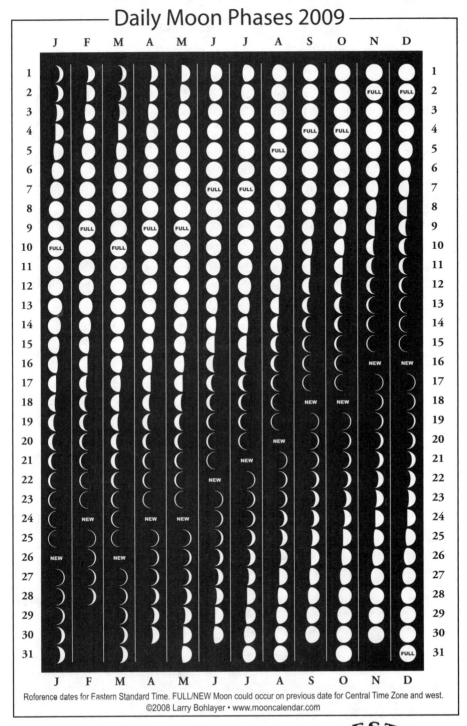

Reference dates for Eastern Standard Time. FULL/NEW Moon could occur on previous date for Central Time Zone and west.
©2008 Larry Bohlayer • www.mooncalendar.com

Larger moon calendars, cards, imprints and astronomy items available from Celestial Products.

800-235-3783

www.mooncalendar.com

Visibility of The Planets, 2009

VENUS is a brilliant object in the evening sky from the beginning of the year until the second half of March, when it comes too close to the Sun for observation. It reappears at the beginning of April as a morning star and can be seen until the start of December, when again it comes too close to the Sun for observation. Venus is in conjunction with Mars on April 18 and June 19 and with Saturn on October 13.

MARS is too close to the Sun for observation until the start of February, when it appears in the morning sky in Sagittarius. Its westward elongation gradually increases as it passes through Capricornus, Aquarius, Pisces, Cetus, into Pisces again, Aries, Taurus, Gemini, Cancer and into Leo, where it can be seen for more than half the night. Mars is in conjunction with Mercury on January 26 and March 1, with Jupiter on February 17, and with Venus on April 18 and June 19.

JUPITER can be seen in the evening sky in Sagittarius at the beginning of January, and then passes into Capricornus during the first week of January, remaining here throughout the year. In the second week of January it becomes too close to the Sun for observation and reappears in the morning sky in early February. After mid-May it can be seen for more than half the night. It is at opposition on August 14, when it is visible throughout the night. From mid-November until the end of the year it can be seen only in the evening sky. Jupiter is in conjunction with Mars on February 17 and with Mercury on Feb. 24.

SATURN rises shortly before midnight in Leo at the beginning of the year, and is at opposition on March 8, when it can be seen throughout the night. From mid-June to early August it is visible only in the evening sky, and then becomes too close to the Sun for observation. In early October it reappears in the morning sky in Virgo and remains in the morning sky until late December. Saturn is in conjunction with Mercury on August 18 and with Venus on October 13.

MERCURY can only be seen low in the east before sunrise, or low in the west after sunset. It is visible in the mornings between January 27 to March 22, May 28 to July 6, and September 28 to October 23. The planet is brighter at the end of each period. It is visible in the evenings between January 1 to 15, April 9 to May 9, July 22 to September 14, and November 22 to December 30. The planet is brighter at the beginning of each period.

Visibility of Planets in Morning and Evening Twilight

	Morning			Evening		
VENUS	April 1	–	December 1	January 1	–	March 24
MARS	February 1	–	December 31			
JUPITER	February 7	–	August 14	January 1	–	January 11
				August 14	–	December 31
SATURN	January 1	–	March 8	March 8	–	August 31
	October 6	–	December 31			

RADIO TELEPHONE INFORMATION – VHF SYSTEM

GENERAL INFORMATION ON THE 156-162 MHz VHF-FM MARINE RADIO

Calling Guidelines: Avoid excessive calling. Make calls as brief as possible. Give name of called vessel first, then "This is (name of your vessel)," your call sign (if you have a Station License), and the word "Over." If station does not answer, delay your repeat call for 2 minutes. At the end of your message, sign off with "This is (your vessel's name)," your call sign, and "Out."

Range and Power: Operation is essentially line-of-sight. Since the elevation of antennas at both communications points extends the "horizon," range may be 20 to 50 miles on a 24-hour basis between a boat and a land station. Effective range between boats will be less because of lower antenna heights. 25 watts is the maximum power permitted.

Interference factor: Most VHF-FM equipment has 6 or more channels, so it is possible to shift to a clear channel. Like the FM in your home radio, the system is practically immune to interference from ignition noise, static, etc., except under unusual conditions.

Channelization: A minimum of 3 channels is required by the FCC. Two are mandatory: Channel 16 (156.800 MHz), the International Distress frequency; and Channel 06 (156.300 MHz), the Intership Safety Frequency. The Coast Guard *strongly recommends* that you have Channel 22A as your third channel.

Channel	Purpose and Comments
16 156.800 MHz	**Distress and Safety**: Ship to Shore and Intership. Guarded 24 hours by the Coast Guard. No routine messages allowed other than to establish the use of a working channel. *See page 239* for Distress calling procedure. **Calling**: Ship to Shore and Intership. Use Channel 16 to establish contact, then switch to a working channel (see below). Calling Channel: New England waters. Commercial and pleasure.
09 156.450 MHz	**Boater Calling:** Commercial and Non-Commercial
06 156.300 MHz	**Intership Safety:** No routine messages allowed. 06 is limited to talking with the Coast Guard and others at the scene of an emergency, and to information on the movement of vessels.
22A 157.100 MHz (**21** in Canada) 161.65 MHz	**Maritime Safety Information** channel. Not guarded by the CG, but after a vessel makes contact with the CG for non-distress calls on Channel 16, they will tell you to switch to and use *only* 22A for communicating. Channel 22A is also used for CG weather advisories and Notices to Mariners. *Times* of these broadcasts given on Channel 16.
12, 14, 20A, 65A, 66A, 73, 74, 77	**Ship to Shore and Intership:** Port operations, harbormasters, etc. (Your electronics dealer should have local frequencies.)
08, 67, 88A	**Commercial (intership only):** For ocean vessels, dredges, tugs, etc.
07A, 10, 11, 18A, 19A, 79A, 80A	**Commercial only**
13	**Intership Navigation Safety:** (bridge to bridge). Ships > 20 m length maintain a listening watch on this channel in US waters.
68, 69, 71, 72, 78A	**Ship to Shore and Intership, Pleasure craft only:** Shore stations, marinas, etc. The best channels for general communication.
70 156.525	**Digital Selective Calling (DSC).** Special equipment required.

MARINE EMERGENCY AND DISTRESS CALLS

Speak Slowly and Clearly. Use: Ch. 16 (156.800 MHz) or 2182 kHz or CB Radio Ch. 9

I. DISTRESS SIGNAL (top priority)

If you are in DISTRESS (i.e. when threatened by grave and imminent danger) transmit the International DISTRESS call on either 2182 kHz or 156.800 MHz (Channel 16) — "MAYDAY MAYDAY MAYDAY, THIS IS (Your vessel's name and call sign repeated three times)"

IF CALLING FROM A VESSEL IN TROUBLE — give:

1. WHO you are (Your vessel's name, registration number or call sign).
2. WHERE you are (Your vessel's position in latitude/longitude or true bearing and distance in nautical miles from a known geographical point. Local names known only in the immediate vicinity are confusing).
3. WHAT is wrong (Nature of Distress or Difficulty, if not in Distress).
4. Kind of assistance desired.
5. Number of persons aboard and condition of any injured.
6. Present seaworthiness of your vessel.
7. Description of your vessel (length, type cabin, masts, power, color of hull, superstructure and trim).
8. Your listening frequency and schedule.

IF CALLING WHILE OBSERVING ANOTHER VESSEL IN DIFFICULTY — give:

1. Your position and the bearing and distance of the vessel in difficulty.
2. Nature of Distress or Difficulty.
3. Description of the vessel in Distress or Difficulty, (see item 7 above).
4. Your intentions, course and speed, etc.
5. Your radio call sign, name of your vessel, listening frequency and schedule.

If there is no immediate response, repeat appropriate messages above; if still no response, you may send on any other available frequency until you make contact.

IF YOU HEAR A MAYDAY CALL — Immediately discontinue any transmission. Note details in your radio log right away. Do not make any transmission on this DISTRESS channel until MAYDAY condition is lifted by the Coast Guard, unless you are in a position to be of assistance.

II. URGENCY SIGNAL (second in priority)

If you have an URGENT message to send (threat to a vessel's safety or to someone on board, overboard or within sight), use the same procedure as above but say the word "PAN" three times. "PAN" (pronounced "PAWN") is also used as a warning signal that a Distress Signal may be sent out at a later stage. Radio telegraph signal is – (T) – (T) – (T)

III. SAFETY SIGNAL (third priority)

If you wish to report navigation or weather warnings (ice, derelicts, tropical storms, etc.) use the same procedure as above but say the word "SECURITY" (pronounced SAY-CUR-I-TAY) three times. Radio telegraph signal is – • • – (X) – • • – (X) – • • – (X)

OTHER DISTRESS SIGNALS — See p. 6.. Flag Hoist "NC"; Flying ensign upside down; Hoisting any square flag with a ball above or below it; Flames in a bucket; Red flare or rocket; Orange smoke; Dye marker; Flashlight • • • (S) – – – (O) • • • (S); Firing a gun repeatedly; Fog horn sounded continuously; Raising and lowering arms slowly. NOTE: 3 aerial or hand-held flares now mandatory for night use for all vessels over 16'.

U.S. COAST GUARD STATIONS – MAINE TO KEY WEST
(monitoring Ch. 16, 2182 kHz, 156.800 MHz)

1st District – Boston – (800) 848-3942

Eastport, ME (207) 853-2845
Jonesport, ME (207) 497-2134
Southwest Harbor, ME (207) 244-4250
Rockland, ME (207) 596-6667
Boothbay Hbr., ME (207) 633-2644
S. Portland, ME (207) 767-0364
Portsmouth, NH (603) 436-4415
Merrimac-Newburyport, MA (978) 465-0731
Gloucester, MA (978) 283-0704
Boston, MA (617) 223-3224
Point Allerton-Hull, MA (781) 925-0166
Scituate, MA (781) 545-3801
Cape Cod Canal-E. Entr. (508) 888-0020
Provincetown, MA (508) 487-0077
Chatham, MA (508) 945-3830
Brant Pt.-Nantucket, MA (508) 228-0388
Woods Hole, MA (508) 457-3219
Menemsha, MA (508) 645-2662
Castle Hill, Newport, RI (401) 846-3676
Point Judith, RI (401) 789-0444
New London, CT (860) 442-4471
New Haven, CT (203) 468-4498
Sector NY, NY (718) 354-4353
Fire Island, NY (631) 661-9101
New York Station, NY (718) 354-4149
Eatons Neck, NY (631) 261-6959
Kings Point, NY (516) 466-7135
Jones Beach, NY (516) 785-2995
Moriches, NY (631) 395-4400
Shinnecock, NY (631) 728-0078
Montauk, NY (631) 668-2773
Sandy Hook, NJ (732) 872-3429

5th District – Portsmouth, VA (757) 398-6486

Manasquan Inlet, NJ (732) 899-0887
Shark River, NJ (732) 776-6730
Barnegat, NJ (609) 494-2661
Atlantic City, NJ (609) 637-2230
Cape May, NJ (609) 898-6995
Great Egg, NJ (609) 399-0144
Indian River Inlet, DE (302) 227-2440
Ocean City, MD (410) 289-1905

5th District, cont.

Chincoteague, VA (757) 336-2874
Little Creek-Norfolk, VA (757) 464-9371
Wachapreague, VA (757) 787-9526
Portsmouth, VA (757) 483-8527
St. Inigoes, MD (301) 872-4344
Crisfield, MD (410) 968-0323
Annapolis, MD (410) 267-8108
Oxford, MD (410) 226-0581
Curtis Bay-Baltimore, MD (410) 576-2625
Stillpond, MD (410) 778-2201
Cape Charles, VA (757) 331-2000
Milford Haven, VA (804) 725-3732
Oregon Inlet, NC (252) 441-6260
Hatteras Inlet, NC (252) 986-2176
Hobucken, NC (252) 745-3131
Ocracoke, NC (252) 928-4731
Fort Macon, NC (252) 247-4581
Elizabeth City, NC (252) 335-6086
Wrightsville Beach, NC (910) 256-4224
Emerald Isle, NC (252) 354-2719
Oak Island, NC (910) 278-1133

7th District – Miami, FL (305) 415-6751

Georgetown, SC (843) 546-2052
Charleston, SC (843) 724-7629
Tybee, GA (912) 786-5440
Brunswick, GA (912) 267-7999
Mayport, FL (904) 564-7592
Ponce de Leon Inlet, FL (386) 428-9085
Cape Canaveral, FL (321) 853-7601
Fort Pierce Inlet, FL (772) 464-6100
Lake Worth Inlet, FL (561) 840-8503
Ft. Lauderdale, FL (954) 927-1611
Miami Beach, FL (305) 535-4368
Islamorada, FL (305) 664-4404
Marathon, FL (305) 743-1945
Key West, FL (305) 292-8862

Canada-Nova Scotia & New Brunswick

Department of Fisheries and Oceans Canada
200 Kent St., 13th fl., Station 13228
Ottawa, Ontario, Canada K1A OE6
(613) 993-0999
Maritimes - Dartmouth, N.S. (902) 426-3760

The Coast Guard Navigation Information Service (NIS)
Access All USCG Districts: Local Notices to Mariners (without enclosures) via the Internet.
If you have problems call NIS Watchstander, 24 hrs. a day - (703) 313-5900
INTERNET
USCG - www.navcen.uscg.gov / Canada - www.notmar.gc.ca/
USCG Suspected Terrorist Incidents Hotline, National Response Center (NRC) 1-800-424-8802
USCG Maritime Safety Line 1-800-682-1796

To update nautical charts see website: http://chartmaker.ncd.noaa.gov/ and click on -Critical Corrections>Chart Updates. This website includes information from NOS (National Ocean Service), NGA (National Geospatial-Intelligence Agency, and the US Coast Guard Local Notice to Mariners. This website must not be viewed as a substitute for either the USCG LNM or the NGA NTM. Aid to navigation changes and other important information published in the USCG and NGA notices are available at http://www.navcen.uscg.gov/lnm/ and http://www.nga.mil/portal/ site/maritime/. Canadian Notices to Mariners are available at http://www.notmar.gc.ca/

Halifax to Key West

Station		Channel	MHz	kHz	
CANADA					**Atlantic Time**
Sydney, NS	VCO	21B	162.4	—	Continuous reports
			162.475		
			162.55		
			—	2749	0340, 1040, 1710, 2040
Halifax, NS	VCS	21B	162.4	—	Continuous reports
			162.475		
			162.55		
			—	2749	0410, 1140, 1610, 2110
UNITED STATES					**Eastern Time**
SW Harbor, ME	NMF-44	22A	157.1	2670	0635, 1835
Portland, ME		22A	157.1	2670	0605, 1805
Boston, MA	NMF	22A	157.1	2670	0535, 1735
Woods Hole, MA		22A	157.1	—	0505, 1705
				2670	1140, 2340
Long Is. Sound, NY		22A	157.1	—	0620, 1820
Activities, NY		22A	157.1	—	0550, 1750
Moriches, NY	NMY-41	22A	157.1	2670	0710, 1910
Atlantic City, NJ		22A	157.1	—	0603, 1803
				2670	0603, 1703
Cape May, NJ		22A	157.1	—	0603, 1803
				2670	0603, 1703
Baltimore, MD		22A	157.1	—	0705, 2030
Eastern Shore, MD		22A	157.1	—	0645, 2100
				2670	0903, 2133
Hampton Roads, VA		22A	157.1	—	0620, 2130
				2670	0833, 2103
Cape Hatteras, NC		22A	157.1	—	0555, 2000
				2670	0803, 2033
Charleston, SC		22A	157.1	—	0700, 1700
				2670	1120, 2320
Fort Macon, GA		22A	157.1	2670	0733, 2003
Mayport, FL		22A	157.1	—	0715, 1715
				2670	0120, 1320
Miami, FL	NCF	22A	157.1	—	0730, 1730
				2670	1050, 2250
Key West, FL	NOK	22A	157.1	—	0700, 1700
BERMUDA					**Bermuda Time**
Bermuda Hbr. Radio		27	—	2582	0035, 0435, 0835, 1235, 1635, 2035
		02	162.4		Continuous reports

NOAA RADIOFAX SERVICE

National Weather Service provides a Radiofacsimile service for weather and charts prepared by the U.S. Coast Guard Communication Station, Boston (NMF). For more specific information, contact the **Ocean Prediction Ctr.**, www.opc.ncep. noaa.gov/. For information on receiving radiofax charts via email go to http://weather.noaa.gov/fax/marine.shtml

MARINE COMMUNICATIONS

Emergencies: The Coast Guard is required to monitor Channel 16; they are not required to answer the telephone. In an emergency, use your VHF radio to call the Coast Guard on Channel 16 (156.80 MHz). Digital Selective Calling (DSC) is on Channel 70 on your VHF. The Coast Guard urges, in the strongest terms possible, that you take the time to interconnect your GPS and DSC-equipped radio. Doing so may save your life in a distress situation!

DSC: As part of the Global Maritime Distress and Safety System (GMDSS), Rescue 21 is the Coast Guard system that provides the emergency response made possible by DSC-equipped VHF radios. It has been active for a while, and if you don't yet have a DSC-VHF radio, you need to know the significant advantages it offers.

Rescues initiated by DSC-equipped radios are far quicker and more successful. Why? With the push of one button, an automated digital distress alert is sent to other DSC-equipped vessels and rescue facilities. This transmission includes your vessel's unique, 9-digit MMSI (Marine Mobile Service Identity) number, which contains your vessel's description for easier identification by response teams. If connected to a compatible GPS, the signal will give your vessel's latitude and longitude for faster and more efficient assistance or rescue. For more information go to: www. navcen.uscg.gov/marcomms/

Domestic users (non-commercial) who do not travel outside of the US can be issued an MMSI number without applying for an FCC Station License. You can register for an MMSI at online at www.BoatUS.com, or www.seatow.com/boating_safety/.

Non-emergency: Near shore (range will vary) a cell phone can be used successfully for non-emergency calls. The usable distance assumes line-of-sight, so an antenna which is higher may help communicate farther. That distance may be less where there are fewer cell towers.

MARINE WEATHER FORECASTS

VHF-FM, NOAA All-Hazards Weather Radio - Continuous broadcasts 24 hours a day are provided by the National Weather Service with taped messages repeated every 4-6 minutes. These are updated every 3-6 hours and include weather and radar summaries, wind observations, visibility, sea conditions and detailed local forecasts. NOAA VHF-FM broadcasts can be received 20-40 miles from transmitting site.

	MHz		MHz
WX-1	162.550	WX-5	162.450
WX-2	162.400	WX-6	162.500
WX-3	162.475	WX-7	162.525
WX-4	162.425		

It is recommended that a separate NOAA Weather Radio receiver be carried aboard so that mariners may maintain a simultaneous watch on NOAA Weather radio and marine VHF channels.

TIME SIGNALS

Bureau of Standards Time Signals: WWV, Ft. Collins, Col., every min. on 2500, 5000, 10000, 15000, 20000, 25000 kHz. **Canadian Time Signals:** CHU, (frequently easier to get than WWV) 45° 17' 47" N, 75° 45' 22" W. Continuous transmission on 3330 kHz, 7335 kHz and 14670 kHz. For more information on time visit the following websites. http://tf.nist.gov/timefreq/, http://nist.time.gov/

Omission of a tone indicates the 29th second of each minute. The new minute is marked by the full tone *immediately* following the voice announcement. Five sets of two short tones mark the first five seconds of the next minute. The hour is identified by a pulse of one full second followed by 12 seconds of silence.

HYPOTHERMIA
and Cold Water Immersion
What You Need To Know

It is not uncommon for a boater to fall off a boat or dock. Most are rescued immediately. However, when rescue is delayed and conditions are present which threaten survival, all who go boating should know what to do.

Hypothermia is a state of low body core temperature - specifically below 95° F. This loss of body heat may be caused by exposure to cold air or cold water. Since water conducts heat away 25 times more quickly than air, time is critical for rescue. There are many variables beyond water temperature which combine to determine survival time: whether a life jacket is on, body size and composition, type of clothing, movement in the water, etc. Wearing a Personal Floation Device (PFD) greatly extends survival time by keeping your head above water and by allowing you to float without expending energy.

What a person in the water should do:
1. If at all possible, get out of the water, or at least grab hold of anything floating. If the boat is swamped, stay with it and crawl as far out of the water as possible.
2. Do not try to swim, unless a boat or floating object is very nearby and you are certain you can get to it.
3. Control heat loss by keeping clothing on as partial insulation. In particular, keep the head out of water. To protect the groin, sides, and chest from heat loss, use the H.E.L.P. (heat escape lessening position), a fetal position with hands clasped around the legs, which extends survival time.
4. Conserve energy by remaining as still as possible. Physical effort promotes heat loss. Swimming, or even treading water, reduces survival time.

The states of hypothermia:
1. Mild: the victim feels cold and exhibits violent shivering, lethargy and slurred speech
2. Medium: loss of some muscle control, incoherence, stupor, exhaustion
3. Severe: unconsciousness, respiratory distress, possible cardiac arrest

What a rescuer should do:
1. Move the victim to a warm place, position on his/her back, and check breathing and heartbeat.
2. Start CPR if necessary (see pp.244-245).
3. Carefully remove wet clothing, cutting it away if necessary.
4. Take steps to raise the body temperature gradually: cover the victim with blankets or a sleeping bag, and apply warm moist towels to the neck, chest, and groin.
5. Provide warm oral fluids and sugar sources after uncontrolled shivering stops and the patient shows evidence of rewarming.

What NOT to do:
1. Do not give alcohol, coffee, tea, or nicotine. If the victim is not fully conscious, do not attempt to provide food or water.
2. Do not massage arms or legs or handle the patient roughly, as this could cause cold blood from the periphery to circulate to the body's core, which needs to be warm first.

see **Emergency First Aid, pp. 244-245**

EMERGENCY FIRST AID

These are guidelines to be used only when professional help is not readily available.

Good Samaritan laws were enacted to encourage people to help others in emergency situations. Laws vary from state to state, but all require that the caregiver use common sense and a reasonable level of skill.

Before giving care to a conscious victim you must first get consent. If the victim does not give consent call 911. Consent may be implied if a victim is unconscious, confused, or seriously ill.

Prevent disease transmission by avoiding contact with bodily fluids, using protective equipment such as disposable gloves and thoroughly washing hands after giving care.

PRIMARY ASSESSMENT
Check for: 1. Unresponsiveness 2. Breathing - Look, listen and feel. 3. Pulse - If pulse and breathing are present, check for and control any severe bleeding.

* If no pulse or signs of breathing, begin CPR for 60 seconds, then call for help.

* If pulse is present but no breathing, begin Rescue Breathing.

* If airway is obstructed, do Heimlich to clear airway. Do not use Heimlich if drowning is suspected; go to Rescue Breathing.

CPR* - Use only when there is no sign of breathing and no pulse. Roll victim onto back as a unit, being careful to keep spine in alignment, and open airway. Tilt head back and lift chin. Look, listen and feel for breath for 3-5 seconds. If no breath, keep head tilted back, pinch nose shut, seal your lips tight around victim's mouth, GIVE 2 FULL BREATHS for 1 to 1 ½ seconds each checking for chest rise. Feel for pulse at side of neck for 5-10 seconds. If no pulse, locate notch at lower end of breast bone, place heel of other hand on breastbone next to fingers, remove hand from notch and put it on top of other hand, keeping fingers off of chest. Position shoulders over hands and compress breastbone 1 ½ to 2 inches. GIVE 30 COMPRESSIONS in approx. 20 seconds. GIVE 2 FULL BREATHS. Do 4 cycles of 30 compressions and 2 breaths. Recheck pulse after 1 minute. If no pulse, give 2 full breaths and continue CPR. Phone for help.

* To perform CPR you should be trained. Courses are available through the American Red Cross and the American Heart Association. If you are unable to do Rescue Breathing the American Heart Association suggests that Chest Compressions alone can be effective in helping to circulate oxygenated blood through the body.

RESCUE BREATHING - no obstruction. Pulse present, unresponsive, no breathing. Roll victim onto back and open airway. Tilt head back and lift chin. Look, listen and feel for breath for 3-5 seconds. If no breath, keep head tilted back, pinch nose shut, seal your lips tight around victim's mouth, GIVE 2 FULL BREATHS, GIVE NORMAL BREATH over 1 second until chest rises for 1 to 1 ½ seconds each. Feel for pulse at side of neck for 5-10 seconds. If pulse present, begin Rescue Breathing for 1 minute. Phone for help. Keep head tilted back, pinch nose. Give 1 breath every 5 seconds. Look, listen and feel for breath between breaths. RECHECK PULSE EVERY MINUTE. If victim has pulse but is not breathing, continue rescue breathing. If victim has no pulse or breath, go to CPR.

OBSTRUCTED AIRWAY - If victim cannot cough, breathe, or speak, use HEIMLICH. If drowning suspected, use Rescue Breathing. Do not try to clear water from lungs. Roll to side if vomiting occurs so victim won't choke.

HEIMLICH - If victim is conscious, stand behind him. Wrap your arms around victim's waist. Place your fist (thumbside) against the victim's stomach in the midline, just above the navel and well below the rib margin. Grasp your fist with other hand. Press into stomach with a quick upward thrust. If victim is unconscious, lay victim on back, do finger sweep on adult (on child only if you can see object). Attempt rescue breathing. If airway remains blocked, give 6-10 abdominal thrusts and repeat as necessary.

BLEEDING - Apply pressure directly over wound with a dressing, until bleeding stops or until EMS rescuers arrive. If possible, press edges of a large wound together before using dressing and bandage. If bleeding continues, apply additional bandages and continue to maintain pressure. If possible, elevate wounded area, apply ice wrapped in cloth to wound and keep the patient warm.

SHOCK - Confused behavior, rapid pulse and breathing, cool moist skin, blue tinge to lips and nailbeds, weakness, nausea and vomiting, etc. Keep patient lying down with legs elevated. Remove wet clothing. Maintain normal body temperature. Do not give victim food or drink.

BURNS, SCALDS - No open blisters: Use cool water, then cover with a moist sterile dressing. Open blisters - Heat: Cover with dry sterile dressing. Do not put water on burn or remove clothing sticking to burn. Treat for shock. Open blisters - Chemical: Flush all chemical burns with water for 15 to 30 minutes. Remove all clothing on which chemical has spilled. Cover with dry sterile dressing and treat for shock. Eyes: Flush with cool water only for 15 minutes.

FRACTURES - Do not move victim or try to correct any deformity. Immobilize the area. If bone penetrates the skin use a sterile dressing and control bleeding before splinting. Splint a broken arm to the trunk or a broken leg to the other leg. A padded board or pole can be used along the side, front or back of a broken limb. A pillow or a rolled blanket can be used around the arm or leg. For an injured shoulder put a pillow between the arm and chest and bind arm to body. For an injured hip, place pillow between legs and bind legs together.

HEAD, NECK and **SPINE INJURIES** - Do not move victim or try to correct any deformity. Stabilize head and neck as you found them.

POISONING - Call for help immediately. Contact Poison Control Center 1-800-222-1222. Have poison container available. Keep syrup of ipecac and activated charcoal available, but do not administer unless advised to do so. Antidotes listed on label may be wrong.

HEAT PROSTRATION - Strip victim. Move to shaded area. Wrap in cool, wet sheet. Treat for shock.

EXPOSURE TO COLD - Provide a warm dry bunk and warm drink, not coffee, tea or alcohol. Frostbite: Rewarm slowly, beginning with the body core rather than the extremities. Elevate and protect affected area. Do not rub frozen area, break blisters or use dry heat to thaw. Treat for shock. *see* Hypothermia and Cold Water Immersion article p. 243.

SUNBURN – Treat heat prostration if present. Take the heat out of the skin by using a cool damp cloth laid over the area. Do not apply ice as this may damage the skin further. Painkillers like acetaminophen (Tylenol) or ibuprofen may be used for pain. Use topical lotions to keep the skin moist and reduce dehydration. Those containing Aloe work well. If the skin is blistering, prevent secondary infection by keeping the area clean and by applying an antibacterial cream. Rest, keep hydrated and seek medical help if area does not improve.

FORECASTING
with Wind Direction and Barometric Pressure

Wind Dir.	Pressure	Trend	Likely Forecast
SW to NW	30.1-30.2	Steady	Fair, little temp. change
	30.1-30.2	Rising rapidly	Fair, perhaps warmer with rain
	30.2+	Steady	Fair, no temp. change
	30.2+	Falling	Fair, gradual rise in temp.
S to SW	30.0	Rising slowly	Clearing, then fair
S to SE	30.2	Falling rapidly	Increasing wind, rain to follow
S to E	29.8	Falling rapidly	Severe NE gale, heavy rain or snow
SE to NE	30.1-30.2	Falling slowly	Rain
	30.1-30.2	Falling rapidly	Increasing wind and rain
	30.0	Falling slowly	Rain continuing
	30.0	Falling rapidly	Rain, high wind, then clearing and cooler
E to NE	30.0+	Falling slowly	Rain with light winds
	30.1	Falling rapidly	Rain or snow, increasing wind
Shifting W	29.8	Rising rapidly	Clearing and cooler

KEYS TO PREDICTING THE WEATHER

Note: A few of our readers may miss the original article, called Fitzroy's Barometer Instructions, but in an effort to save space and make its wisdom more readable, the Editors offer the revision below. We hope Admiral Fitzroy, a pioneer in weather observation and forecasting, would forgive the changes in the light of improvements to the clarity of the content.

The most important point to remember about barometric pressure is that the trend (up, down, or steady) and the rate of change are far more predictive of coming weather than the position of the pointer at any one time. Tap your barometer periodically; the pointer's direction should indicate the pressure trend. Pay little attention to the words Stormy, Rain, Change, Fair, Very Dry; they are traditional, decorative, and often inaccurate.

In addition to changes in barometric pressure, the state of the air (cool, dry, warm, moist) and the appearance of the sky foretell coming weather. See Weather Signs on the next page.

A handy saying has much truth in it: Long foretold – long last; short notice – soon past. Slow changes last longer; sudden changes are quickly over. A steady barometer with dry air indicates continuing fine weather. A rapid rise or fall of barometric pressure indicates unsettled or stormy weather for a short period of time.

WEATHER SIGNS IN THE SKY

Signs of Good Weather

- A gray sky in the morning or a "low dawn" – when the day breaks near the horizon, with the first streaks of light very low in the sky – brings fair weather.
- Light, delicate tints or colors with soft, undefined clouds accompany fine weather.
- Seabirds flying out early and far to sea suggest moderate wind and fair weather.
- A rosy sky at sunset, whether clear or cloudy: "Red sky at night, sailor's delight."

Signs of Bad Weather

- "Red sky at morning, sailor take warning." Poor weather, wind and maybe rain.
- A "high dawn" – when the first streaks of daylight appear above a bank of clouds – often precedes a turn for worse weather.
- Light scud clouds driving across higher, heavy clouds show wind and rain.
- Hard-edged, inky clouds foretell rain and strong wind.
- Seabirds hanging over the land or headed inland suggest coming wind and rain.
- Remarkable clearness of atmosphere near the horizon, when distant hills or vessels are raised by refraction, are signs of an Easterly wind and indicate coming wet weather.

Signs of Wind

- Soft-looking, delicate clouds indicate light to moderate wind.
- Wind is suggested by hard-edged, oily-looking, ragged clouds, or a bright yellow sky at sunset.
- A change in wind is indicated by high clouds crossing the sky in a different direction from that of lower clouds.
- Increasing wind and possibly rain are preceded by greater than usual twinkling of stars, indistinctness of the moon's horns, "wind dogs" (fragments of rainbows) seen on detached clouds, and the rainbow.
- "First rise after very low, indicates a stronger blow."
- "When the wind shifts against the sun, trust it not, for back it will run."

Dew Point and Humidity Afloat

Some sage said that there is no place more damp than a boat - in the driest of weather. Boats are surrounded by water, leak water, carry water and ice, and whatever corners remain dry for a while soon succumb to the circulation of damp air down below. The only thing dry on the boat might be the wit of the skipper proclaiming during a monsoon that "It's a lovely day."

So why bother with humidity and dew point? Simple: it's wise to know your enemy. These two related weather factors, if ignored, can ruin your day. Understanding them can increase your boating enjoyment and make you look wise.

Relative humidity (RH) is measure of the air's capacity to hold water vapor at a certain temperature. At higher temperatures the air can hold more moisture. A 50%RH at 60°F is pleasant, but 50%RH at 80°F is unpleasant because the air is holding far more moisture. If you don't have a hygrometer (humidity indicator) or sling psychrometer, listen to the marine forecast. High humidity can make everyone uncomfortable.

Dew Point is the temperature to which air must be cooled for suspended water vapor to condense into (visible) water. Fog and rain are examples. Dew on the deck in the morning means the night temperatures were low enough that the air could no longer hold all its daytime moisture. If the dew point and air temperature are quite close, then fog is more likely, and you might wait before heading to your next destination. Dew points under 55°F are comfortable, but those above 64°F are sticky to oppressive.

How to measure dew point? Marine weather forecasts often give this figure. There are some handheld digital instruments for under $100 which measure temperature, humidity, and dew point. If the outside temperature is 75°F and the dew point is in the 50s, go boating. If there's little spread between air temperature and dew point, you might want to postpone your voyage.

The table below is for those who have a sling psychrometer to measure dry-bulb and wet-bulb temperatures. A dry bulb reading of 70°F and a wet-bulb reading that is 4°F lower yields a dew point of 64°F, which means uncomfortably damp with possible fog. Periodic measurements which show an increase in the difference between dry- and wet-bulb readings mean fog should dissipate and visibility increase.

Sling Psychrometers and Hygrometers are available at www.robertwhite.com.

DEW POINT

Dry-bulb temp. F	Difference between dry-bulb and wet-bulb temperatures														Dry-bulb temp. F
	1°	2°	3°	4°	5°	6°	7°	8°	9°	10°	11°	12°	13°	14°	
+50	+48	+46	+44	+42	+40	+37	+35	+32	+29	+25	+21	+17	+12	+5	+50
52	50	48	46	44	42	40	37	35	32	29	25	21	17	11	52
54	52	50	49	47	44	42	40	37	35	32	28	25	21	16	54
56	54	53	51	49	47	45	42	40	37	35	32	28	25	21	56
58	56	55	53	51	49	47	45	43	40	38	35	32	28	25	58
+60	+58	+57	+55	+53	+51	+49	+47	+45	+43	+40	+38	+35	+32	+28	+60
62	60	59	57	55	54	52	50	48	45	43	41	38	35	32	62
64	62	61	59	57	56	54	52	50	48	46	43	41	38	35	64
66	64	63	61	60	58	56	54	52	50	48	46	44	41	39	66
68	67	65	63	62	60	58	57	55	53	51	49	46	44	42	68
+70	+69	+67	+66	+64	+62	+61	+59	+57	+55	+53	+51	+49	+47	+45	+70
72	71	69	68	66	64	63	61	59	58	56	54	52	50	47	72
74	73	71	70	68	67	65	63	62	60	58	56	54	52	50	74
76	75	73	72	70	69	67	66	64	62	61	59	57	55	53	76
78	77	75	74	72	71	69	68	66	65	63	61	59	57	55	78
+80	+79	+77	+76	+74	+73	+72	+70	+68	+67	+65	+64	+62	+60	+58	+80
82	81	79	78	77	75	74	72	71	69	67	66	64	62	61	82
84	83	81	80	79	77	76	74	73	71	70	68	67	65	63	84
86	85	83	82	81	79	78	76	75	74	72	70	69	67	66	86
88	87	85	84	83	81	80	79	77	76	74	73	71	70	68	88
+90	+89	+87	+86	+85	+84	+82	+81	+79	+78	+76	+75	+73	+72	+70	+90
92	91	89	88	87	86	84	83	82	80	79	77	76	74	73	92
94	93	92	90	89	88	86	85	84	82	81	79	78	76	75	94
96	95	94	92	91	90	88	87	86	84	83	82	80	79	77	96
98	97	96	94	93	92	91	89	88	87	85	84	82	81	80	98
+100	+99	+98	+96	+95	+94	+93	+91	+90	+89	+87	+86	+85	+83	+82	+100

BEAUFORT SCALE

Wind Speed in knots (33' above sea level)	Wind Condition	Sea Conditions	Approx. Wave ht. (in ft.)	Force
0-1	**Calm**	Smooth, mirror-like sea.	–	**0**
1-3	**Light Air**	Scale-like ripples; no foam crests.	¼	**1**
4-6	**Light Breeze**	Short wavelets; glassy crests; non-breaking.	½	**2**
7-10	**Gentle Breeze**	Large wavelets; glassy crests, some breaking; occasional white foam.	2	**3**
11-16	**Moderate Breeze**	Small waves becoming longer; frequent white foam crests.	3½	**4**
17-21	**Fresh Breeze**	Moderate waves with more pronounced long form; many white foam crests; some spray.	6	**5**
22-27	**Strong Breeze**	Large waves form; white foam crests everywhere; probably more spray.	9½	**6**
28-33	**Near Gale**	Sea heaps up; white foam from breaking waves is blown in streaks with the wind.	13½	**7**
34-40	**Gale**	Moderately high waves with greater length; edges of crests break into spindrift; foam is blown in well-defined streaks with the wind.	18	**8**
41-47	**Strong Gale**	High waves; crests of waves start to topple and roll over; spray may affect visibility.	23	**9**
48-55	**Storm**	Very high waves; overhanging crests; resulting foam is blown in dense white patches with the wind; sea surface takes on a whiter look; the tumbling of the sea becomes heavy and shock-like; visibility affected.	29	**10**
56-63	**Violent Storm**	Exceptionally high waves; sea covered with long white patches of foam blown in direction of wind; all wave crests are blown into froth; visibility affected.	37	**11**
64+	**Hurricane**	Air is filled with spray and foam; sea is completely white with driving spray; visibility seriously affected.	45	**12**

Note: In enclosed waters or near land with an offshore wind, wave heights will be less, the waves steeper and not so long. In many tidal waters wave heights are apt to increase considerably in a very short time and conditions can be more dangerous near land than in the open sea.

WIND CHILL FACTOR

Wind Speed (mph)	Temperature in Fahrenheit					
	+30	+20	+10	0	-10	-20
10	+21	+ 9	- 4	- 16	- 28	- 41
20	+17	+ 4	- 9	- 22	- 35	- 48
30	+15	+ 1	- 12	- 26	- 39	- 53
40	+13	- 1	- 15	- 29	- 43	- 57
50	+12	- 3	- 17	- 31	- 45	- 60

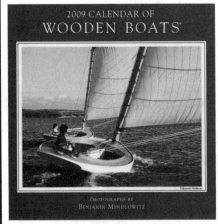

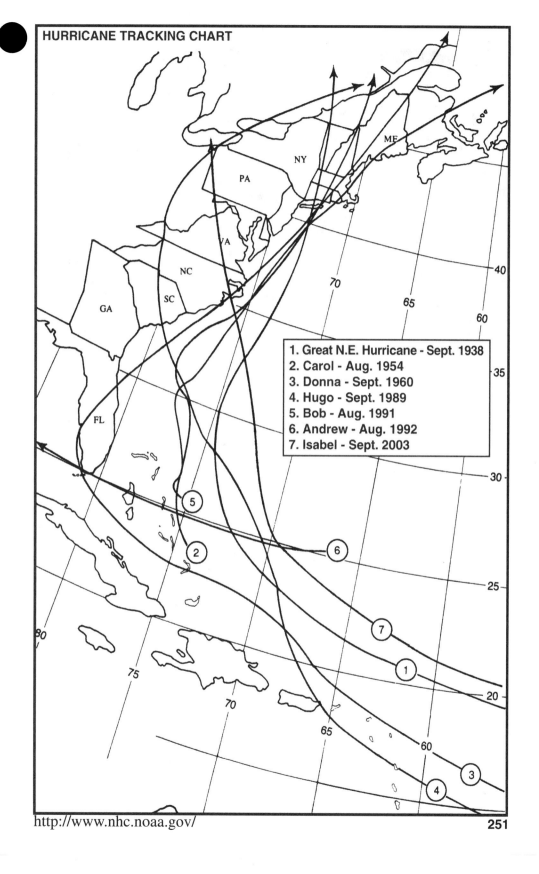

HURRICANE TRACKING CHART

1. Great N.E. Hurricane - Sept. 1938
2. Carol - Aug. 1954
3. Donna - Sept. 1960
4. Hugo - Sept. 1989
5. Bob - Aug. 1991
6. Andrew - Aug. 1992
7. Isabel - Sept. 2003

Announcing a STORY CONTEST

For the 2010 ELDRIDGE we are accepting entries of stories, fictional or factual, in which tide or current plays a role. The winning entry will be published in our 2010 edition, and the author will receive $200. Submitted material becomes the property of the Publishers and will not be returned.

Here are the rules:

1. Submit your entry in writing (typed) by mail to:
   Eldridge Tide and Pilot Book
   Story Contest
   P.O. Box 775
   Medfield, MA 02052
2. Deadline: August 1, 2009
3. Winner Notification: October 1, 2009
4. Limit: 600 words
5. Authors must provide their name, address, daytime phone number and whether the piece is fictional or factual.

HURRICANES

Over the years Atlantic Coast hurricanes have taken a great toll in lives and property. An understanding of their characteristics may help you avoid some of the dangers.

Hurricanes affecting the North Atlantic usually move W or NW from the Eastern Caribbean, veering to the N and E, and losing force as they spin out to sea. They are areas of low barometric pressure; the sharper and deeper the drop in pressure, the more violent the winds will generally be. The "eye" or storm center may travel forward from 5 to 50 m.p.h. The winds around it travel COUNTER-CLOCKWISE, spiraling inwards and **accelerating** as they approach the eye. See diagram below.

In the right or more dangerous semicircle, where the spiral winds are heading in the same general direction as the eye, the two velocities are combined. Winds of 100 m.p.h., combined with a forward speed of 20 m.p.h., yield a velocity of 120 m.p.h. in the right semicircle. In the left or less dangerous semicircle, the velocity would be 80 m.p.h.

If the eye is moving directly toward you, the wind direction will remain fairly constant, and the velocity will increase until the eye arrives. Then the wind will drop, perhaps to a dead calm, and the sky will suddenly clear. Then the eye has passed, the wind will suddenly rise, stronger than ever, but from the opposite direction. This treacherous shift and the greater velocity make the vicinity of the eye most dangerous.

In the diagram, vessels A, B, and C are at positions A1, B1, and C1 as the storm approaches them. By the time the storm has passed, the vessels will be at positions A2, B2, and C2 respectively. Each vessel will have experienced an entirely different series of wind directions and velocities. See illustration:

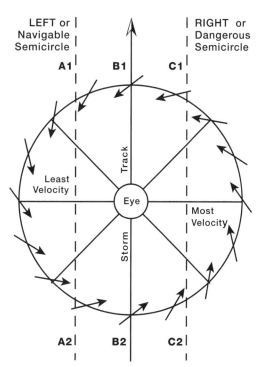

The vessel at A1 has the wind from the NE, backing to N, NW, and at A2, from W. The least velocity is from the N.

At B1 the vessel has wind from a constant direction, ENE, but increasing in velocity until the eye arrives. After the calm of the eye, the wind rises violently from the WSW and stays from that direction, with decreasing velocity, through position B2.

The vessel at C1 has wind from the E, veering to SE, S, and finally to SW at C2. The greatest velocity is from the S.

If space and time permit, try to escape by proceeding at right angles to the storm track. Vessels to the left of the track, or near the center of the track, should try to head to the left of the track. On the right side of the track, it may be better to head farther to the right. Which way to go will depend on several factors which the mariner must weigh: how far away the storm is, the speed of the vessel, and sea conditions or sea room on either side of the expected storm path.

Hurricane Precautions Alongshore

Extremely high tides accompany hurricanes. If the storm arrives anywhere near the usual time of high water, low-lying areas will be flooded. Especially high tides will occur in all bays or V-mouth harbors if they are facing the wind direction. High water in all storm areas will remain a much longer time.

At or near the coastline, pull small craft well above the high water mark, dismast sailboats, remove outboard motors, and remove or lash down loose objects.

Seek the most protected anchorage possible, considering possible wind direction reversals, extreme tides, and other vessels. If on a mooring or at anchor, use maximum scope, allowing for room to swing clear of other boats. In a real blow it is easy to slack off, but not to shorten scope. Use as much chain in your anchor rode as possible. Another piece of chain or a weight, attached halfway along your mooring or anchor line, will help absorb sudden strains. Use chafing gear liberally at bits and bow chocks to minimize fraying of lines. Rig fenders to minimize damage from/to other boats.

Shut off gas, stove tanks, etc. Douse any fires in heating stoves. Secure all portholes, skylights, ventilators, hatch covers, companionways, etc. Pump the boat dry.

At a wharf or pier, use fenders liberally. If possible, rig one or more anchors abreast of the boat in the event the tide rises above pilings.

Boats are replaceable: don't wait for the last moment to get ashore!

Hurricane Precautions Offshore

Monitor storm reports on your radio. The U.S. Coast Guard warns all vessels offshore to seek shelter at least 72 hours ahead of a hurricane.

But, if caught offshore with no chance to reach shelter, watch the wind most carefully. First, note that if you face the wind, the eye is about 10 points or 112° to your right. If the wind "backs" (moves counter-clockwise), you are already on the less dangerous side of the storm track. If the wind "veers" (moves clockwise), you are on the right or more dangerous side. If the wind direction is constant, the hurricane's eye is headed directly at you, so make haste to get to the left side of the track.

Use your radiotelephone to advise the Coast Guard and other vessels of your position. Have a liferaft and safety equipment (flares, flashlights, EPIRB, etc.) ready. Don PFDs. If it is impossible to hold your intended course, head your powerboat directly into the wind and sea, using only enough power to maintain steerageway. If power fails, rig a sea anchor or drogue to keep the bow to the wind.

Sailing vessels heaving to should consider doing so on the starboard tack (boom to port) in the more dangerous semicircle, or on the port tack (boom to starboard) in the less dangerous semicircle, to keep the wind drawing aft.

NOTE: This information is necessarily very general, the diagram (p. 253) is over-simplified, and the suggestions assume a straight storm track. If the storm track curves to the right, vessels A and B will have an easier time of it, but C may wind up on the track. The best advice: monitor weather reports continuously and seek shelter well ahead of time.

National Weather Service – www.nws.noaa.gov
National Hurricane Center – www.nhc.noaa.gov
NOAA Hurricane Research – www.aoml.noaa.gov/hrd

WEATHER NOTES
From Maine to the Chesapeake

Sea Fog

There is always invisible moisture in the air, and the warmer the air, the more moisture it can contain invisibly. But when such a mass of moist air is cooled off, as it does when passing over a body of cooler water, the moisture often condenses into visible vapor, or fog. The fog clears when the air temperature rises, from the sun or a warm land mass, or by a warm, dry wind.

To predict fog accurately, you can use a "sling psychrometer." This instrument uses two thermometers side by side, one of which has a wick fastened to the bulb end. After wetting the wick on the "wet bulb" thermometer, the user swings the instrument in a circle for 60-90 seconds. This causes water to evaporate from the wet bulb thermometer, lowering its reading. The dry bulb thermometer simply tells air temperature. The difference in readings between the dry bulb and wet bulb thermometers determines the relative humidity of the air, and - especially valuable for determining the likelihood of fog - the dew point. The dew point is the (lower) temperature to which air must be cooled for condensation, or fog, to occur. (see Dew Point Table, p. 248)

Eastport, Maine to Cape Cod

Cold water (48°-55°) off the northern New England coast often causes heavy fog conditions in the spring and summer, when a warm moist southwesterly flow of air passes over it. East of Portland to the Bay of Fundy, fog is not apt to occur when the dew point is under 55°, unless there is a very warm moist wind. The effect of the cold water on the warm air is reduced if the winds become brisk, as they are apt to do in the afternoon. Visibility should then improve.

Long Island Sound and the New Jersey Coast

Summertime warm water (in the 70s) in this area rarely cools down any warm air mass enough to produce fog. This is not the case farther offshore, where cooler water temperatures (in the 50s) can produce fog.

On the south coast of Long Island, when the southwest wind blows toward the shore at the same time as the ebb tide, inlets can become dangerous with short, steep seas. Also, offshore swells can become very high near the mouths of inlets.

On the New Jersey shore, prevailing winds in summer are southerly, increasing in mid-morning to rarely more than 20 knots and usually dying down at dusk. Occasional summer thunderstorms can be expected. Any brisk winds from the east, northeast, or southeast can produce dangerous conditions along this lee shore and at the mouths of inlets. When the wind is from offshore like this, inlets should be entered on a flood tide.

At the mouth of Delaware Bay, seas can build up to a hazardous degree when there is a southeast wind at the same time as an ebb current at the mouth of the Bay.

Chesapeake Bay

There is little chance of fog in this region because of the warmth of the water. The Bay has quirks of its own in weather and sea conditions. It is a narrow and fairly shallow body of water, and winds tend to blow up or down it. Sharp seas can result, depending on the direction of the current and the wind. Opposing forces make for rough water.

Prevailing winds in spring and summer are southerly, freshening in the afternoon after a morning of calm. Summer thunderstorms occur frequently in afternoon and early evening, usually from the west. In the fall, after a cold front passes through, the winds will shift into the north or northeast, usually for three days, and increase in velocity, causing seas to build up. Calm follows for a day or so until the wind shifts to the southwest.

CLEAN VESSEL ACT

Congress passed the Clean Vessel Act in 1992 (CVA) to help reduce pollution from vessel sewage discharges. The Act established a five-year federal grant program administered by the U.S. Fish and Wildlife Service and authorized $40 million from the Sport Fish Restoration Account of the Aquatic Resources Trust Fund for use by the States. Federal funds can constitute up to 75% of all approved projects with the remaining funds provided by the States or marinas. Reauthorized in 1998, Congress extended the pumpout grant program through 2003, providing $50 million to continue to provide alternatives to overboard disposal of recreational boater sewage. In 2005 the Safe, Accountable, Flexible, and Efficient Transportation Equity Act (SAFETEA) was signed into law and provided for the reauthorization of the Sport Fish Restoration Act including the CVA Program through 2009. The amendment eliminated the Aquatic Resources Trust Fund and established the Sport Fish Restoration and Boating Trust Fund.

The Clean Vessel Act provides a portion of its total funding for educational outreach regarding the effects of boater sewage and the means by which boaters can avoid improper sewage disposal. NOAA will mark pumpout and dump station locations on its nautical charts.

The Act was created to provide a viable alternative to the overboard disposal of recreational boater sewage. All recreational vessels must have access to pumpouts funded under the Clean Vessel Act. The Act made grants available to the States on a competitive basis for the construction and/or renovation, operation and maintenance of pumpout and portable toilet dump stations. States may sub-grant to public and private marinas to install pumpouts. Since the Act's passage in 1992, grants have been awarded to install 2,200 pumpout stations and 1,400 dump stations. A maximum fee of $5.00 may be charged for use of pumpout facilities constructed or maintained with grant funds.

Contact: Steve_Jose@fws.gov

For Information on pumpout station locations in your state, see p. 261.

U.S. Fish and Wildlife Service
Wildlife and Sport Fish Restoration Program
4401 North Fairfax Drive, MBSP 4020
Arlington, Virginia 22203
(703) 358-2156 (voice), (703) 358-1837 (fax)
http://wsfrprograms.fws.gov/Subpages/GrantPrograms/CVA/CVA.htm

VESSEL SAFETY CHECK PROGRAM

The Vessel Safety Check (VSC) program is a win-win proposition for boaters who wish to have a trained examiner inspect their boats, at no cost, to ensure they have all the equipment required by regulations.

The VSC program involves the U.S. Coast Guard Auxiliary, the U.S. Power Squadron, and the National Association of Boating Law Administrators to provide free safety examinations of recreational boats. Examiners are available to check boats for meeting federal and state requirements for the equipment that they are required to carry. They also check for general safety conditions.

Owners of boats that meet the VSC requirements receive the VSC decal to indicate that their boat was examined and passed. If a boat does not meet all of the VSC requirements, the owner gets a listing of what is needed. No tickets are ever issued and no reports are made to law enforcement authorities.

If you are interested in having a VSC go to: http://safetyseal.net/GetVSC/ and enter your zipcode for the nearest examiner. You may email the examiner from the site to arrange for a VSC, or contact either of the following organizations:

U.S. Power Squadron: http://www.usps.org/
U.S. Coast Guard Auxiliary: http://www.cgaux.org/

U.S. COAST GUARD BOARDINGS

The U.S. Coast Guard has the authority to enforce federal laws by making inquiries, examinations, inspections, searches, seizures, and arrests on the waters over which the United States has jurisdiction.

U.S. Coast Guard personnel are armed and may use necessary force to compel compliance. They are charged with enforcement of laws dealing with safety, water pollution, drug smuggling, illegal immigration, and the 200 mile fishery conservation zone. In nearly half the boardings, they find some kind of noncompliance with regulations. A civil penalty of up to $500 may be imposed for failure to comply with equipment or numbering regulations, navigation rules, accident reporting procedures, etc.

A boat underway that is hailed by a U.S. Coast Guard vessel or patrol boat is required to follow the boarding officer's instructions, which may be to stop, to continue at reduced speed, or to maneuver in such a way as to permit boarding. Instructions will depend on sea conditions. The Coast Guard follows a standard procedure before boarding, and the boarding team will provide an explanation before the actual boarding. If the boarding party has full cooperation from you, the inspection will be completed quickly.

The editors wish to thank the U. S. Power Squadron (USPS) for permission to reprint this article from their Seamanship Manual.

TIPS FOR ENVIRONMENTALLY RESPONSIBLE BOATING

Waste Disposal

Become familiar with U.S. Coast Guard pollution regulations (p. 260). Know what you can dispose of, and where. When in doubt, carry it out. Learn where the No Discharge Areas are in your waters. If your Marine Sanitation Device has a holding tank, know where you can find a pumpout station (see p. 261 for a listing of agency phone numbers and websites by state).

Cleaning

When washing a boat's deck and hull surfaces, people often use products that contain toxic ingredients such as chlorine, phosphates, and ammonia. Just as these chemicals act as degreasers on the boat, they also act as degreasers on fish, drying the natural oil fish need for their gills to take in oxygen. To reduce your need for toxic products:

- Rinse your boat only with fresh water.
- Use baking soda, vinegar, lemon juice, borax, and "elbow grease."

Sanding

Sanding and scraping your boat releases noxious particles in the air and onto the ground and water around you. Always sand and scrape on shore, away from the water. Use a vacuum sander to collect and store the harmful dust before it can get into the water or your eyes and lungs.

Painting

Anti-fouling paints usually have toxic metals such as copper, mercury, arsenic, or tributyltin. All have severe impact on human health and the underwater ecosystem. There are state laws regulating bottom paints. Contact your local marine supply store or state boating agency to find out more.

Fueling

Take precautions not to overfill your fuel tank. If there is overflow onto the boat or dock, wipe it up with a rag. Do not hose it into the water. If you do spill fuel or oil into the water, do not disperse it with detergent or soap. That sends the problem down to the bottom where it becomes more toxic and more difficult to clean up. If the spill is large or if it discolors the surface of the water, you are required to report it to the National Response Center at 800-424-8802 or to the U.S. Coast Guard on VHF channel 16. Failure to do so is illegal and can cost you civil penalties and/or criminal sanctions.

For More Information:

GOOD MATE Recreational Boating & Marina Manual
The Ocean Conservancy 1300 19th Street, NW, 8th floor, Washington, DC 20036
tel. 202-429-5609, 800-519-1541
Download PDF at http://www.auxmdept.org/Good%20Mate%20Manual.pdf

COAST GUARD POLLUTION REGULATIONS

The Damage Caused by Pollution

Sewage is not just a repulsive visual pollutant. The microorganisms in sewage, including pathogens and bacteria, degrade water quality by introducing diseases like hepatitis, cholera, typhoid fever and gastroenteritis, which can contaminate shellfish beds. Shellfish are filter feeders that eat tiny food particles filtered through their gills into their stomachs, along with bacteria from sewage. Nearly all waterborne pathogens can be conveyed by shellfish to humans. Stormwater runoff and drainage from fertilized lawns contain chemical products and nutrients. Although nutrients are necessary for waterborne plants, when too abundant they can stimulate algae blooms. This process leads to oxygen depletion, which can harm and kill aquatic life.

Federal Regulations for Waste Disposal

Prohibited in all waters: The discharge of plastic or garbage mixed with plastic, including synthetic ropes, fishing nets and plastic bags.

Prohibited within 25 n.m. of land. The discharge of dunning, lining, and packing materials that float.

Prohibited within 12 n.m. of land: The discharge of unground garbage larger than 1 inch, including food waste, paper, rags, glass, metal, bottles, crockery and similar refuse.

Prohibited within 3 n.m. of land: The discharge of any garbage, including ground food waste, paper, rags, glass, etc.

Marine Sanitation Devices

Vessels under 65' may install type I, II or III MSD. Vessels over 65' must install a type II or III MSD. All installed MSD's must be U. S. Coast Guard certified.

- **Type I** MSDs are allowed only on vessels under 65'. They treat sewage with disinfectant chemicals before discharge. The discharge must not show any visible floating solids, and must have a fecal coliform bacterial count not greater than 1000 per 100 milliliters of water.

- **Type II** MSDs provide a higher level of treatment than Type I, using greater levels of chemicals to create effuent having less than 200 per 100 milliliters and suspended solids not greater than 150 milligrams per liter.

- **Type III** MSDs do not allow discharge of sewage, except through a Y-valve to discharge at a pumpout facility, or overboard when outside the 3 nautical miles. They include holding tanks, recirculating and incinerating units.

- **Portable toilets** or "porta-potties" are not considered installed toilets and are not subject to MSD regulations. They are, however, subject to the disposal regulations which prohibit the disposal of raw sewage within the three-mile limit or territorial waters of the U.S.

No Discharge Areas (NDAs) are water bodies where the Environmental Protection Agency (EPA) and local communities prohibit the discharge of all vessel sewage. Chesapeake Bay and Narragansett Bay, among others, have been designated as NDAs.

When operating vessel in NDAs, the operator must secure each Type I or Type II MSD in a manner which prevents discharge of treated or untreated sewage. Acceptable methods of securing the MSD include: closing the seacock and removing the handle, padlocking the seacock in the closed position, using a non-releasable wire-tie to hold the seacock in the closed position, or locking the door to the space enclosing the toilets with a padlock or door handle key lock.

Type III MSDs, or holding tanks, must also be secured in a manner that prevents discharge of sewage. Acceptable methods of securing the device include: closing each valve leading to an overboard discharge and removing the handle, padlocking each valve leading to an overboard discharge in the closed position, or using a non-reusable wire-tie to hold each valve leading to an overboard discharge in a closed position. Sewage held in Type III MSDs can be removed by making arrangements with landside pumpout stations or pumpout boats. Call Harbormaster for details.

HOW TO FIND PUMPOUT STATIONS

Please be sure to call or radio in advance for rates and availability. While we have taken all possible care in compiling this list, changes may have occurred and we cannot guarantee accuracy. For more current information check the state website or call the agency listed.

Look for Clean Vessel Act (CVA):
http://wsfrprograms.fws.gov/Subpages/GrantPrograms/CVA/CVA.htm

Most major harbors now have a pumpout boat. Contact the local Harbormaster. Many monitor VHF channel 09.

MAINE: ME Dept. of Environ. Protection, 207-287-7905
www.maine.gov/dep/blwq/topic/vessel/pumpout/

NEW HAMPSHIRE: NH Environ. Serv., 603-271-3414
www.des.nh.gov/wmb/cva/dir_map.htm

MASSACHUSETTS: MA Fish & Wildlife, 617-626-1520
www.mass.gov/dfwele/dmf/programsandprojects/cvabig.htm
MA Coastal Zone Mgmt., 617-626-1200
www.mass.gov/czm/nda/pumpouts/index.htm

RHODE ISLAND: RI Environ. Mgmt., 401-222-6800
www.dem.ri.gov/programs/benviron/water/shellfsh/pump/index.htm

CONNECTICUT: CT Environ. Protection, 860-424-3000
www.ct.gov/dep/cwp/view.asp?a=2705&q=323750

NEW YORK: Sea Grant NY Boating and Marine Facility, 315-312-3042
www.nysgextension.org/pumpouts/pumpouts-search.html

NEW JERSEY: NJ Fish & Wildlife, 609-748-2056, 609-748-2020
www.state.nj.us/dep/fgw/cvahome.htm

DELAWARE: Delaware Fish & Wildlife, 302-739-9911
www.dnrec.state.de.us/fw/pumpout.htm

MARYLAND: MD Natural Resources, 410-260-8770
www.dnr.state.md.us/boating/pumpout/

VIRGINIA: VA Dept. of Health, 804-864-7468
www.vdh.state.va.us/EnvironmentalHealth/Wastewater/MARINA/
pumpoutdata/index.htm

NORTH CAROLINA: NC Div. of Coastal Management, 919-733-2293
http://dcm2.enr.state.nc.us/Marinas/pumplist.htm

SOUTH CAROLINA: SC Dept. of Health & Env. Control, 843-953-9062
www.scdhec.gov/environment/ocrm/outreach/cva.htm
A Guide to Marine Sewage Disposal Stations in Coastal South Carolina
(www.scdhec.gov/environment/ocrm/pubs/general.htm)

GEORGIA: Contact local marinas

FLORIDA: Dept. of Environ. Protection, 850-245-2100
www.dep.state.fl.us/cleanmarina/CVA/default.htm

TABLE FOR CONVERTING SECONDS TO DECIMALS OF A MINUTE

From many sources, including charts, Light Lists, and Notices to Mariners, positions are in degrees, minutes, and seconds. These are written either 34° 54' 24" or 34-54-24

However, for navigating with GPS, Loran, chart plotters, and celestial calculators, it can be useful to convert the last increment – seconds – to either tenths or hundredths of a minute. The numbers above become 34° 54.40' or 34-54.4

Secs.	Tenths	Hundredths	Secs.	Tenths	Hundredths	Secs.	Tenths	Hundredths
1	.0	.02	21	.4	.35	41	.7	.68
2	.0	.03	22	.4	.37	42	.7	.70
3	.1	.05	23	.4	.38	43	.7	.72
4	.1	.07	24	.4	.40	44	.7	.73
5	.1	.08	25	.4	.42	45	.8	.75
6	.1	.10	26	.4	.43	46	.8	.77
7	.1	.12	27	.5	.45	47	.8	.78
8	.1	.13	28	.5	.47	48	.8	.80
9	.2	.15	29	.5	.48	49	.8	.82
10	.2	.17	30	.5	.50	50	.8	.83
11	.2	.18	31	.5	.52	51	.9	.85
12	.2	.20	32	.5	.53	52	.9	.87
13	.2	.22	33	.6	.55	53	.9	.88
14	.2	.23	34	.6	.57	54	.9	.90
15	.3	.25	35	.6	.58	55	.9	.92
16	.3	.27	36	.6	.60	56	.9	.93
17	.3	.28	37	.6	.62	57	1.0	.95
18	.3	.30	38	.6	.63	58	1.0	.97
19	.3	.32	39	.7	.65	59	1.0	.98
20	.3	.33	40	.7	.67	60	1.0	1.00

TABLE OF EQUIVALENTS
and other useful information

Length

English	Metric
1 inch	2.54 centimeters
1 foot	.30 meters
1 fathom	1.61 meters
1 statute mile	1.61 kilometers
1 nautical mile	1.85 kilometers

Metric	English
1 meter	39.37 inches
"	3.28 feet
"	.55 fathoms
1 kilometer	.62 statute miles
"	.54 nautical miles

Nautical	Terrestrial
1 fathom	6 feet
1 cable	608 feet
1 nautical mile	6076 feet
"	1.15 statute miles
1 knot	1.15 mph
7 knots	8 mph approx.

Capacity

English	Metric
1 quart	.95 liters
1 gallon	3.78 liters

Metric	English
1 liter	1.06 quarts
"	.26 US gallons

Weight

English	Metric
1 ounce	28.35 grams
1 pound	.45 kilograms
1 US ton	.907 metric tons
"	.893 long tons

Metric	English
1 gram	.035 ounces
1 kilogram	2.20 pounds
1 metric ton	2204.6 pounds

Weight of 1 US Gallon

Gasoline	6 pounds
Diesel fuel	7 pounds
Fresh water	8.3 pounds
Salt water	8.5 pounds

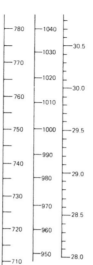

Barometric Pressure: millimeters, millibars,

Temperature:
$$C^\circ = (F^\circ - 32) \times 5/9$$
$$F^\circ = C^\circ \times 9/5 + 32$$

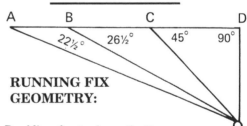

RUNNING FIX GEOMETRY:

Doubling the Angle on the Bow
1. Angle DCO = 45°; Angle CDO = 90°; True distance run (CD) = distance DO.
2. Angle DAO = 22½°; Angle DCO = 45°; True distance run (AC) = Distance CO.

Other Useful Bow Bearings
3. Angle DBO = 26½°; Angle DCO = 45°; True distance run (BC) = distance DO.
4. Example 3 also works with angles of 25° and 41°; 32° and 59°; 35° and 67°; 37° and 72° when distance run will be distance DO.

CRUISING CHARTS

DESIRABLE CRUISING CHARTS FROM CAPE BRETON I. TO KEY WEST, FL.

Numbers listed to the left are general coastal charts. Indented numbers refer to harbor charts. (C) indicates Loran-C lines of position on the chart. 1:80(000), 1:40(000), etc. indicates scale.

USA, NOAA: http://www.nauticalcharts.noaa.gov, **CANADA:** www.charts.gc.ca

Canada

4013	Halifax to Sydney 1:350
4279	Bras d'Or Lake 1:60
4447	Pomquet and Tracadie Harbours 1:25
4385	Chebucto Hd. to Betty Is. 1:39
4321	Cape Canso to Liscomb Is. 1:108.8
4227	Country Hbr. to Ship Hbr. 1:50
4320	Egg Is. to W. Ironbound Is. 1:145
4012	Yarmouth to Halifax 1:300
4386	St. Margaret's Bay 1:39.4
4381	Mahone Bay 1:38.9
4384	Pearl Is. to Cape LaHave 1:39
4211	Cape LaHave to Liverpool Bay 1:37.5
4230	Little Hope Is. to Cape St. Mary's 1:50
4240	Liverpool Hbr. to Lockeport Hbr. 1:60
4241	Lockeport to Cape Sable 1:60
4242	Cape Sable to Tusket Is. 1:60
4243	Tusket Is. to Cape St. Mary's 1:60
4010	Bay of Fundy (inner portion) 1:200
4011	Appr. to Bay of Fundy 1:300
4118	St. Mary's Bay 1:60
4396	Annapolis Basin 1:24
4141	St. John to Evandale 1:30
4116	Appr. to St. John 1:60
4340	Grand Manan 1:60

U.S. East Coast

13325 (C) Quoddy Narrows to Petit Manan Is. 1:80
13312 (C) Frenchman & Blue Hill Bays & apprs. 1:80
13302 (C) Penobscot Bay and apprs. 1:80
    13315 Deer Is. Thoro. and Casco Pass. 1:20
    13308 Fox Islands Thorofare 1:15
13288 (C) Monhegan Is. to Cape Elizabeth 1:80
    13290 Casco Bay 1:40
13286 (C) Cape Elizabeth to Portsmouth 1:80
    13283 Cape Neddick Hbr. to Isles of Shoals 1:20, Portsmouth Hbr. 1:10
13278 (C) Portsmouth to Cape Ann 1:80, Hampton Harbor 1:30
    13281 Gloucester Hbr. and Annisquam R. 1:10
13267 (C) Massachusetts Bay 1:80
    13275 Salem and Lynn Harbors 1:25, Manchester Harbor 1:10
    13276 Salem, Marblehead & Beverly Hbrs. 1:10
    13270 Boston Harbor 1:25
13246 (C) Cape Cod Bay 1:80
    13253 Plymouth, Kingston and Duxbury Hbrs. 1:20, Greens Hbr. 1:10

    13236 Cape Cod Canal and approaches 1:20
13237 (C) Nantucket Sound and approaches 1:80
    13241 Nantucket Island 1:40
    13242 Nantucket Harbor 1:10
13218 (C) Martha's Vineyard to Block Island 1:80
    13230 Buzzards Bay 1:40, Quicks Hole 1:20
    13233 Martha's Vineyard 1:40, Menemsha Pond 1:20
    13221 Narragansett Bay 1:40
13205 (C) Block Island Sound and apprs. 1:80
    13214 Fishers Island Sound 1:20
    13212 Approaches to New London Hbr. 1:20
    13219 Point Judith Harbor 1:15
    13217 Block Island 1:15
    13211 North Shore of Long Is. Sd.-Niantic Bay & Vicinity 1:20
    13213 New London Harbor and Vicinity 1:10, Bailey Point to Smith Cove 1:5
    13209 Block Is. Sd. & Gardiners Bay, Long Is., 1:40
12354 (C) Long Island Sound - eastern part 1:80
    12375 Connecticut R. -Long Is. Sd. to Deep R. 1:20
    12374 Duck Island to Madison Reef 1:20
    12373 Guilford Hbr to Farm R. 1:20
    12371 New Haven Harbor 1:20
    12370 Housatonic R. and Milford Hbr. 1:20
    12362 Port Jefferson & Mt. Sinai Hbrs. 1:10
12363 (C) Long Island Sound - western part 1:80
    12369 Stratford to Sherwood Pt. 1:20
    12368 Sherwood Pt. to Stamford Hbr. 1:20
    12367 Greenwich Pt. to New Rochelle 1:20
    12366 L.I. Sd. and East R., Hempstead Hbr. to Tallman Is. 1:20
    12365 L.I. Sd. S. Shore, Oyster and Huntington Bays 1:20
12353 (C) Shinnecock Light to Fire Island Light 1:80
    12352 Shinnecock B. to E. Rockaway In. 1:20; 1:40
    12339 East R. - Tallman I. to Queensboro Br. 1:10
    12331 Raritan Bay and Southern Part of Arthur Kill 1:15

12327 New York Harbor 1:40
12335 Hudson & E. Rs. -Governors I. to
67 St. 1:10
12326 (C) Appr. to N.Y., Fire I. to Sea Girt
1:80
12350 Jamaica Bay and Rockaway In.
1:20
12323 (C) Sea Girt to Little Egg In. 1:80
12324 Sandy Hook to Little Egg Harbor
1:40
12318 (C) Little Egg In. to Hereford In. 1:80,
Absecon In. 1:20
12316 Little Egg Harbor to Cape May
1:40
12304 (C) Delaware Bay 1:80
12311 Delaware R.- Smyrna R. to
Wilmington 1:40
12312 Wilmington to Philadelphia 1:40
12277 Chesapeake and Delaware Canal,
Salem R. Ext. 1:20
12214 (C) Cape May to Fenwick I. 1:80
12211 (C) Fenwick I. to Chincoteague In.1:80,
Ocean City In. 1:20
12210 (C) Chincoteague In. to Great Machipongo
In. 1:80, Chincoteague In. 1:20
12221 (C) Chesapeake Bay Entrance 1:80
12222 Cape Charles to Norfolk Hbr. 1:40
12224 Cape Charles to Wolf Trap 1:40
12228 Pocomoke and Tangier Sds. 1:40
12231 Tangier Sd.-northern part 1:40
12225 (C) Wolf Trap to Smith Point 1:80
12230 (C) Smith Point to Cove Point 1:80
12263 (C) Cove Point to Sandy Point 1:80
12273 (C) Sandy Point to Susquehanna River
1:80
12233 Chcsapcakc Bay to Pincy Pt. 1:40
12274 Head of Chesapeake Bay 1:40
12278 Appr. to Baltimore Harbor 1:40
12286 Piney Pt. to Lower Cedar Pt. 1:40
12288 Lower Cedar Pt. to Mattawoman
Cr. 1:40
12289 Mattawoman Cr. to Georgetown
1:40; Washington Hbr. 1:20
12282 Severn and Magothy Rs. 1:25
12253 Norfolk Hbr. and Elizabeth R. 1:20
12254 Cape Henry to Thimble Shoal
Lt. 1:20
12256 Chesapeake Bay-Thimble Shoal
Channel 1:20

12245 Hampton Roads 1:20
12207 (C) Cape Henry to Currituck Bch. Lt.
1:80
12205 Cape Henry to Pamlico Sd.
incl. Albemarle Sd. 1:40; 1:80
12204 (C) Currituck Beach Lt. to Wimble
Shoals 1:80
11555 (C) Cape Hatteras-Wimble Shoals to
Ocracoke In. 1:80
11548 (C) Pamlico Sd.-western part 1:80
11550 Ocracoke In. and N. Core
Sd. 1:40
11545 Beaufort In. and S. Core
Sd. 1:40, Lookout Bight 1:20
11544 (C) Portsmouth I. to Beaufort
incl. Cape Lookout Shoals 1:80
11543 (C) Cape Lookout to New R. 1:80
11539 (C) New R. In. to Cape Fear 1:80
11536 (C) Appr. to Cape Fear R. 1:80
11535 (C) Little R. In. to Winyah Bay Entr.1:80
11531 Winyah Bay to Bulls Bay 1:80
11532 Winyah Bay 1:40
11521 (C) Charleston Hbr. & Appr. 1:80
11513 (C) St. Helena Sd. to Savanna
R. 1:80
11509 (C) Tybee I. to Doboy Sd. 1:80
11502 (C) Doboy Sd. to Fernandina 1:80
11488 (C) Amelia I. to St. Augustine 1:80
11486 (C) St. Augustine Lt. to Ponce
de Leon In. 1:80
11484 (C) Ponce de Leon In. to Cape
Canaveral 1:80
11476 (C) Cape Canaveral to Bethel
Shoal 1:80
11474 (C) Bethel Shoal to Jupiter In. 1:80
11466 (C) Jupiter In. to Fowey Rocks
1:80, Lake Worth In. 1:10
11469 Straits of FL.Fowey Rks., Hillsboro
Inlet to Bimini Is. Bahamas 1:100
11462 (C) Fowey Rocks to Alligator
Reef 1:80
11452 (C) Alligator Reef to Sombrero
Key 1:80
11442 (C) Sombrero Key to Sand Key 1:80
11439 (C) Sand Key to Rebecca Shoal 1:80
11438 Dry Tortugas 1:30

To find your nearest NOAA Chart Agent: http://www.naco.faa.gov/Agents.asp

Print-on-Demand Nautical Charts for up to date NOAA charts: http://www.OceanGrafix.com

NOAA has posted all 1000+ of its US Nautical charts on the internet. The charts can be viewed using any internet browser. Each chart is up-to-date with the most recent Notices to Mariners. Use these online charts as a ready reference or planning tool, then use one of NOAA's printed or digital charts for actual navigation. Online charts can be viewed at: http://ocsdata.ncd.noaa.gov/OnLineViewer/

For US, USCG Local Notices to Mariners for Critical Chart updates: http://www.navcen.uscg.gov/lnm/

Ferry Service Information

*Vehicle reservations may be required.

MAINE

-- *For all Maine Ferry Service information:* www.exploremaine.org/ferry/index.html

Bar Harbor/Portland*, Yarmouth, N.S., The Cat (877) 359-3760

Maine State Ferry Service, Rockland ME, (207) 596-2202. General Schedule Information and Daily Operations Update (800) 491-4883. www.state.me.us/mdot/opt/ferry/ferry.htm

   Frenchboro Ferry* between Frenchboro and Bass Harbor, Bass Harbor (207) 244-3254

   Islesboro Ferry* between Islesboro (207) 734-6935 and Lincolnville (207) 789-5611

   Matinicus Island Ferry* between Matinicus Island and Rockland (207) 596-2202

   North Haven Ferry* between North Haven (207) 867-4441 and Rockland (207) 596-2202

   Vinalhaven Ferry* between Vinalhaven (207) 863-4421 and Rockland (207) 596-2202

Monhegan Island Ferry between Monhegan Island and Port Clyde (207) 372-8848. www.monheganboat.com

MASSACHUSETTS

Steamship Authority*, www.steamshipauthority.com/ssa/

   Between Woods Hole, MA, (508) 548-3788 (information), (508) 477-8600 (car reservation) and Martha's Vineyard (508) 693-9130

   Between Hyannis and Nantucket. Hyannis (508) 771-4000, Fast Ferry (508) 495-3278 (passenger reservation), Nantucket (508) 228-0262

Island Queen Ferry between Falmouth and Oak Bluffs, M.V. (508) 548-4800. www.islandqueen.com

MV Express between New Bedford and Martha's Vineyard (866) 683-3779. www.mvexpressferry.com/

Hy-Line Cruises, Nantucket, Martha's Vineyard, Hyannis loop (800) 492-8082. www.hy-linecruises.com/

-- *For more Martha's Vineyard ferry information:* www.mvol.com/directory/transportation/Ferries/

Freedom Cruise Line between Harwichport and Nantucket (508) 432-8999. www.nantucketislandferry.com

Cuttyhunk Is. Ferry between Cuttyhunk and New Bedford (508) 992-0200. www.cuttyhunkferryco.com/

RHODE ISLAND

Vineyard Fast Ferry between Quonsett, RI and M.V. (401) 295-4040. www.vineyardfastferry.com

Block Island Ferry* and High Speed Ferry between Block Island and Jerusalem (Pt. Judith) (866) 783-7996. www.blockislandferry.com/

CONNECTICUT

Orient Point, L. I. NY* between New London and Orient Pt., Cross Sound Ferry, New London, CT (860) 443-5281, Orient Pt. (631) 323-2525. www.longislandferry.com

Block Is. High Speed Ferry between New London (860) 444-4624 and Block Is. (401) 466-2212. www.goblockisland.com

Fishers Island Ferry* between Fishers Island, NY (631) 788-7744 and New London, CT (860) 442-0165 (car reservations). www.fiferry.com

Port Jefferson, L.I., NY Ferry* between Port Jefferson, NY (631) 473-0286 and Bridgeport, CT (888) 443-3779. www.portjeffersonferry.com

NEW YORK

Long Island *is served by two year-round ferry lines that cross Long Island Sound connecting Port Jefferson to Bridgeport, CT and Orient Point to New London, CT. For more information for ferries in Long Island Sound (Fire Island, Shelter Island, etc.):* www.webscope.com/li/ferries.html#si

NEW JERSEY

Lewes, DE Ferry* between Lewes, DE and Cape May, NJ (800) 643-3779. www.capemaylewesferry.com

NORTH CAROLINA

-- *To request all NC ferry routes and schedules (877) 368-4968 or download at* www.ncferry.org

Hatteras Inlet Ferry* between Hatteras (800) 368-8949 and Ocracoke (800) 345-1665. www.ncferry.org

Cedar Island and Swan Quarter Ferry between Cedar Island (800) 856-0343 and Swan Quarter (800) 773-1094. www.ncferry.org

Where To Buy The Eldridge Tide and Pilot Book

CANADA

NOVA SCOTIA
Halifax
Binnacle Yachting Equip.
ONTARIO
Toronto
Nautical Mind Bookstore
QUEBEC
Montreal
McGill Maritime

UNITED STATES

Available at most:
Boater's World
West Marine

**For store locations in each
state see website:**
BoatersWorld.com
WestMarine.com

ME
Bar Harbor
Sherman's Book Store
Bath
Maine Maritime Museum
Blue Hill
Blue Hill Books
North Light Books
Boothbay
Sherman's Book Store
Brooklin
The Wooden Boat Store
Brooksville
Buck's Harbor Marine
Brunswick
Brunswick Bookland
Gulf of Maine Books
Camden
Owl and the Turtle
Sherman's of Camden
Castine
Four Flags
Damariscotta
Maine Coast Book Shop
Kittery
Jackson Hardware
Little Deer Isle
Buck's Harbor Marine
Northeast Harbor
F.T. Brown Co.
Portland
Chase Leavitt
Hamilton Marine
Rockland
Rockland Boat
Searsport
Hamilton Marine
South Freeport
*Brewer's Yacht Yard
Southwest Harbor
Henry Hinckley Co.
Stonington
Billings Diesel & Marine
Yarmouth
Landing Boat Supply
York Harbor
York Harbor Marine Services

NH
Concord
Contoocook River Canoe Co.
Keene
Toadstool Bookshop
Litttleton
Village Bookstore

Portsmouth
New England Marine
West Marine
Seabrook
West Marine
Wolfboro
The Country Bookseller

MA
Bass River
Ship Shops, Inc.
Beverly Farms
The Bookshop
Boston
Borders Books and Music
Boston Hbr. Sailing Club
Boston Sailing Center
Boxell's Chandlery
*Robert E. White Instruments
Brewster
Brewster Bookstore
Buzzards Bay
Red Top Sporting Goods
Cambridge
Globe Corner Bookstore
Porter Sq. Books, Inc.
Cataumet
*Kingman Yachting Center
*Parker's Boat Yard
Charlestown
Constitution Marina
Chatham
Cabbages & Kings
Mayflower Shop
Stage Harbor Marine
Yellow Umbrella Books
Cohasset
Buttonwood Books
Concord
Concord Book Shop
Cotuit
Peck's Boats
Cuttyhunk
Island Market
Danvers
Boater's World
West Marine
Dedham
West Marine
Duxbury
Bayside Marine Corp.
Edgartown
Edgartown Books
Edgartown Marine Outfitters
Everett
Book Stall
Fairhaven
Fairhaven Shipyard
Falmouth
Booksmith-Falmouth Plaza
Eastman's Sport & Tackle
Falmouth Bait and Tackle
Falmouth Marine & Yachting
*MacDougalls'
Framingham
Recreational Equip., Inc. (REI)
Gloucester
The Bookstore
Brown's Yacht Yard
Building Center of Gloucester
Enos Marine
Gosnold
Island Market and Chandlery
Hanover
Sylvester Co.
Harwichport
*Allen Harbor Marine Serv.
Cape Yachts

Hingham
Old Salt Outfitters
RNR Marine
Hull
Hull Lifesaving Museum Store
Hyannis
Hyannis Marina
Sports Port
Kingston
Viking Marine and Photo
Manchester
Crocker's Boatyard
New England Small Craft, Inc.
Mansfield
Book Ends
Marblehead
The Forepeak
Hugo Books
F. L. Woods
Lynn Marine Supply Co.
Marblehead Outfitters
Spirit of '76 Bookstore
Marion
Book Stall
*Burr Bros. Boats
Mashpee
Booksmith
Bosun's Marine
Market St. Bookshop
Nantucket
Mitchell's Book Corner
*Nantucket Boat Basin
Nantucket Ship Chandlery
Natick
Natick Outdoor Store
New Bedford
Bay Fuels, Inc.
C.E. Beckman Co.
CMS Enterprises Inc.
The Gear Locker
I.M.P. Fishing Gear, LTD
Lighthouse Marine Supply Store
Luzo Fishing Gear
New Bedford Ship Supply
Newburyport
Book Rack
Newton
Charles River Canoe & Kayak
North Dartmouth
Baker Books
North Falmouth
Mad Fish Outfitters
N. Falmouth Hardware & Marine
Orleans
Booksmith/Musicsmith
Goose Hummock Shop
Nauset Marina
Osterville
Books by the Sea
Oyster Harbors Marine, Inc
Peabody
West Marine
Plymouth
Boater's World
West Marine
Provincetown
Land's End Mar. Supply
Quincy
*Marina Bay
Raynham
Slip's Capeway Marine
Reading
Recreational Equip., Inc. (REI)
Rockport
Toad Hall Bookstore
Salem
Derby Sq. Bookstore
Nautical Traders

Sandwich
Sandwich Ship Supply
Scituate
Front St. Bookshop
Seekonk
Boater's World
West Marine
South Dartmouth
Cape Yachts
*Concordia Co.
Navigator Shop
South Yarmouth
Riverview Bait & Tackle
Swansea
Newsbreak, Inc.
Vineyard Haven
Bunch of Grapes Book Store
Gannon & Benjamin Marine Rwy.
Martha's Vineyard Fuel & Ice
*Martha's Vineyard Shipyard
Wakefield
Boats and Motors
Watertown
Associated Yachts
West Dennis
Sportsman's Landing
Westport
Partners Village Store
Weymouth
Monahan's Marine
Winthrop
Woodside Hardware
Woburn
West Marine

RI
Barrington
Barrington Books
*Brewer Cove Haven Marina
Block Island
Ned Phillips Co.
Bristol
Herreshoff Museum
Charlestown
Breachway Bait and Tackle
East Greenwich
Norton's Shipyard & Marina
Jamestown
*Conanicut Marine
Dutch Harbor Boatyard
*Jamestown Boat Yard
Narragansett
R.I. Engine Co.
Wilcox Marine Supply of R.I.
Newport
Armchair Sailor Bookstore
Newport Nautical Supply
North Kingston
R.I. Mooring Service
Portsmouth
Hinckley Yacht Service
Ship's Store & Rigging
Providence
The Map Center
New England Marine Supply Co.
Wakefield
Ram Point Marina
Snug Harbor Marina
Wickford
The Hour Glass
Kayak Centre

CT
Branford
Birbarie Marine Sales
Clinton
Riverside Basin Marina
Danbury
Boarders Books & Music

Deep River
*Brewer Deep River Marina
East Norwalk
Coastwise Boatworks
Norwalk Cove Marina
Essex
The Chandlery at Essex
Fairfield
West Marine
Greenwich
Outdoor Traders
Guilford
Breakwater Books
Madison
R. J. Julia Booksellers
Milford
Ship's Store
at Milford Boat Works
Mystic
Bank Sq. Books
*Brewer Yacht Yard
Mystic Seaport Stores
New London
Crocker's Boatyard
Noank
Noank Shipyard
Spicer's Marinas
Norwalk
Small Boat Shop
Old Lyme
Kellog Marine Supply
Old Lyme Marina
Old Saybrook
Emerson & Cook Book Co.
North Cove Outfitters, Inc.
River's End Bait & Tackle
Pawcatuck
Sail the Sounds
Portland
Wm. J. Petzold Inc.
Portland Boat Works
South Norwalk
Rex Marine Center
South Windsor
Fish and Boat Center
Stamford
*Brewer Yacht Haven
Hathaway Reiser & Raymond
Landfall Navigation
Stonington
Dodson's Boat Yard
Don's Dock
Wilcox Marine Supply
Waterford
Defender Industries
Hillyer's Tackle Shop
Wilton
Outdoor Sports Center

NY
Albany
Book House of Stuyvesant Plaza
Aquebogue
Larry's Lighthouse Marina
Babylon
West Marine
Bohemia
Borders Books & Music
Brooklyn
Bernie's Fishing Tackle
City Island, Bronx
Bridge Marine Supply
Cold Spring
Hudson Valley Outfitters
Connelly
Rondout Yacht Basin
East Hampton
Seacoast Enterprises
Three Mile Harbor Boat Yard

Fisher's Island
Pirate's Cove Marine
Freeport
Fred Chall Marine
Freeport Marine Supply
Garden City
West Marine
Glen Cove
*Brewer Glen Cove Marina
Greenport
*Brewer Stirling Harbor Marina
*S.T. Preston
White's Hardware
Hampton Bays
Modern Yachts
Huntington
Book Revue
Compass Rose Marine
Coney's Marine Corp.
West Shore Marina
Island Park
West Marine
Mamaroneck
Boater's World
Montauk
Montauk Marine Basin
New Rochelle
Post Marine Supply
West Harbor Yacht Serv.
New York City
Hagstrom Map & Travel Ctr.
New York Kayak Co.
*New York Nautical
Northport
Tidewater Marine
Oyster Bay
Nobman's Marine Hdwre.
Oyster Bay Marine Supply
Seawanhaka Boat Yard
Patchoque
B. Sack
Port Jefferson
West Marine
Port Washington
*Brewer Capri Marina
Riverhead
West Marine
Sag Harbor
Emporium Hardware, Inc.
Henry Persan & Sons
Sag Harbor Yacht Yard
Saugerties
Atlantic Kayak Tours
Shelter Island
Coecles Harbor Marina
Southold
Wego Bait and Tackle
Staten Island
Nautical Chart Supply
Upper Nyack
Julius Peterson
Westhampton Beach
Chesterfield Assoc.
West Islip
West Marine

NJ
Atlantic Highlands
West Marine
Bayonne
Ken's Marina Services
True World Tackle
Belfort
Mariner's Mart
Blackwoods
Blackwell's Book Services
Brick
West Marine
Bridgewater
Baker & Taylor

Cape May
  Sea Gear Marine Supply
  *South Jersey Marina
  Tony's Marine Supply
Eatontown
  West Marine
Hoboken
  Carter Craft
Lodi
  West Marine
Mays Landing
  Boater's World
Mt. Laurel
  Boater's World
  West Marine
Paramus
  Ramsey Outdoor
Perth Amboy
  West Marine
Somers Point
  West Marine
South Amboy
  Lockwood Boat Works
  Morgan Marina
Toms River
  Boaters World
  West Marine
Weehawken
  Port Imperial Marina

PA
Pittsburgh
  West Marine
Phildelphia
  Boater's World
  Pilot House Nautical Books

DE
Bear
  West Marine
Newark
  Boater's World
  Newark News Stand
Rehoboth Beach
  West Marine
Wilmington
  Borders Books & Music

MD
Annapolis
  Boater's World
  Fawcett Boat Supplies
  West Marine
Baltimore
  Maryland Nautical Sales
  West Marine
Easton
  Boater's World
  West Marine
Edgewater
  West Marine
Georgetown
  Georgetown Yacht Basin
Glen Burnie
  Boater's World
  West Marine
Middle River
  Boater's World
Ocean City
  West Marine
Oxford
  Hinkley Yacht Services
Pasadena
  West Marine
Salisbury
  Boater's World
Solomons
  West Marine
Stevensville
  Boater's World

Tracey's Landing
  West Marine

Washington, DC
  Washington Marina

VA
Alexandria
  West Marine
Deltaville
  West Marine
Glen Allen
  West Marine
Hampton
  Boater's World
  West Marine
Norfolk
  Boater's World
  W.T. Brownley Co.
  West Marine
Portsmouth
  Tidewater Yacht Agency
Richmond
  Boater's World
Virginia Beach
  Boater's World
  West Marine
Woodbridge
  West Marine

NC
Atlantic Beach
  Boater's World
Beaufort
  N.C. Maritime Museum
  Scuttlebutt
Charlotte
  West Marine
Cornelius
  West Marine
Morehead City
  Dee Gee's
  West Marine
Nags Head
  West Marine
New Bern
  West Marine
Pineville
  Boater's World
Raleigh
  Boater's World
  West Marine
Southport
  Boater's World
Wilmington
  Boater's World
  West Marine

SC
Bluffton
  Boater's World
Charleston
  Boater's World
  City Marina Ship's Shop
  West Marine
Columbia
  Boater's World
Hilton Head
  West Marine
Mt. Pleasant
  Boater's World
Murrells Inlet
  Boater's World
Myrtle Beach
  Boater's World

GA
Brunswick
  Boater's World

Duluth
  Boater's World
Savannah
  Boater's World
  West Marine
Smyrna
  Boater's World

FL
Daytona
  West Marine
Delray
  West Marine
Ft. Lauderdale
  Bluewater Books & Charts
  Boater's World
  West Marine
Fort Pierce
  West Marine
Hollywood
  West Marine
Jacksonville
  Pier "17" Marina
  West Marine
Key Largo
  Boater's World
  West Marine
Key West
  Boater's World
Lake Worth
  Boater's World
Largo
  Boater's World
Marathon
  Boater's World
  West Marine
Miami
  West Marine
North Palm Beach
  West Marine
St. Augustine
  West Marine
Stuart
  West Marine
Vero Beach
  West Marine
West Palm Beach
  West Marine

CO
Fort Collins
  Geomart

IL
Momence
  Baker & Taylor

Please check with the following:
Amazon.com

Available at most East Coast:
Boater's World
West Marine

For store locations in each state see website:
BoatersWorld.com
WestMarine.com

Advertisers in book.
Refer to page 271.

For a current list of
ELDRIDGE dealers
please visit:
www.eldridgetide.com
www.robertwhite.com

269

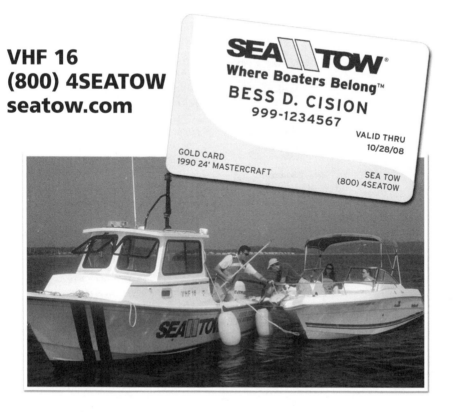

It helps to say, "I saw it in the **ELDRIDGE TIDE BOOK.**"

INDEX TO ADVERTISERS

For more information about **ELDRIDGE** advertisers and links to their websites visit: **www.eldridgetide.com**

271